Automata and Languages

Automata and Languages

JOHN M. HOWIE

Regius Professor of Mathematicas
University of St Andrews

CLARENDON PRESS · OXFORD
1991

Oxford University Press, Walton Street, Oxford OX2 6DP
Oxford New York Toronto
Delhi Bombay Calcutta Madras Karachi
petaling Jaya Singapore Hong Kong Tokyo
Nairobi Dar es Salaam Cape Town
Melbourne Auckland
and associated companies in
Berlin Ibadan

Oxford is a trade mark of Oxford University Press

Published in the United States
by Oxford University Press, New York

© John M. Howie, 1991

A catalogue record for this book
available from the British Library

Library of Congress Cataloging in Publication Data

ISBN 0–19–853424–8 (h/b)
ISBN 0–19–853442–6 (p/b)

Printed and bound in
Great Britain by Biddles Ltd,
Guildford and King's Lynn

To Dorothy

Preface

Like many another book, this one has grown out of a course given to university students over a number of years. The growing importance of discrete mathematics is making its mark on undergraduate and postgraduate syllabuses, and I expect that many university teachers besides myself have felt hampered in giving a course on Automata and Languages by the lack of a suitable textbook. Books do exist, of course, but being on the whole aimed at fairly sophisticated readers they can be a bit discouraging for undergraduates and even for junior postgraduates. My hope is that this book will stimulate interest in a lively and fascinating area by providing a readable and not too difficult introduction.

While my main aim has been to write for mathematics students, I hope that my readers will include students of computer science. Such students will be of widely varying mathematical backgrounds, and it seemed appropriate to include an introductory chapter in which the relevant mathematical ideas and notations are discussed. Many readers will of course be able to skate very lightly over this chapter.

Whether such light skating is possible depends very much on the precise mathematical background of the reader, and this raises an interesting point, for it seems that the structure and content of mathematics service courses in most universities are far from satisfactory from the point of view of computer science. A first course in abstract algebra is the most obviously useful preliminary to a course on theoretical computer science, but mathematics departments tend to expect their students to have the sophistication that comes from a solid background in calculus and linear algebra before being let loose on abstract algebra, and this can be something of a disincentive. Even a standard first course in abstract algebra is far from ideal. Such a course very properly concentrates on groups, rings and fields, but it is the humble semigroup and the scarcely less humble monoid that make the biggest impact on the theory of automata. I hope that Chapter 1 provides something close to a geodesic path through the most relevant ideas. Most of the ideas are in fact not needed until Chapter 3, and from the teaching point of view it can be useful to plunge straight into Chapter 2, developing mathematical ideas as and when they are required.

In no sense does the book set out to be comprehensive, and doubtless many readers will be disappointed by the omission of this or that favourite topic. I remember, however, how much in my own student days I appreciated the slim volumes of the Oliver and Boyd *University Mathematical Texts* series, not only for their accessible prices (still less than ten shillings in those far off times), but also for the fact that one felt able actually to read them right through. So I have tried to be interesting, encouraging and persuasive rather than comprehensive. The reader whose appetite is whetted can turn to the more advanced books mentioned in the Bibliography.

It would, I think, take a course of 40 or 50 hours to cover all the material in the book. A shorter course with an algebraic flavour could deal with Chapters 2, 3 and 7, while Chapters 2, 4, 5 and 6 would constitute a course with a stronger emphasis on languages and machines.

It is a pleasure to record thanks to the University of St Andrews for the sabbatical leave that made this book possible, to Northern Illinois University for taking me in during part of that leave, to Robert McFadden, Helena Sezinando, Jack Jones and Tom Hall, who commented on various drafts, and above all to the St Andrews students who were the guinea pigs as the ideas for the book took shape.

St Andrews   J. M. H.
January 1991

Contents

1 Mathematical preliminaries   1
1.1 Sets   1
1.2 Functions   7
1.3 Equivalence relations   12
1.4 Semigroups and monoids   15
1.5 Morphisms and congruences   22
1.6 Transformation monoids   28
1.7 Free semigroups and monoids   29
Exercises 1   33

2 Automata   39
2.1 Finite state automata   40
2.2 Initial and final states; the behaviour of an automaton   43
2.3 Incomplete and non-deterministic automata   48
2.4 Finite sets   55
2.5 Rational sets and Kleene's Theorem   57
Exercises 2   64

3 The syntactic monoid   67
3.1 The syntactic congruence   67
3.2 The minimal automaton   71
3.3 Reduced automata   75
3.4 The transformation monoid of an automaton   82
3.5 The calculation of the syntactic monoid   86
Exercises 3   88

4 Languages   92
4.1 Phrase structure grammars   92
4.2 Regular and rational languages   96
4.3 Context-free languages   103
4.4 Chomsky Normal Form   106
Exercises 4   117

5 Pushdown automata   119
5.1 Pushdown automata   121
5.2 Examples   123
5.3 Recognition by empty memory stack   129
5.4 Pushdown automata and context-free languages   133

5.5 The Pumping Lemma: non-context-free languages   143
5.6 Closure properties of context-free languages   147
Exercises 5   154

6 Turing machines   157
6.1 Definitions and examples   158
6.2 Turing machines as text-processors   164
6.3 Alternative Turing machines   168
6.4 Non-deterministic Turing machines   174
6.5 Turing machines and context-free languages   179
6.6 Turing machines and phrase structure grammars   181
6.7 The Chomsky hierarchy   189
6.8 Sets that are not TM-recognizable   190
Exercises 6   193

7 Varieties   196
7.1 Some more algebra   196
7.2 Varieties and F-varieties   201
7.3 The variety theorem   207
7.4 Star-free languages and aperiodic monoids   217
Exercises 7   226

Solutions to exercises   229

References   287

Notations used in the text   289

Index   291

1 Mathematical preliminaries

Introduction

Most of the mathematical equipment we shall need is quite elementary, and many readers will be able to skip lightly over this chapter. The algebraic ideas in Sections 1.3 to 1.7 are required primarily in Chapters 3 and 7, and some further bits of algebra, required only in Chapter 7, are deferred until Section 7.1.

1.1 Sets

When 'New Mathematics' invaded school syllabuses throughout the Western World in the 1960s, parents of small children were bewildered by what seemed to be a strange and esoteric language, and set theory acquired (in the popular imagination at least) a totally unjustified aura of mystery. (The small children themselves were much less bothered.) To be sure, there are some subtleties and some pitfalls in *axiomatic* set theory, but these will not concern us at all. For our purposes a *set A* is simply a collection of objects of some kind. The objects are called the *members* or the *elements* of the set A; if a is an element of A we write '$a \in A$' and read 'a belongs to A'. Thus, for example, if R is the set of all readers of this book, and if Susan is a reader of this book, then we can write 'Susan $\in R$'.

As with many other mathematical symbols, we can use an oblique stroke to negate the meaning. So if George is *not* a reader of this book we may write 'George $\notin R$'.

Two sets A and B are said to be *equal* if they consist of the same elements. We write $A = B$.

Many of the sets we shall be concerned with are sets of numbers, such as the set N of *natural numbers,* whose elements are $1, 2, 3, \ldots$, the set N^0 of *non-negative integers,* whose elements are $0, 1, 2, 3, \ldots$, and the set Z of all *integers,* whose elements are

$$0, 1, -1, 2, -2, 3, -3, \ldots.$$

These are all *infinite* sets; for an example of a *finite* set consider D_{30}, the set of all positive divisors of 30, consisting of the elements

$$1, 2, 3, 5, 6, 10, 15, 30.$$

We can specify a set, especially a finite set, simply by listing its elements. The standard convention is to list the elements between curly brackets $\{, \}$. Thus

$$D_{30} = \{1, 2, 3, 5, 6, 10, 15, 30\}.$$

The other way of specifying a set is by means of a defining property. Again we use curly brackets, but this time the convention is different. For example, we write

$$D_{30} = \{n : n \in N, \ n \text{ divides } 30\}$$

(and read 'the set of n such that n belongs to N and n divides 30') or

$$D_{30} = \{n \in N : n \text{ divides } 30\}$$

(and read 'the set of n belonging to N such that n divides 30'). To give another example,

$$\{n : n \in N, \ n < 7\} \ = \ \{n \in N : n < 7\} \ = \ \{1, 2, 3, 4, 5, 6\}.$$

Listing is of course less useful in the case of infinite sets, but we do allow ourselves to write

$$\{2, 4, 6, \ldots\}$$

instead of $\{n \in N : 2 \text{ divides } n\}$. Another alternative expression for this same set is $\{2n : n \in N\}$.

Sometimes a set has just one element and we call it a *singleton* set. For example, the set

$$\{n \in N : n > 1 \text{ and } n \text{ divides } 7\}$$

has the single element 7 (since 7 is a prime number); we write the set as $\{7\}$. There is a logical distinction between the number 7 and the set $\{7\}$, but usually the difference is not important.

Sometimes sets are even smaller. For example,

$$\{n \in \mathbf{Z} : n^2 < 0\}$$

contains no elements at all, since $n^2 \geq 0$ for every n in $\mathbf{Z}$, whether positive or negative. We have a special symbol $\emptyset$ for the *empty* (or *null*, or *void*) set containing no elements. Thus

$$\{n \in \mathbf{Z} : n^2 < 0\} = \emptyset,$$

$$\{n \in \mathbf{N} : 1 < n < 11, \ n \text{ divides } 11\} = \emptyset,$$

and so on.

The number of elements in a finite set A, sometimes called the *cardinality* of A, is denoted by $|A|$. Thus

$$|D_{30}| = 8, \ |\{7\}| = 1, \ |\emptyset| = 0.$$

If A and B are two sets, then the *union* of A and B, denoted by $A \cup B$, is defined by the rule that $x \in A \cup B$ if and only if $x \in A$ *or* $x \in B$. The *intersection* of A and B, denoted by $A \cap B$, is defined by the rule that $x \in A \cap B$ if and only if $x \in A$ *and* $x \in B$. Thus

$$\{1, 2, 3, 4\} \cup \{3, 4, 5, 6\} = \{1, 2, 3, 4, 5, 6\},$$
$$\{1, 2, 3, 4\} \cap \{3, 4, 5, 6\} = \{3, 4\}.$$

The sets A and B are called *disjoint* if $A \cap B = \emptyset$. Thus, for example, the sets

$$A = \{n \in \mathbf{N} : n > 1, \ n \text{ divides } 12\}$$
$$\text{and } A = \{n \in \mathbf{N} : n > 1, \ n \text{ divides } 35\}$$

are disjoint.

It is sometimes useful to extend the notations $\cup$ and $\cap$ so as to cope with more than two sets. Notations such as $A \cap B \cap C$ or $A_1 \cup A_2 \cup \ldots A_n$ are easy enough to interpret, but it can be convenient at times to replace the latter by

$$\bigcup_{i=1}^{n} A_i \quad \text{or} \quad \bigcup_{i \in \{1,2,\ldots,n\}} A_i.$$

To be more formal about it, let I be a non-empty set (either finite or infinite) and for each i in I let A_i be a set. We say that $\{A_i : i \in I\}$ is an *indexed family* of sets, indexed by I. Then we define the sets

$$\bigcup_{i \in I} A_i \quad \text{and} \quad \bigcap_{i \in I} A_i$$

by the rules that

$$x \in \bigcup_{i \in I} A_i$$

if and only if $x \in A_i$ for *some* i in I, and

$$x \in \bigcap_{i \in I} A_i$$

if and only if $x \in A_i$ for *all* i in I.

We can be more formal still. If we write $(\exists i \in I)\ x \in A_i$ to mean 'there exists i in I such that $x \in A_i$' and $(\forall i \in I)\ x \in A_i$ to mean 'for all i in I, $x \in A_i$' then

$$\bigcup_{i \in I} A_i = \{x : (\exists i \in I)\ x \in A_i\},$$

$$\bigcap_{i \in I} A_i = \{x : (\forall i \in I)\ x \in A_i\}.$$

The symbols $\exists$ and $\forall$, which are standard in mathematical logic, are called *quantifiers*.

If A and B are sets, we say that A is a *subset* of B, and write either $A \subseteq B$ or $B \supseteq A$, if every element of A is an element of B. Thus, for example, $\mathbf{N} \subseteq \mathbf{Z}$, $\emptyset \subseteq \mathbf{Z}$, $\mathbf{Z} \subseteq \mathbf{Z}$, and

$$\{n \in \mathbf{N} : n \text{ divides } 30\} \subseteq \{n \in \mathbf{N} : n \text{ divides } 60\}.$$

If $A \subseteq B$ and $A \neq B$ then we shall say that A is a *proper subset* of B and write either $A \subset B$ or $B \supset A$.

Note 1.1.1. This is the first of our notations that is not absolutely standard. In algebra the standard notation is usually as I have given it, but in some other parts of mathematics the notation $A \subset B$ is used with the meaning I have given to $A \subseteq B$.

Notice that the statement $A = B$ is equivalent to $A \subseteq B$ and $B \subseteq A$.

From time to time we shall find it convenient to employ the symbols $\Rightarrow$ (implies) and $\Leftrightarrow$ (implies and is implied by). So $A \subseteq B$ can be defined by

$$x \in A\ \Rightarrow\ x \in B,$$

and $A = B$ by

$$x \in A\ \Leftrightarrow\ x \in B.$$

If A is a subset of a set S, then the *complement of A in S* is defined as

$$\{x \in S : x \notin A\}.$$

Our usual notation for this will be $S\backslash A$, but if the context makes it clear what S is then we can get away with the simpler notation A'.

It is frequently useful to visualize $A \cup B$, $A \cap B$ and $A' = S\backslash A$ by means of a *Venn diagram*. The 'universal' set S is represented by a rectangle and the subsets by circles inside the rectangle. In the following three diagrams the shaded areas represent $A \cup B$, $A \cap B$ and A' respectively.

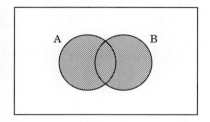
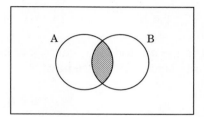

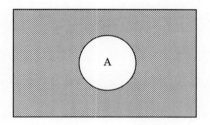

The three operations $\cup$, $\cap$ and $\backslash$ (union, intersection and complementation) are often referred to as the *Boolean* operations (after George Boole (1815–1864)). They give rise to an *algebra* of subsets, some of whose rules of operation, such as

$$(A \cup B) \cup C = A \cup (B \cup C) \quad \text{the } \textit{associative } \text{law}$$

and

$$A \cup (B \cap C) = (A \cup B) \cap (A \cup C) \quad \text{the } \textit{distributive } \text{law},$$

are reminiscent of ordinary algebra, while others have no obvious counterparts. We explore this aspect of set theory in Exercise 1.1, but for future reference we record here the so-called *De Morgan laws*, (after Augustus De Morgan (1806–1871)): for all subsets A and B of a fixed set S,

$$(A \cup B)' = A' \cap B', \quad (A \cap B)' = A' \cup B'.$$

In the more general notation,

$$S\backslash(A \cup B) = (S\backslash A) \cap (S\backslash B), \quad S\backslash(A \cap B) = (S\backslash A) \cup (S\backslash B). \quad (1.1.2)$$

If A and B are sets, then the *Cartesian product* $A \times B$ (called after René Descartes (1596–1650)) is defined as the set of all ordered pairs (a, b) with $a \in A$, $b \in B$:

$$A \times B = \{(a, b) : a \in A,\ b \in B\}.$$

For example, if $A = \{a_1, a_2\}$ and $B = \{b_1, b_2, b_3\}$, then

$$A \times B = \{(a_1, b_1),\ (a_1, b_2),\ (a_1, b_3),\ (a_2, b_1),\ (a_2, b_2),\ (a_2, b_3)\}.$$

It is not an accident that $A \times B$ has $6 = 2 \times 3$ elements: in fact for all finite sets A and B we have

$$|A \times B| = |A|\,|B|.$$

More generally, if $A_1, A_2, \ldots, A_n$ are sets, then the *Cartesian product* of $A_1, A_2, \ldots, A_n$ is defined by

$$A_1 \times A_2 \times \cdots \times A_n = \{(a_1, a_2, \ldots, a_n) : a_1 \in A_1,\ a_2 \in A_2, \ldots, a_n \in A_n\}.$$

If $A_1 = A_2 = \cdots = A_n = A$ then we denote the product by A^n. Thus

$$A^n = \{(a_1, a_2, \ldots, a_n) : a_1, a_2, \ldots, a_n \in A\}.$$

For example, if $A = \{1, 2\}$ then A^3 consists of the eight triples

$$(1, 1, 1),\quad (1, 1, 2),\quad (1, 2, 1),\quad (1, 2, 2)$$
$$(2, 1, 1),\quad (2, 1, 2),\quad (2, 2, 1),\quad (2, 2, 2).$$

In general, we have
$$|A^n| = |A|^n$$

for every finite set A and for $n = 2, 3, \ldots$.

The Cartesian product is familiar in elementary mathematics in the shape of the coordinate plane

$$\mathbf{R}^2 = \{(x, y) : x, y \in \mathbf{R}\}$$

and its three-dimensional analogue. (Here $\mathbf{R}$ denotes the set of all *real* numbers.) It can be useful to visualize Cartesian products as if the sets A and B were intervals in $\mathbf{R}$.

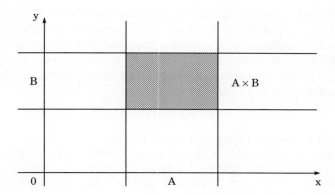

The elements of a set may themselves be sets. Thus, for example, given a set A we may wish to consider the *set of all subsets of* A. This frequently called the *power set* of A, and we shall denote it by $\mathcal{P}(A)$. If

$$A = \{a_1, a_2, \ldots, a_n\},$$

a set of cardinality n, then we can specify a subset B of A by considering each of $a_1, a_2, \ldots, a_n$ in turn and stating whether or not it does or does not belong to B. If we write down 0 for a 'no' and 1 for a 'yes' then B is specified by means of a sequence of length n of 0's and 1's. (For example, the subset $\{1, 3, 5\}$ of the set $\{1, 2, 3, 4, 5\}$ is specified by the sequence 10101.) Thus the number of different subsets of A is the number of sequences of length n of 0's and 1's. We deduce:

Theorem 1.1.3. *For every finite set A,*

$$|\mathcal{P}(A)| = 2^{|A|}. \qquad\qquad \blacksquare$$

Notice that the sequence $00 \ldots 0$ gives the *empty* subset $\emptyset$, while the sequence $11 \ldots 1$ gives the set A itself.

In the case where A is an infinite set we shall sometimes want to consider $\mathcal{F}(A)$, the set of all *finite* subsets of A.

1.2 Functions

Let A and B be sets. A *relation* $\mathbf{P}$ *from* A *to* B is defined as a subset of the Cartesian product $A \times B$. We say that a and b are *related* (by $\mathbf{P}$) if $(a, b) \in \mathbf{P}$.

This is a very general definition, and we shall want to specialize it in two different directions. First, we shall say that a relation φ from A to B (that is, a subset φ of $A \times B$) is a *function*, or a *map*, or a *mapping*, from A to B if for every a in A there is a unique b in B for which $(a, b) \in \varphi$. It can be useful to write '$\exists!$' as an abbreviation for 'there exists a unique'; so symbolically we have

$$(\forall a \in A)(\exists! \, b \in B) \, (a, b) \in \varphi. \tag{1.2.1}$$

The unique b corresponding to a given a is called the *image* of a under the function φ, or the *value* of the function φ at the point a. It is denoted by $\varphi(a)$. The set A is called the *domain* of φ and is sometimes written $\operatorname{dom} \varphi$. The set $\{\varphi(a) : a \in A\}$ (which may well be a proper subset of B) is called the *image* (or *range*) of φ and is sometimes written $\operatorname{im} \varphi$.

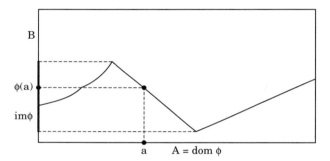

Two functions φ_1 and φ_2 from A to B are said to be *equal* if they take the same value for every element in their common domain. This amounts to saying that φ_1 and φ_2 are equal as subsets of $A \times B$.

To indicate that φ is a function we shall write $\varphi : A \to B$ (reading 'φ from A to B') rather than $\varphi \subseteq A \times B$. We can usefully think of φ as a 'rule' which determines the value $\varphi(a)$ for each a in the domain A.

The familiar functions of analysis are of course instances of the notion we have been describing: for example,

$$\varphi(x) = \sin x \quad (x \in \mathbf{R}),$$
$$\varphi(x) = \sqrt{x} \quad (x \in \mathbf{R}, \ x \geq 0),$$

and so on. In the first case the domain is $\mathbf{R}$ and the image is the closed interval $[-1, 1] \left(= \{x \in \mathbf{R} : -1 \leq x \leq 1\}\right)$. In the second case

$$\operatorname{dom} \varphi = \operatorname{im} \varphi = [0, \infty) \left(= \{x \in \mathbf{R} : x \geq 0\}\right).$$

A function $\varphi : A \to B$ is said to be *one-one*, or *injective*, or an *injection*, if

$$(\forall a_1, a_2 \in A) \, \varphi(a_1) = \varphi(a_2) \implies a_1 = a_2.$$

Equivalently, we have the 'contrapositive' version of the same thing:

$$(\forall a_1, a_2 \in A)\; a_1 \neq a_2 \;\Rightarrow\; \varphi(a_1) \neq \varphi(a_2).$$

In words, φ is *injective* if distinct elements of A have distinct images under φ.

The function $\sigma : \mathbf{R} \to \mathbf{R}$ given by

$$\sigma(x) = \sin x \quad (x \in \mathbf{R})$$

is not injective, since (for example)

$$\sin 0 = \sin \pi \; (= 0).$$

The function $\kappa : \mathbf{R} \to \mathbf{R}$ given by

$$\kappa(x) = x^3 \quad (x \in \mathbf{R}) \tag{1.2.2}$$

is injective; for

$$\begin{aligned}
\kappa(x) = \kappa(y) &\Rightarrow x^3 - y^3 = 0 \\
&\Rightarrow (x - y)(x^2 + xy + y^2) = 0 \\
&\Rightarrow x = y,
\end{aligned}$$

since

$$x^2 + xy + y^2 = (x + \tfrac{1}{2}y)^2 + \tfrac{3}{4}y^2 > 0$$

unless $x = y = 0$.

A function $\varphi : A \to B$ is said to be *onto,* or *surjective,* or a *surjection,* if $\mathrm{im}\,\varphi = B$, that is, if

$$(\forall b \in B))(\exists a \in A)\; \varphi(a) = b.$$

For example, the function $\tau : \mathbf{R} \to \mathbf{R}$ given by

$$\tau(x) = x^2 \quad (x \in \mathbf{R})$$

is *not* surjective, since $\mathrm{im}\,\tau = [0, \infty) \subset \mathbf{R}$. The cube function κ, however, given by (1.2.2), is surjective.

A function φ is said to be *one-one onto,* or *bijective,* or a *bijection,* if it is both injective and surjective, that is, if

$$(\forall b \in B)(\exists! \, a \in A)\; \varphi(a) = b.$$

The cube function κ provides an example.

If $\varphi : A \rightarrow B$ and $\psi : B \rightarrow C$ are functions, then the *composition* of φ and ψ, written $\psi \circ \varphi$, is a function from A to C defined by

$$(\psi \circ \varphi)(a) = \psi(\varphi(a)) \quad (a \in A).$$

Given

$$\theta : A \rightarrow B, \quad \varphi : B \rightarrow C, \quad \psi : C \rightarrow D,$$

we see that for all a in A

$$\big(\psi \circ (\varphi \circ \theta)\big)(a) = \psi\big((\varphi \circ \theta)(a)\big) = \psi\Big(\varphi(\theta(a))\Big),$$

$$\big((\psi \circ \varphi) \circ \theta\big)(a) = (\psi \circ \varphi)(\theta(a)) = \psi\Big(\varphi(\theta(a))\Big);$$

thus

$$\psi \circ (\varphi \circ \theta) = (\psi \circ \varphi) \circ \theta. \tag{1.2.3}$$

We express this property of functions by saying that the operation $\circ$ is *associative*.

For every set A we have an identity function $\mathrm{id}_A : A \rightarrow A$ defined by

$$\mathrm{id}_A(a) = a \quad (a \in A).$$

Notice that, for every $\varphi : A \rightarrow B$,

$$\varphi \circ \mathrm{id}_A = \varphi = \mathrm{id}_B \circ \varphi. \tag{1.2.4}$$

Given $\varphi : A \rightarrow B$, we say that $\theta : B \rightarrow A$ is an *inverse function* of φ if

$$\theta \circ \varphi = \mathrm{id}_A, \quad \varphi \circ \theta = \mathrm{id}_B.$$

In fact we may talk of *the* inverse function of φ, because if we have another function $\theta' : B \rightarrow A$ with the property that

$$\theta' \circ \varphi = \mathrm{id}_A, \quad \varphi \circ \theta' = \mathrm{id}_B,$$

then by (1.2.4) and (1.2.3)

$$\theta' = \theta' \circ \mathrm{id}_B = \theta' \circ (\varphi \circ \theta) = (\theta' \circ \varphi) \circ \theta = \mathrm{id}_A \circ \theta = \theta.$$

We write the unique inverse of $\varphi : A \rightarrow B$ as $\varphi^{-1} : B \rightarrow A$. It is defined by the properties

$$\varphi^{-1} \circ \varphi = \mathrm{id}_A, \quad \varphi \circ \varphi^{-1} = \mathrm{id}_B. \tag{1.2.5}$$

Of course not every function $\varphi : A \rightarrow B$ has an inverse. In fact, if φ has an inverse then it must be bijective. For the existence of φ^{-1} satifying (1.2.5) gives that for all a_1, a_2 in A

$$\varphi(a_1) = \varphi(a_2) \Rightarrow \varphi^{-1}\big(\varphi(a_1)\big) = \varphi^{-1}\big(\varphi(a_2)\big) \Rightarrow a_1 = a_2,$$

and so φ is injective. Also, for all b in B,

$$b = \varphi\big(\varphi^{-1}(b)\big) \in \mathrm{im}\,\varphi,$$

and so φ is surjective.

We have proved half of the following theorem:

Theorem 1.2.6. *A function $\varphi : A \to B$ has an inverse if and only if it is bijective.*

Proof. Suppose now that φ is bijective. For each b in B there is a unique a in A such that $\varphi(a) = b$. Define $\theta : B \to A$ by taking $\theta(b)$ to be this a for each b in B. Then, simply by virtue of our definition,

$$\varphi\big(\theta(b)\big) = b \quad (b \in B).$$

Also, again by the definition,

$$\theta\big(\varphi(a)\big) = a \quad (a \in A).$$

Thus

$$\theta \circ \varphi = \mathrm{id}_A, \quad \varphi \circ \theta = \mathrm{id}_B,$$

and so θ is the inverse of φ. ∎

Example 1.2.7. Consider $\varphi : \mathbf{R} \to \mathbf{R}$, defined by

$$\varphi(x) = 2x - 3 \quad (x \in \mathbf{R}).$$

The inverse function φ^{-1} is given by

$$\varphi^{-1}(x) = \frac{1}{2}(x + 3) \quad (x \in \mathbf{R}).$$

It can be useful to employ the notation φ^{-1} even when φ is not a bijection. Given an arbitrary function $\varphi : A \to B$ and a subset C of B, we define

$$\varphi^{-1}(C) = \{a \in A : \varphi(a) \in C\}.$$

If $C = \{c\}$, a singleton subset of B, then we write $\varphi^{-1}(c)$ rather than $\varphi^{-1}(\{c\})$: thus

$$\varphi^{-1}(c) = \{a \in A : \varphi(a) = c\}.$$

Notice that $\varphi^{-1}(c)$ is a *subset* of A, which may contain more than one element, or — if $c \notin \mathrm{im}\,\varphi$ — may be empty. If φ happens to be a bijection then the two meanings for $\varphi^{-1}(c)$ do not conflict, for in this case

$$\varphi^{-1}(\{c\}) = \{\varphi^{-1}(c)\}$$

for every c in $B = \mathrm{im}\,\varphi$.

Example 1.2.8. Consider the function $\varphi : \mathbf{R} \to \mathbf{R}$ defined by

$$\varphi(x) = x^2 - 2x \quad (x \in \mathbf{R}).$$

The graph of the function is given as follows:

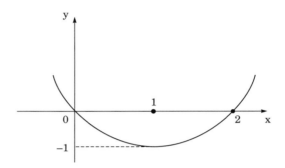

The function, with image $[-1, \infty] \subset \mathbf{R}$, is not surjective. It is not injective either, since (for example) $\varphi(0) = \varphi(2) = 0$. If $y < -1$ then $\varphi^{-1}(y) = \emptyset$. If $y > -1$ then

$$\varphi^{-1}(y) = \{x \in \mathbf{R} : x^2 - 2x - y = 0\}.$$

Finding the elements of this set amounts to solving the quadratic equation $x^2 - 2x - y = 0$; thus

$$\varphi^{-1}(y) = \{1 + \sqrt{y+1}, \ 1 - \sqrt{y+1}\}.$$

If $y = -1$ then $\varphi^{-1}(y) = \{1\}$.

1.3 Equivalence relations

Let us now return to the original definition of a relation $\mathbf{P}$ from A to B as a subset of $A \times B$, and specialize to the case where $B = A$. We say that $\mathbf{P}$ is a *relation on* A. If $(a, a') \in \mathbf{P}$ we say that a and a' are *related* (by $\mathbf{P}$) and sometimes write $a \, \mathbf{P} \, a'$. The relation $\mathbf{P}$ is said to be an *equivalence* on A if

$$(\forall a \in A) \ a \, \mathbf{P} \, a \quad \text{(the \textit{reflexive} law)}, \tag{1.3.1}$$

$$(\forall a, b \in A) \ a \, \mathbf{P} \, b \ \Rightarrow \ b \, \mathbf{P} \, a \quad \text{(the \textit{symmetric} law)}, \tag{1.3.2}$$

$(\forall a, b, c \in A)$ $a\,\mathbf{P}\,b$ and $b\,\mathbf{P}\,c$ $\Rightarrow$ $a\,\mathbf{P}\,c$ (the *transitive* law).   (1.3.3)

The most obvious example of an equivalence relation is the equality relation '=', defined on any set A. Regarded as a subset of $A \times A$, it consists of all pairs (a, a) (where $a \in A$). Since this can be visualized as the diagonal

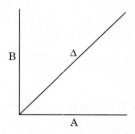

in a Cartesian diagram, equality is often called the *diagonal* relation and we shall sometimes want to denote it by Δ (or by Δ_A if the context makes it necessary):

$$\Delta = \Delta_A = \{(a, a) : a \in A\}. \qquad (1.3.4)$$

Another, less trivial, classical example of an equivalence relation, when $A = \mathbf{Z}$, is the *arithmetical congruence* $\mathbf{K}_n$ $(n \geq 2)$, where

$$a\,\mathbf{K}_n\,b \text{ if and only if } a \equiv b\,(\mathrm{mod}\,n),$$
$$\text{i.e., if and only if } n \text{ divides } b - a.$$

It is routine to check that this relation satisfies (1.3.1)–(1.3.3).

Let $\mathbf{P}$ be an equivalence on a set A and let $a \in A$. We define the *equivalence class,* or $\mathbf{P}$-*class,* containing a by

$$a\mathbf{P} = \{x \in A : x\,\mathbf{P}\,a\}.$$

For example, if we consider the arithmetical congruence $\mathbf{K}_5$ we see that

$$2\mathbf{K}_5 = \{2, 7, 12, \ldots\} \cup \{-3, -8, -13, \ldots\};$$

that is, $2\mathbf{K}_5$ consists of all integers leaving remainder 2 when divided by 5. The equivalence $\mathbf{K}_5$ in fact partitions $\mathbf{Z}$ into disjoint subsets

$$0\mathbf{K}_5 = \{0, \pm 5, \pm 10, \ldots\},$$
$$1\mathbf{K}_5 = \{1, 6, 11, 16, \ldots\} \cup \{-4, -9, -14, \ldots\},$$
$$2\mathbf{K}_5 = \{2, 7, 12, 17, \ldots\} \cup \{-3, -8, -13, \ldots\},$$
$$3\mathbf{K}_5 = \{3, 8, 13, 18, \ldots\} \cup \{-2, -7, -12, \ldots\},$$
$$4\mathbf{K}_5 = \{4, 9, 14, 19, \ldots\} \cup \{-1, -6, -11, \ldots\}.$$

This, or something very like it, is what happens in general. The set of all **P**-classes gives a *partition* of A, in the sense that:

every element of A belongs to exactly one **P**-class. (1.3.5)

To see this, notice first that $a \in a\mathbf{P}$ by the reflexive law (1.3.1), so that every element of A belongs to at least one **P**-class. The uniqueness follows from the fact that distinct **P**-classes are disjoint. To see that this is so, suppose that $x \in a\mathbf{P} \cap b\mathbf{P}$, so that

$$x \, \mathbf{P} \, a, \quad x \, \mathbf{P} \, b.$$

For all y in $a\mathbf{P}$ we thus have

$$y \, \mathbf{P} \, a, \quad a \, \mathbf{P} \, x, \quad x \, \mathbf{P} \, b,$$

giving $y \, \mathbf{P} \, b$ by the transitive law; thus $y \in b\mathbf{P}$. We have shown that $a\mathbf{P} \subseteq b\mathbf{P}$, and we may equally well prove that $b\mathbf{P} \subseteq a\mathbf{P}$. Hence $a\mathbf{P} = b\mathbf{P}$. We have shown that if two **P**-classes are not disjoint then they are identical, which is exactly what we require.

Example 1.3.6. Let **P** be the relation on the set **R** of real numbers defined by the rule that $a \, \mathbf{P} \, b$ if and only if $b - a \in \mathbf{Z}$. It is easily seen that **P** is an equivalence. A real number x has a unique expression as $n + y$, where n is an integer and $0 \leq y < 1$. The integer n is called the *integral part* of x, and y is called the *fractional part* of x. It is easy to see that $a \, \mathbf{P} \, b$ if and only if a and b have the same fractional part. Each **P**-class is expressible as $y\mathbf{P}$, where $0 \leq y < 1$, and

$$y\mathbf{P} = \{n + y : n \in \mathbf{Z}\}.$$

Suppose now that we have a partition π of A, that is, a collection $X_1, X_2, \ldots$ of subsets of A with the property:

every element of A belongs to exactly one X_j (1.3.7)

Then we can define an equivalence **P** on A by the rule that $a \, \mathbf{P} \, b$ if and only if a and b are in the same X_j. The **P**-classes are then precisely the sets X_j.

The two notions, of *equivalence* and *partition,* are in fact completely interchangeble, and we shall feel free to use whichever is more convenient at the time.

Given an equivalence **P** on A, we shall denote the set of **P**-classes — it is a set of subsets — by $A/\mathbf{P}$. It is called the *quotient set* of A by the equivalence **P**. For example,

$$\mathbf{Z}/\mathbf{K}_5 = \{0\mathbf{K}_5, 1\mathbf{K}_5, 2\mathbf{K}_5, 3\mathbf{K}_5, 4\mathbf{K}_5\},$$

a set with 5 elements.

Example 1.3.8. Let S be the relation on the set $A = (\mathbf{R}\backslash\{0\})^2$ defined by

$$(x_1, y_1) \ \mathbf{S} \ (x_2, y_2) \ \text{if and only if} \ x_1 y_2 = x_2 y_1.$$

Then S is an equivalence. (This is easily seen if we rewrite $x_1 y_2 = x_2 y_1$ as $y_1/x_1 = y_2/x_2$.) For all $k \neq 0$ the S-class $(1, k)\mathbf{S}$ is given by

$$(1, k)\mathbf{S} = \{(x, y) \in A : y/x = k\},$$

and can be visualized as the line through $(1, k)$ with gradient k (and with the point $(0, 0)$ 'pinched out').

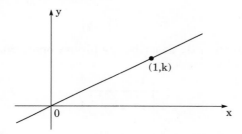

The quotient set $A/\mathbf{S}$ is the set of all lines through $(0, 0)$ other than the x- and y-axes, with $(0, 0)$ pinched out. Notice that the S-classes form a partition of A

We mention now an important connection between functions and equivalences. If $\varphi : A \to B$ is a function then the relation

$$\{(a, a') \in A \times A : \varphi(a) = \varphi(a')\}$$

is an equivalence on A. We call it the *kernel equivalence* of φ and write it as $\ker \varphi$. Notice that $\ker \varphi = \Delta_A$ if and only if φ is one-one.

1.4 Semigroups and monoids

Many of the sets encountered in mathematics have an associated algebraic structure, in the sense that we are able to add, multiply, subtract and divide their elements. These familiar arithmetical operations are all examples of what are called *binary operations* on a set. Formally, if S is a non-empty set, then a *binary operation* β on S is defined as a map $\beta : S \times S \to S$.

For example, consider the operation α of addition on the set $\mathbf{N}$ of natural numbers. We have a map $\alpha : \mathbf{N} \times \mathbf{N} \to \mathbf{N}$ given by

$$\alpha(x, y) = x + y \quad ((x, y) \in \mathbf{N} \times \mathbf{N}).$$

Under the terms of our definition, subtraction is *not* a binary operation on N, for $x - y$ may fall outside N. (Subtraction *is* a binary operation on Z, since $x - y$ is certainly an integer whenever x and y are integers.)

The notation $\beta(x, y)$ is cumbersome, and we usually prefer to use notations that derive from the familiar examples already mentioned. Usually we shall employ the multiplicative notation $x.y$ or xy, and we shall often refer to the operation as 'multiplication', though it may be far removed from multiplication in the ordinary arithmetical sense.

Example 1.4.1. Let $S = \{1, 2, 3, 4\}$, and define a 'multiplication' on S by

$$xy = \max\{x, y\}.$$

Then we can conveniently describe the binary operation by means of a *Cayley table*

	1	2	3	4
1	1	2	3	4
2	2	2	3	4
3	3	3	3	4
4	4	4	4	4

where we find the value of xy by looking in the row labelled x and the column labelled y:

$$y$$
$$\vdots$$
$$x \quad \cdots \quad xy$$

(Tables of this kind are named after Arthur Cayley (1821–1895), though we are in fact using his device in a more general setting than Cayley envisaged.)

A non-empty set S with a binary operation is called a *semigroup* if the operation is *associative,* that is, if (in multiplicative notation) for all a, b, c in S

$$(ab)c = a(bc).$$

If we want to draw particular attention to the binary operation that makes S into a semigroup, we refer to 'the semigroup $(S, .)$'. An important example (one where we do *not* write the operation multiplicatively) is the semigroup $(N, +)$, where N as usual denotes the set of natural numbers. Another example is $(N, .)$, the set of natural numbers under multiplication.

We say that a semigroup $(S, .)$ is a *semigroup with identity*, or *monoid*, if there exists an element 1 in S such that, for all a in S,

$$1a = a1 = a.$$

The element 1 is called the *identity element* of S. There can be at most one such element: if we also have $ea = ae = a$ for every a in S, then

$$e = 1e \quad \text{(since 1 is an identity element)}$$
$$= 1 \quad \text{(since } e \text{ is an identity element).}$$

Example 1.4.1 provides an example of a finite monoid, with identity element 1.

If we want to draw particular attention to the binary operation and also to the identity element 1 that makes S into a monoid, we refer to 'the monoid $(S, ., 1)$'. Thus we have the monoid $(\mathbf{N}, ., 1)$ consisting of the natural numbers under multiplication. Anther important example, not written multiplicatively, is the monoid $(\mathbf{N}^0, +, 0)$.

Let $(S, .)$ be a semigroup, with $|S| \geq 2$. An element 0 of S is called a *zero* element if, for all a in S,

$$0a = a0 = 0.$$

If such an element exists then it is unique, for if we also have $za = az = z$ for every a in S then

$$z = 0z \quad \text{(since } z \text{ is a zero element)}$$
$$= 0 \quad \text{(since 0 is a zero element).}$$

The monoid of Example 1.4.1 does have a zero element, namely 4.

Example 1.4.2. Let $S = \{a, b, c\}$ and let multiplication be given by the table

	a	b	c
a	a	a	a
b	b	b	b
c	c	c	c

Then S is a semigroup. It has no identity element and no zero element.

For subsets A and B of a semigroup S it is frequently useful to define

$$AB = \{ab : a \in A, \ b \in B\}.$$

Notice carefully that the logic of this notation implies that

$$A^2 = \{aa' : a, a' \in A\},$$

and that this may well be different from

$$\{a^2 : a \in A\}.$$

(Notice that we have a potential source of confusion here, since A^2 may also mean the Cartesian product $A \times A$; the context should normally make it clear which of the two meanings applies.) If one or other of the sets A, B is a singleton we simplify the notation: if A, B are subsets of S and a, b are elements of S, then

$$Ab = \{xb : x \in A\}, \quad aB = \{ay : y \in B\}.$$

It is easy to see that the associativity of the semigroup operation implies that

$$(AB)C = A(BC)$$

for all subsets A, B, C of S. Consequently, just as notations such as abc are unambiguous in a semigroup, so we may unambiguously use notations such as AbA, xSy, etc.

A non-empty subset T of a semigroup S is called a *subsemigroup* of S if it is *closed* with respect to the multiplication, i.e., if we have the implication

$$a, b \in T \implies ab \in T.$$

Using the notation of the last paragraph, we can express this implication very succinctly by writing

$$T^2 \subseteq T.$$

If S is a monoid then we say that a subsemigroup T of S is a *submonoid* of S if $1 \in T$. In Example 1.4.2 above, $\{a\}$ and $\{a, b\}$ are both subsemigroups of S. In Example 1.4.1, $\{1, 2, 3\}$ is a submonoid of S.

Notice carefully that it is possible for a subset to be both a subsemigroup and (in its own right) a monoid, but to fail to be a submonoid: in Example 1.4.1 the subset $\{2, 3, 4\}$ is a subsemigroup and a monoid (with identity element 2) but is *not* a submonoid of S, since it does not contain the identity element 1 of S.

A monoid G is called a *group* if

$$(\forall a \in G)(\exists x \in G)\, ax = xa = 1. \tag{1.4.3}$$

The element x is in fact uniquely determined by a: if we also have $ax' = x'a = 1$ then

$$x' = x'1 = x'(ax) = (x'a)x = 1x = x.$$

Accordingly, we always write the element a associated in this way with a as a^{-1}, and call it the *inverse* of a in G. It is defined by the property that

$$aa^{-1} = a^{-1}a = 1.$$

Neither of the semigroups in Examples 1.4.1 and 1.4.2 is a group. For an interesting example of a group, let us consider:

Example 1.4.4. Let $G = \{i, a, b, f, g, h\}$, where i, a, b, f, g and h are functions defined on $\mathbf{R}\backslash\{0,1\}$ by

$$i(x) = x, \quad a(x) = \frac{1}{1-x}, \quad b(x) = 1 - \frac{1}{x},$$

$$f(x) = \frac{1}{x}, \quad g(x) = 1 - x, \quad h(x) = \frac{x}{x-1}.$$

The operation is $\circ$ (composition of functions) and it is a straightforward exercise in applying the rule for composition of functions to obtain the Cayley table

	i	a	b	f	g	h
i	i	a	b	f	g	h
a	a	b	i	h	f	g
b	b	i	a	g	h	f
f	f	g	h	i	a	b
g	g	h	f	b	i	a
h	h	f	g	a	b	i

For example, with $x \neq 0, 1$,

$$(a \circ f)(x) = a(1/x) = 1/\left(1 - \frac{1}{x}\right) = x/(x-1) = h(x),$$

and so $a \circ f = h$;

$$(f \circ g)(x) = f(1-x) = 1/(1-x) = a(x),$$

and so $f \circ g = a$.

We saw earlier (1.2.1) that composition of functions is associative. Hence $(G, \circ)$ is a semigroup. It is clearly a monoid, with i as identity element, and in fact is a group, with

$$i^{-1} = i, \; a^{-1} = b, \; b^{-1} = a, \; f^{-1} = f, \; g^{-1} = g, \; h^{-1} = h.$$

In a semigroup S we are often interested in elements a with the property that $a^2 = a$. Such elements are called *idempotents* and evidently have the property that

$$a = a^2 = a^3 = \cdots.$$

Let e be an idempotent in a semigroup S, and let

$$H_e = \{a \in S : ae = ea = a \text{ and } (\exists a' \in S) \; aa' = a'a = e\}. \qquad (1.4.5)$$

Certainly H_e is non-empty, since $e \in H_e$ (with $e' = e$). In fact we have:

Theorem 1.4.6. *Let e be an idempotent in a semigroup S, and let H_e be defined as above. Then H_e is a subsemigroup of S and is a group with identity element e.*

Proof. If $a, b \in H_e$ then

$$(ab)e = a(be) = ab, \quad e(ab) = (ea)b = ab,$$

and

$$(ab)(b'a') = a(bb')a' = aea' = aa' = e,$$
$$(b'a')(ab) = b'(a'a)b = b'eb = b'b = e.$$

Thus $ab \in H_e$, and so H_e is a subsemigroup of S. To show that H_e is a group, notice first that e is an identity element for H_e. Let $a \in H_e$ and let $a' \; (\in S)$ be such that $aa' = a'a = e$. If we now define $a^{-1} = ea'e$, then from the idempotent property of e it is easy to verify that

$$a^{-1}e = ea^{-1} = a^{-1}, \quad aa^{-1} = a^{-1}a = e.$$

Thus $a^{-1} \in H_e$ and is the required inverse of a in H_e. ∎

A subsemigroup of a semigroup S that happens to be a group in its own right will be called a *subgroup* of S.

In Examples 1.4.1 and 1.4.2 above, every element is idempotent and $H_e = \{e\}$ in all cases. For a more interesting example, see Exercise 1.4. That exercise also illustrates the potentially treacherous point that, while a group is a special case of a monoid, it is perfectly possible for a subgroup of a monoid not to be a submonoid.

Let S be a finite semigroup and let a be an element of S. The elements

$$a, a^2, a^3, \ldots$$

cannot all be distinct, since S is finite. Let a^m ($m \geq 1$) be the first member of the sequence to repeat, and let the first repetition occur at a^{m+r} (with $r \geq 1$). Thus

$$a^m = a^{m+r}, \tag{1.4.7}$$

and the elements $a, a^2, \ldots, a^{m+r-1}$ are all distinct. From (1.4.7) it is easy to deduce that

$$a^m = a^{m+kr}$$

for all $k \geq 0$. If $n \geq m$ we may write

$$n = m + kr + p,$$

with $k \geq 0$ and $0 \leq p \leq r - 1$. Hence

$$a^n = a^{m+kr} a^p = a^{m+p}.$$

The powers of a form a subsemigroup of S, denoted by $\langle a \rangle$, and we have now shown that

$$\langle a \rangle = \{a, a^2, \ldots, a^{m+r-1}\}.$$

The subset $H_a = \{a^m, a^{m+1}, \ldots, a^{m+r-1}\}$ is clearly a subsemigroup of $\langle a \rangle$ (and hence of S). It is in fact a sub*group*. The identity element is a^{m+p}, where

$$\left(a^{m+p}\right)^2 = a^{m+p},$$

i.e., where

$$2m + 2p \equiv m + p \ (\text{mod } r),$$

i.e., where

$$p \equiv -m \ (\text{mod } r).$$

To find the inverse of a typical element a^{m+u} we choose v so that

$$a^{m+u}.a^{m+v} = a^{m+p},$$

i.e., so that

$$v \equiv p - u - m \equiv -u - 2m \ (\text{mod } r).$$

For example, if $m = 2$ and $r = 6$, then

$$\langle a \rangle = \{a, a^2, \ldots, a^7\}$$

and $a^2 = a^8$. The elements of H_a multiply in accordance with the table

	a^2	a^3	a^4	a^5	a^6	a^7
a^2	a^4	a^5	a^6	a^7	a^2	a^3
a^3	a^5	a^6	a^7	a^2	a^3	a^4
a^4	a^6	a^7	a^2	a^3	a^4	a^5
a^5	a^7	a^2	a^3	a^4	a^5	a^6
a^6	a^2	a^3	a^4	a^5	a^6	a^7
a^7	a^3	a^4	a^5	a^6	a^7	a^2

The identity is a^{2+p}, where $p \equiv -2 \pmod 6$, i.e., $p = 4$. Thus a^6 is the identity element. The inverse of (say) $a^5 = a^{2+3}$ is a^{2+v}, where $v \equiv -3 - 4 \pmod 6$, i.e., $v = 5$. Thus a^7 is the inverse of a^5. Similarly, a^2 and a^4 are mutually inverse, while each of a^3 and a^6 is its own inverse.

Perhaps the most important conclusion from our analysis of powers in a finite semigroup is:

Theorem 1.4.8. *Every element in a finite semigroup has a power which is an idempotent.*

Proof. For each such element a the identity of the group H_a is the required idempotent. ■

We end this section by mentioning briefly the idea of *generators* in a semigroup. If S is a semigroup and X is a subset of S then we write $\langle X \rangle$ for the smallest subsemigroup of S containing X, and we refer to $\langle X \rangle$ as the subsemigroup of S *generated by* X. If $\langle X \rangle = S$ we say that X is a *set of generators* for S, or that S is *generated by* X. If $U = \{u_1, u_2, \ldots, u_k\}$ then we write $\langle u_1, u_2, \ldots, u_k \rangle$ rather than $\langle \{u_1, u_2, \ldots, u_k\} \rangle$.

Example 1.4.9. In the group G considered in Example 1.4.4, $\langle a, f \rangle = G$, since

$$a \circ a = b, \quad f \circ f = i, \quad a \circ f = h, \quad f \circ a = g.$$

The generating set is not unique: for example, we also have $\langle f, g \rangle = G$, since

$$f \circ g = a, \quad g \circ f = b, f \circ f = i, \quad f \circ g \circ f = h.$$

1.5 Morphisms and congruences

Let S and T be semigroups. A map $\varphi : S \to T$ is called a *morphism* (or a *homomorphism*) if, for all a, b in S,

$$\varphi(ab) = \varphi(a)\varphi(b).$$

A slightly different definition applies to monoids: if S and T are monoids, with identity elements 1_S, 1_T respectively, then $\varphi : S \to T$ is called a *morphism* if, for all a, b in S,

$$\varphi(ab) = \varphi(a)\varphi(b)$$

and

$$\varphi(1_S) = 1_T.$$

To emphasize the distinction between the two concepts of morphism, we shall sometimes refer to *semigroup morphisms* and to *monoid morphisms*.

A morphism $\alpha : S \to T$ which is both injective and surjective is called an *isomorphism*. Since it is bijective there is an inverse map $\alpha^{-1} : T \to S$, which is also bijective. (See Theorem 1.2.4.) In fact α^{-1} is an isomorphism, since, for all x, y in T,

$$\alpha^{-1}(xy) = \alpha^{-1}\big(\alpha(a)\alpha(b)\big), \text{ where } a = \alpha^{-1}(x), \ b = \alpha^{-1}(y),$$
$$= \alpha^{-1}\big(\alpha(ab)\big) = ab = \alpha^{-1}(x)\alpha^{-1}(y);$$

and, if we are dealing with monoids

$$\alpha^{-1}(1_T) = 1_S.$$

We shall sometimes want to say that S is *isomorphic* to T, and to write $S \simeq T$, without specifying the precise isomorphism $\alpha : S \to T$. It is easy to check that the relation $\simeq$ is an equivalence.

An equivalence relation ρ on a semigroup or monoid S is called a *congruence* on S if it has the property that

$$(a, b) \in \rho \text{ and } c \in S \ \Rightarrow \ (ca, cb) \in \rho, \ (ac, bc) \in \rho. \tag{1.5.1}$$

Alternatively, the equivalence ρ is a *congruence* if

$$(a, b), \ (c, d) \ \in \ \rho \ \Rightarrow \ (ac, bd) \in \rho. \tag{1.5.2}$$

It is a routine matter to show that the two definitions are equivalent. (See Exercise 1.9.)

Let ρ be a congruence on a semigroup or monoid S, and consider the quotient set S/ρ, consisting of all the ρ-classes $a\rho$ ($a \in S$). The congruence property enables us to define a binary operation on S/ρ by the rule that

$$(a\rho)(b\rho) = (ab)\rho. \tag{1.5.3}$$

This definition looks innocent enough, but in fact by writing a ρ-class as $a\rho$ we are arbitrarily choosing a representative of the ρ-class, and it is conceivable that the ρ-class obtained on the right-hand side depends on the choices we make for representatives of the two ρ-classes on the left.

So suppose that we now choose different representatives a' and b' of the ρ-classes on the left, obtaining $(a'b')\rho$ on the right. Then $a \rho a'$ and $b \rho b'$, and so, by (1.5.2), $ab \rho a'b'$. Thus

$$(a'b')\rho = (ab)\rho.$$

We have shown that the ρ-class on the right is independent of the choices we make on the left. Thus (1.5.3) does indeed define a binary operation on the set S/ρ. We say that the operation given by (1.5.3) is 'well-defined'. It is certainly associative, since, for all $a\rho$, $b\rho$, $c\rho$ in S/ρ,

$$[(a\rho)(b\rho)](c\rho) = [(ab)\rho](c\rho) = [(ab)c]\rho$$
$$= [a(bc)]\rho = (a\rho)[(bc)\rho] = (a\rho)[(b\rho)(c\rho)].$$

Thus S/ρ is a semigroup, and if S is a monoid then so is S/ρ, since

$$(1\rho)(a\rho) = (1a)\rho = a\rho = (a1)\rho = (a\rho)(1\rho)$$

for every a in S.

The map $\rho^\natural : S \to S/\rho$ defined by

$$\rho^\natural(a) = a\rho \quad (a \in S) \tag{1.5.4}$$

is a surjective morphism, and its kernel equivalence — see the final paragraph of Section 1.2 — is the congruence ρ:

$$\ker \rho^\natural = \rho. \tag{1.5.5}$$

We have seen that the consideration of a congruence ρ on a semigroup S leads in a natural way to a morphism $\rho^\natural$. There is an equally natural connection in the other direction: if $\varphi : S \to T$ is a morphism, then $\ker \varphi$ is a congruence on S; for

$$(a,b), \ (c,d) \ \in \ \ker \varphi$$
$$\Rightarrow \varphi(a) = \varphi(b), \ \varphi(c) = \varphi(d)$$
$$\Rightarrow \varphi(ac) = \varphi(a)\varphi(c) = \varphi(b)\varphi(d) = \varphi(bd)$$
$$\Rightarrow (ac, bd) \ \in \ \ker \varphi.$$

The following result, stated for monoids, in fact holds very generally in abstract algebra, and is a cornerstone of the subject:

Theorem 1.5.6. *Let S, T be monoids, let $\varphi : S \to T$ be a morphism, and let $\rho = \ker \varphi$. Then there exists an injective morphism $\alpha : S/\rho \to T$ such that* $\operatorname{im} \alpha = \operatorname{im} \varphi$ *and such that the diagram*

commutes.

Proof. We have not used the notion of a commutative diagram before, but its meaning is clear enough. By saying that the given diagram commutes, we are merely saying that the two routes from S to T amount to the same thing, i.e., that $\alpha \circ \rho^\natural = \varphi$.

To prove the result, define $\alpha : S/\rho \to T$ by

$$\alpha(s\rho) = \varphi(s) \quad (s\rho \in S/\rho).$$

If we choose a different representative s' of the ρ-class $s\rho$, then

$$(s, s') \in \rho = \ker \varphi$$

and so

$$\alpha(s'\rho) = \varphi(s') = \varphi(s) = \alpha(s\rho).$$

Thus the map α is well-defined. It is injective, since, for all $x\rho$, $y\rho$ in S/ρ,

$$\begin{aligned}
\alpha(x\rho) = \alpha(y\rho) &\Rightarrow \varphi(x) = \varphi(y) \\
&\Rightarrow (x, y) \in \ker \varphi = \rho \\
&\Rightarrow x\rho = y\rho.
\end{aligned}$$

It is a morphism, since, for all $x\rho$, $y\rho$ in S/ρ,

$$\begin{aligned}
\alpha\big((x\rho)(y\rho)\big) = \alpha\big((xy)\rho\big) &= \varphi(xy) \\
&= \varphi(x)\varphi(y) = \alpha(x\rho)\alpha(y\rho),
\end{aligned}$$

and

$$\alpha(1_S\rho) = \varphi(1_S) = 1_T.$$

It is clear that $\operatorname{im} \alpha = \operatorname{im} \varphi$, and we complete the proof by observing that for all s in S

$$(\alpha \circ \rho^\natural)(s) = \alpha(s\rho) = \varphi(s). \qquad \blacksquare$$

Next, we have a theorem dealing with a slightly more general situation:

Theorem 1.5.7. *Let S, T, U be monoids and let $\varphi : S \to T$, $\psi : S \to U$ be morphisms such that φ is surjective and $\ker \varphi \subseteq \ker \psi$. Then there is a morphism $\beta : T \to U$ such that $\operatorname{im} \beta = \operatorname{im} \psi$ and the diagram*

commutes.

Proof. Let $t \in T$. Since φ is surjective, there exists s in S (not in general unique) such that $\varphi(s) = t$. Define $\beta(t)$ to be $\psi(s)$. The value of $\beta(t)$ is in fact independent of the choice of s, since if $\varphi(s') = \varphi(s) = t$ then

$$(s', s) \in \ker \varphi \subseteq \ker \psi,$$

and so $\psi(s') = \psi(s)$. So we have a well-defined map $\beta : T \to U$. To show that β is a morphism, let t_1, $t_2 \in T$ and choose s_1, s_2 in S so that $\varphi(s_1) = t_1$, $\varphi(s_2) = t_2$. Then

$$\varphi(s_1 s_2) = \varphi(s_1)\varphi(s_2) = t_1 t_2$$

and so

$$\beta(t_1 t_2) = \psi(s_1 s_2) = \psi(s_1)\psi(s_2) = \beta(t_1)\beta(t_2).$$

Also, $\varphi(1_S) = 1_T$, and so

$$\beta(1_T) = \psi(1_S) = 1_U.$$

It is clear that $\operatorname{im} \beta \subseteq \operatorname{im} \psi$. To show the opposite inclusion, notice that if $u = \psi(s) \in \operatorname{im} \psi$ then $u = \beta\big(\varphi(s)\big) \in \operatorname{im} \beta$. Hence $\operatorname{im} \beta = \operatorname{im} \psi$.

Finally, since the definition of β makes it clear that

$$\beta\big(\varphi(s)\big) = \psi(s)$$

for all s in S, the given diagram commutes. ∎

Next, we have what amounts to a different version of the same result:

Theorem 1.5.8. *Let ρ amd σ be congruences on a monoid S such that $\rho \subseteq \sigma$. Then on the monoid S/ρ there is a congruence, denoted by σ/ρ, with the property that*

$$S/\sigma \simeq (S/\rho)/(\sigma/\rho).$$

Proof. The natural surjective morphisms

$$\rho^\natural : S \to S/\rho, \quad \sigma^\natural : S \to S/\sigma$$

are such that

$$\ker \rho^\natural = \rho, \quad \ker \sigma^\natural = \sigma.$$

Hence, by Theorem 1.5.7 there is a surjective morphism $\beta : S/\rho \to S/\sigma$ such that $\beta \circ \rho^\natural = \sigma^\natural$. The kernel equivalence of β is a congruence, which we may denote by σ/ρ, on S/ρ. By Theorem 1.5.6 there is an isomorphism

$$\alpha : (S/\rho)/(\sigma/\rho) \to S/\sigma. \qquad \blacksquare$$

Notice in passing that the congruence σ/ρ is given by the formula

$$\sigma/\rho = \{(x\rho, y\rho) : (x, y) \in \sigma\}. \tag{1.5.9}$$

We end this section with two further ideas that will prove useful later. If A is a subset of a monoid S and ρ is a congruence on S, we say that ρ *respects* A if A is a union of ρ-classes. Equivalently, ρ *respects* A if, for all (x, y) in ρ, either both x and y are in A or both x and y are outside A.

Example 1.5.10. Consider the monoid $(\mathbf{N}^0, +, 0)$ and the congruence ρ on $\mathbf{N}^0$ given by the rule that $x \rho y$ if and only if $x \equiv y \pmod 4$. Then ρ respects the set

$$2\mathbf{N}^0 = \{0, 2, 4, 6, \ldots\},$$

since $2\mathbf{N}^0$ is the union of the two ρ-classes 0ρ and 2ρ. On the other hand, ρ does not respect the set

$$3\mathbf{N}^0 = \{0, 3, 6, 9, \ldots\},$$

since, for example,

$$(3, 7) \in \rho, \quad 3 \in 3\mathbf{N}^0, \quad 7 \notin 3\mathbf{N}^0.$$

Finally, if S and T are monoids, we say that S *divides* T, and write $S \mid T$, if there exists a submonoid U of T and a surjective morphism $\varphi : U \to S$. In other word, S *divides* T if S is a *morphic image of a submonoid* of T. This will turn out to be an important notion in Chapter 3.

1.6 Transformation monoids

If Q is a finite set and $(S, ., 1)$ is a monoid, we say that S *acts on* Q if there is a map $\alpha : Q \times S \to Q$ with the properties

$$\alpha(q, 1) = q \quad (q \in Q), \tag{1.6.1}$$
$$\alpha(q, st) = \alpha\big(\alpha(q, s), t\big) \quad (q \in Q, \ s, t \in S). \tag{1.6.2}$$

It is usual to write $\alpha(q, s)$ more compactly as qs, whereupon the rules (1.6.1) and (1.6.2) take the more memorable form

$$q1 = q \quad (q \in Q), \tag{1.6.3}$$
$$q(st) = (qs)t \quad (q \in Q, \ s, t \in S). \tag{1.6.4}$$

Such 'actions' are familiar in many parts of mathematics. For example, if we think of the elements of Q as 'vectors' and those of S as 'scalars', then what we have is something quite like multiplication of a vector by a scalar in a vector space, though here we have a lot less structure than in linear algebra.

We now define a relation τ on S by

$$\tau = \{(s, s') \in S \times S : (\forall q \in Q) \, qs = qs'\}. \tag{1.6.5}$$

Then we have:

Theorem 1.6.6. *Let Q be a finite set acted upon by a monoid S, and let τ be the relation defined by (1.6.5). Then τ is a congruence on S.*

Proof. It is clear that τ is an equivalence. If $a \, \tau \, b$ and $c \in S$ then, for all q in Q,

$$q(ac) = (qa)c = (qb)c = q(bc)$$

and so $ac \, \tau \, bc$. Also,

$$q(ca) = (qc)a = (qc)b = q(cb)$$

for all q in Q, since $a \, \tau \, b$ implies in particular that $q'a = q'b$ for all elements q' of the form qc. Thus $ca \, \tau \, cb$. ∎

As before, let Q be a finite set, acted upon by a monoid S. For each element s of S we have a map $\rho_s : Q \to Q$ defined by

$$\rho_s(q) = qs \quad (q \in Q).$$

One useful way of regarding the congruence τ is to remark that $a \, \tau \, b$ if and only if the two maps ρ_a and ρ_b are equal.

If $\tau = \Delta$, the diagonal relation on S described by (1.3.4), we say that the pair (Q, S) is a *transformation monoid*. Thus, in a transformation monoid (Q, S),

$$(\forall\, q \in Q)\; qs_1 = qs_2 \quad \Rightarrow \quad s_1 = s_2. \tag{1.6.7}$$

Here the elements s of S are in one-one correspondence with the maps ρ_s, and one often chooses to regard the elements s as *being* maps from Q to Q. One effect of this restriction is that the finiteness of Q forces the finiteness of S, for $|S|$ cannot be greater than the total number of distinct maps from Q into Q. Thus, if $|Q| = n$, we must have $|S| \le n^n$.

Two transformation monoids (Q, S), (R, T) are said to be *isomorphic* if there exists a bijection $\beta : Q \to R$ and an isomorphism $\theta : S \to T$ such that, for all q in Q and s in S,

$$\beta(qs) = \beta(q)\theta(s).$$

We write $(Q, S) \simeq (R, T)$, or that

$$(\beta, \theta) : (Q, S) \to (R, T)$$

is an isomorphism.

Notice that $(Q, S) \simeq (R, T)$ is a stronger statement than $S \simeq T$. For example, let

$$Q = \{1, 2\}, \quad R = \{1, 2, 3, 4\}, \quad S = \{1_S, s\}, \quad T = \{1_T, t\},$$

with $s^2 = 1_S$, $t^2 = 1_T$. Let

$$1s = 2,\; 2s = 1; \quad 1t = 2,\; 2t = 1,\; 3t = 4,\; 4t = 3.$$

Then (Q, S) and (R, T) are transformation monoids, and $S \simeq T$. However, we cannot have $(Q, S) \simeq (R, T)$, since there can be no bijection between Q and R.

1.7 Free semigroups and monoids

We shall often want to refer to a finite non-empty set A as an *alphabet*. If A is an alphabet, let A^+ consist of all finite sequences

$$(a_1, a_2, \ldots, a_l) \quad (l \ge 1),$$

where $a_1, a_2 \ldots, a_l \in A$. Thus, for example, if $A = \{a, b\}$, the set A^+, which is infinite, consists of

$$(a), (b); \quad (a, a), (a, b), (b, a), (b, b);$$
$$(a, a, a), (a, a, b), (a, b, a), (a, b, b),$$
$$(b, a, a), (b, a, b), (b, b, a), (b, b, b);$$
$$(a, a, a, a), \ldots.$$

In general, if $|A| = n$, there are n^l sequences of length l.

The set A^+ becomes a semigroup if we define

$$(a_1, a_2, \ldots, a_l)(b_1, b_2, \ldots, b_m) = (a_1, a_2, \ldots, a_l, b_1, b_2, \ldots, b_m). \qquad (1.7.1)$$

It is trivial to verify that this is an associative operation. The sequences of length 1 *generate* the semigroup A^+, since every element $(a_1, a_2, \ldots a_l)$ is a finite product $(a_1)(a_2) \ldots (a_l)$ of sequences of length 1. If we now decide, very reasonably, that there is no very useful distinction to be made between an element a of A and the sequence (a) of length 1, we may regard the elements of A^+ as being *words* $a_1 a_2 \ldots a_l$ in the alphabet A. The multiplication given by (1.7.1) then corresponds to simple juxtaposition:

$$(a_1 a_2 \ldots a_l)(b_1 b_2 \ldots b_m) = a_1 a_2 \ldots a_l b_1 b_2 \ldots b_m, \qquad (1.7.2)$$

and the list of elements of $\{a, b\}^+$ may be rewritten as

$$a, b; \quad a^2, ab, ba, b^2;$$
$$a^3, a^2 b, aba, ab^2, ba^2, bab, b^2 a, b^3;$$
$$a^4, \ldots$$

The semigroup A^+ will play an important part in the chapters that follow. It is called the *free semigroup on the set* (or *alphabet*) A.

Frequently we wish to adjoin an identity element 1 to A^+ to form A^*, the *free monoid on the set* A. We may interpret 1 as the *empty sequence* or the *empty word* (with no letters at all) in the alphabet A. Its length is defined as 0.

We shall almost always want to identify the element a with the sequence (a) and to regard A as a subset of A^+. Strictly speaking, there is an injective map $i : A \to A^+$ given by

$$i(a) = (a) \quad (a \in A).$$

It may be worth emphasizing that two words in A^* are equal if and only if they are identical. Thus

$$a_1 a_2 \ldots a_l = b_1 b_2 \ldots b_m$$

if and only if $m = l$ and $a_i = b_i$ $(i = 1, 2, \ldots, l)$.

Algebraically, the importance of the free monoid is given by the following result:

Theorem 1.7.3. *Let A be an alphabet, let $(S, ., 1)$ be a monoid and let $\theta : A \to S$ be a map. Then there is a unique morphism $\varphi : A^* \to S$ such that the diagram*

$$A \overset{\theta}{\longrightarrow} S$$

with maps $i : A \to A^*$ and $\varphi : A^* \to S$

commutes.

Proof. Define $\varphi(1) = 1$ and

$$\varphi(a_1 a_2 \dots a_l) = \theta(a_1)\theta(a_2)\dots\theta(a_l).$$

Then, for all elements $a_1 a_2 \dots a_l, b_1 b_2 \dots b_m$ in A^*,

$$\begin{aligned}
\varphi\big((a_1 a_2 \dots a_l)(b_1 b_2 \dots b_m)\big) &= \varphi(a_1 a_2 \dots a_l b_1 b_2 \dots b_m) \\
&= \theta(a_1)\theta(a_2)\dots\theta(a_l)\theta(b_1)\theta(b_2)\dots\theta(b_m) \\
&= \big(\theta(a_1)\theta(a_2)\dots\theta(a_l)\big)\big(\theta(b_1)\theta(b_2)\dots\theta(b_m)\big) \\
&= \varphi(a_1 a_2 \dots a_l)\varphi(b_1 b_2 \dots b_m),
\end{aligned}$$

and so φ is a morphism. Clearly

$$\varphi\big(i(a)\big) = \theta(a)$$

for every a in A, and so the diagram commutes. The uniqueness of φ is also clear, for if φ' is another morphism satisfying $\varphi'\big(i(a)\big) = \theta(a)$ for all a in A then, for all $a_1 a_2 \dots a_l$ in A^+

$$\begin{aligned}
\varphi'(a_1 a_2 \dots a_l) &= \varphi'\big(i(a_1)i(a_2)\dots i(a_l)\big) \\
&= \varphi'\big(i(a_1)\big)\varphi'\big(i(a_2)\big)\dots\varphi'\big(i(a_l)\big) \\
&= \theta(a_1)\theta(a_2)\dots\theta(a_l) = \varphi(a_1 a_2 \dots a_l).
\end{aligned}$$

Since we also have (by definition of monoid morphisms)

$$\varphi(1) = \varphi'(1) = 1,$$

we thus have $\varphi' = \varphi$. ∎

If we regard A as a subset of A^*, then the commuting diagram condition means that φ coincides with θ on the set A. We say that φ *extends* θ, or that the *restriction* $\varphi|_A$ of φ to A is θ.

An analogous result holds for semigroups, with A^+ replacing A^*.

As a consequence of Theorem 1.7.3 we have the following result for monoids. (A similar theorem can be proved for semigroups.)

Theorem 1.7.4. *Let S be a finite monoid. Then there exists a finite alphabet A and a congruence ρ on A^* such that $A^*/\rho \simeq S$.*

Proof. Let A be a set of generators of S. (If all else fails we can take $A = S$.) Certainly there is a one-one map (the inclusion map) from A into S, and so there is a morphism $\varphi : A^* \to S$ coinciding with the inclusion map on A. The morphim φ is surjective, since every s in S is a product $a_1 a_2 \ldots a_l$ of elements of A, and so is the image under φ of the word $a_1 a_2 \ldots a_l$ in A^*. Defining ρ as the kernel congruence of φ, we now see by Theorem 1.5.6 that $A^*/\rho \simeq S$. ∎

From a combinatorial point of view, the importance of A^* and A^+ lies in the idea of *language,* which in its most general sense is just a subset of some free monoid A^*. The idea of a language as simply a set of words is of course a very primitive one, and in Chapters 2, 3, 4 and 5 we shall be specializing the notion in various ways. At this stage it will be useful to introduce some standard notations.

First, if $A = \{a\}$, a singleton set, we write a^* and a^+ rather than $\{a\}^*$ and $\{a\}^+$. Thus

$$a^* = \{1, a, a^2, \ldots\}, \quad a^+ = \{a, a^2, a^3, \ldots\}.$$

If $w \in A^*$ we shall write $|w|$ for the *length* of w: thus

$$|a_1 a_2 \ldots a_l| = l$$

and, as remarked earlier, $-1- = 0$. Notice that if $w, z \in A^*$ then

$$|wz| = |w| + |z|.$$

If $w \in A^*$ and $a \in A$ then $|w|_a$ denotes the number of occurrences of a in w. Notice that

$$|w| = \sum_{a \in A} |w|_a \quad (w \in A^*),$$

$$|wz|_a = |w|_a + |z|_a \quad (w, z \in A^*, \ a \in A).$$

If $w \in A^*$ then u is called a *left factor* of w if $w = uv$ for some v in A^*. If $w \in A^+$ then u is called a *proper left factor* of w if $u \neq 1$ and there exists v in A^+ such that $w = uv$.

The *content* $C(w)$ of w is the set of letters occurring in w. Thus

$$C(w) = \{a \in A : |w|_a > 0\}.$$

Notice that $C(1) = \emptyset$.

We end this chapter with two important properties of free monoids. A monoid S is called *cancellative* if (for all a, b, c in S)

$$ca = cb \ \Rightarrow \ a = b, \qquad ac = bc \ \Rightarrow \ a = b.$$

Theorem 1.7.5. *For every alphabet A the monoid A^* is cancellative.*

Proof. This almost obvious. If we write

$$a = a_1 a_2 \ldots a_l, \quad b = b_1 b_2 \ldots b_m, \quad c = c_1 c_2 \ldots c_n,$$

then $ca = cb$ means that the words

$$c_1 c_2 \ldots c_n a_1 a_2 \ldots a_l, \quad c_1 c_2 \ldots c_n b_1 b_2 \ldots b_m$$

are identical. It follows that the words $a_1 a_2 \ldots a_l$ and $b_1 b_2 \ldots b_m$ are identical, i.e., that $a = b$. In a similar way one shows that

$$ac = bc \Rightarrow a = b. \qquad \blacksquare$$

A monoid S is called *equidivisible* if (for all x, y, z, t in S) $xy = zt$ implies either that $x = zu$ and $t = uy$ for some u in S, or $z = xv$ and $y = vt$ for some v in S.

Theorem 1.7.6. *For every alphabet A the monoid A^* is equidivisible.*

Proof. This too is almost obvious. If $xy = zt = a_1 a_2 \ldots a_l$ then x and z are left factors of $a_1 a_2 \ldots a_l$. That is,

$$x = a_1 a_2 \ldots a_i, \quad z = a_1 a_2 \ldots a_j,$$

with $0 \le i \le l$, $0 \le j \le l$. If $i \ge j$ then $x = zu$, with $u = a_{j+1} \ldots a_i$, and then

$$t = a_{j+1} \ldots a_l = (a_{j+1} \ldots a_i)(a_{i+1} \ldots a_l) = uy.$$

If $i \le j$ then $z = xv$, with $v = a_{i+1} \ldots a_j$, and

$$y = a_{i+1} \ldots a_l = (a_{i+1} \ldots a_j)(a_{j+1} \ldots a_l) = vt. \qquad \blacksquare$$

Exercises 1

1.1. The laws of Boolean algebra are as follows. (Here A, B and C are subsets of a fixed 'universal' set S, and A' denotes the complement of A in S.)
   1. $(A \cup B) \cup C = A \cup (B \cup C)$, $A \cap B) \cap C = A \cap (B \cap C)$;
   2. $A \cup B = B \cup A$, $A \cap B = B \cap A$;
   3. $A \cup A = A$, $A \cap A = A$;
   4. $A \cup (B \cap C) = (A \cup B) \cap (A \cup C)$,
      $A \cap (B \cup C) = (A \cap B) \cup (A \cap C)$;

5. $A \cup \emptyset = A, \quad A \cap S = A$;
6. $A \cap \emptyset = \emptyset, \quad A \cup S = S$;
7. $(A')' = A$;
8. $A \cup A' = S, \quad A \cap A' = \emptyset$;
9. $(A \cup B)' = A' \cap B', \quad (A \cap B)' = A' \cup B'$.

Define $A + B$ by the formula

$$A + B = (A \cap B') \cup (A' \cap B).$$

Using the laws above, show that
  (i) $A + B = (A \cup B) \cap (A \cap B)'$;
  (ii) $A \cap (B + C) = (A \cap B) + (A \cap C)$;
  (iii) $(A + B) + C = A + (B + C)$.

1.2. It is sometimes useful to write an infinite set in the form of a *list*

$$a_1, \ a_2, \ a_3, \ \dots.$$

It is not always possible to do this: for example, there is no way of listing the set **R** of real numbers in this way. For an example where it is possible, consider the set **Z** of integers, which can be listed

$$0, \ 1, \ -1, \ 2, \ -2, \ 3, \ -3, \ \dots.$$

(i) Write the set $\{x \in \mathbf{Q} : 0 < q \le 1\}$ as a list. [Hint: for $n = 1, 2, 3, \dots$ write down the fractions with denominator n.]

(ii) Write the set $\mathcal{F}$ of all finite subsets of **N** as a list. [Hint: for $n = 1, 2, 3, \dots$ write down the subsets $\{a_1, a_2, \dots, a_r\}$ for which $a_1 + a_2 + \cdots + a_r = n$.]

1.3. Let ρ and σ be equivalence relations on a set X, and recall that $\rho \subseteq \sigma$ means that

$$x \, \rho \, y \ \Rightarrow \ x \, \sigma \, y.$$

(i) Show that $\rho \subseteq \sigma$ if and only if every σ-class is a union of ρ-classes.

(ii) Illustrate this in the case where $X = \mathbf{N}$ and where ρ, σ are defined by

$$a \, \rho \, b \text{ if and only if } a \equiv b \pmod 6,$$
$$a \, \sigma \, b \text{ if and only if } a \equiv b \pmod 3.$$

(iii) Let X be a finite set and let ρ, σ be equivalences on X such that $\rho \subseteq \sigma$. Show that $\rho \subset \sigma$ if and only if $|X/\rho| > |X/\sigma|$.

1.4. If $a_1, a_2, \ldots, a_n \in \{1, 2, \ldots, n\}$, let us write

$$\begin{pmatrix} 1 & 2 & \ldots & n \\ a_1 & a_2 & \ldots & a_n \end{pmatrix}$$

for the map $f : \{1, 2, \ldots, n\} \to \{1, 2, \ldots, n\}$ defined by

$$f(1) = a_1, \ f(2) = a_2, \ldots, f(a_n) = a_n.$$

Write down a Cayley table for the set $S = \{e, a, f, b\}$, where

$$e = \begin{pmatrix} 1 & 2 & 3 & 4 \\ 1 & 2 & 3 & 4 \end{pmatrix} \quad a = \begin{pmatrix} 1 & 2 & 3 & 4 \\ 2 & 1 & 4 & 3 \end{pmatrix},$$
$$f = \begin{pmatrix} 1 & 2 & 3 & 4 \\ 3 & 4 & 3 & 4 \end{pmatrix} \quad b = \begin{pmatrix} 1 & 2 & 3 & 4 \\ 4 & 3 & 4 & 3 \end{pmatrix}.$$

Show that S is a monoid. Show also that S has a subgroup which is not a submonoid.

1.5. Write down a Cayley table for $\{0, e, f, a, b\}$, where 0, e, f, a, b are the maps from $\{0, 1, 2\}$ into itself given by

$$0 = \begin{pmatrix} 0 & 1 & 2 \\ 0 & 0 & 0 \end{pmatrix}, \quad e = \begin{pmatrix} 0 & 1 & 2 \\ 0 & 1 & 0 \end{pmatrix}, \quad f = \begin{pmatrix} 0 & 1 & 2 \\ 0 & 0 & 2 \end{pmatrix},$$
$$a = \begin{pmatrix} 0 & 1 & 2 \\ 0 & 2 & 0 \end{pmatrix}, \quad b = \begin{pmatrix} 0 & 1 & 2 \\ 0 & 0 & 1 \end{pmatrix}.$$

Verify that all the subgroups of this semigroup are trivial.

1.6. A *permutation* σ of $\{1, 2, \ldots, n\}$ is defined as a bijection

$$\sigma : \{1, 2, \ldots, n\} \to \{1, 2, \ldots, n\}.$$

The set of all permutations of $\{1, 2, \ldots, n\}$ forms a group S_n under composition. This group, called the *symmetric group of degree n*, is of order $n!$.

In elementary books on group theory it is shown that every permutation is expressible as a composition of 'disjoint cycles'. (A *cycle*, written $(a_1 a_2 \ldots a_k)$, where $a_1, a_2, \ldots, a_k$ are distinct elements of $\{1, 2, \ldots, n\}$, is a map φ defined by

$$\varphi(a_i) = a_{i+1} \quad (i = 1, 2, \ldots, k-1), \quad \varphi(a_k) = a_1,$$
$$\varphi(x) = x \quad (x \notin \{a_1, a_2, \ldots, a_k\}),$$

and two cycles $(a_1 a_2 \ldots a_k)$ and $(b_1 b_2 \ldots b_l)$ are said to be *disjoint* if the sets $\{a_1, a_2, \ldots, a_k\}$ and $\{b_1, b_2, \ldots, b_l\}$ are disjoint.) A cycle $(a_1 a_2)$ of length 2 is called a *transposition*. To avoid trivialities, suppose that $n \geq 3$.

(i) Show that if $k \geq 3$ then

$$(a_1 a_2 \ldots a_k) = (a_1 a_k) \circ (a_1 a_{k-1}) \circ \cdots \circ (a_1 a_2),$$

and deduce that the group S_n is generated by the set of all transpositions.

(ii) Consider the cycles

$$\tau = (12), \quad \zeta = (12 \ldots n).$$

Show that

$$\zeta^{-1} = \zeta^{n-1}.$$

Show that

$$\zeta \circ \tau \circ \zeta^{-1} = (23),$$

and more generally that

$$\zeta^{i-1} \circ \tau \circ \zeta^{-i+1} = (i \ i+1) \quad (i = 1, 2, \ldots, n-1).$$

Next, show that, for $j = 2, 3, \ldots, n-1$,

$$(j \ j+1) \circ (j-1 \ j) \circ \cdots \circ (23) \circ (12) \circ (23) \circ \cdots \circ (j \ j+1) = (1 \ j+1),$$

and that, for $i = 1, 2, \ldots, n-1$ and $j = 1, 2, \ldots, n-i$,

$$\zeta^{i-1} \circ (1 \ j+1) \circ \zeta^{-i+1} = (i \ i+j).$$

(iii) Deduce that $S_n = \langle \tau, \zeta \rangle$.

1.7. The set of all maps from $\{1, 2, \ldots, n\}$ into itself forms a monoid T_n under composition. It is of order n^n and includes the symmetric group S_n as a subgroup. It is called the *full transformation semigroup* (or the *symmetric semigroup*) of degree n.

Let $n \geq 3$. Let π denote the element of T_n given by

$$\pi(1) = 2, \quad \pi(x) = x \quad (x = 2, 3, \ldots, n),$$

and let ζ, τ be the permutations defined in the previous exercise. For $i \neq j$ in $\{1, 2, \ldots, n\}$, let $\|ij\|$ denote the map φ for which

$$\varphi(i) = j, \quad \varphi(x) = x \quad (x \neq i).$$

Thus, in particular, $\pi = \|12\|$.

(i) Prove the identities

$$(1i) \circ \|12\| \circ (1i) = \|i2\| \quad (i \geq 3),$$
$$(2j) \circ \|12\| \circ (2j) = \|1j\| \quad (j \geq 3),$$
$$(1i) \circ (2j) \circ \|ij\| \circ (2j) \circ (1i) = \|ij\| \quad (i, j \geq 3, \; i \neq j),$$
$$(ij)\|ij\|(ij) = \|ji\| \quad (i, j \geq 1, \; i \neq j).$$

(ii) Let $\varphi \in T_n$ and let $|\operatorname{im}\varphi| = r \leq n - 1$. Let i, j (with $i \neq j$) be such that $\varphi(i) = \varphi(j)$, and let

$$z \in \{1, 2, \ldots, n\} \setminus \operatorname{im}\varphi.$$

Show that

$$\varphi = \hat{\varphi} \circ \|ij\|,$$

where

$$\hat{\varphi}(i) = z, \quad \hat{\varphi}(k) = \varphi(k) \quad (k \neq i).$$

(iii) Deduce that $T_n = \langle \zeta, \tau, \pi \rangle$.

1.8. Show that every submonoid of a finite group is a subgroup.

1.9. Two possible definitions of *congruence* are given in (1.5.1) and (1.5.2). Show that they are equivalent.

1.10. Let F be the free monoid $\{1, a, a^2, \ldots\}$ on one generator, and let U be the submonoid $f \setminus \{a\}$. That is,

$$U = \{1, a^2, a^3, \ldots\}.$$

Show that U is not equidivisible, and deduce by Theorem 1.7.6 that it is not free.

1.11. (i) Let $S = A^+$, a free semigroup. Show that $A = S \setminus S^2$.

(ii) Let $F = \{a, b\}^+$, a free semigroup, and let U be the subsemigroup of F generated by $\{ab, ba, aba, bab\}$. Show that

$$U \setminus U^2 = \{ab, ba, aba, bab\}.$$

Deduce that U is not a free semigroup.

1.12. Let S be a non-empty subset of a free semigroup A^+. Show that

$$\bigcap_{n=1}^{\infty} S^n = \emptyset.$$

1.13. Let S be the free monoid $\{a, b\}^*$ Show that the relation ρ on S defined by the rule that

$$w \,\rho\, z \text{ if and only if } w = z \text{ or } |w|, |z| \geq 3$$

is a congruence on S. List the 8 elements of S/ρ, and draw up a Cayley table for the quotient monoid.

1.14. Let A^* be a free monoid, and let $a \in A$. Let λ, φ_a be defined by

$$\lambda(w) = |w|, \quad \varphi_a(w) = a^{|w|_a} \quad (w \in A^*).$$

Show that $\lambda : A^* \to (\mathbf{N}^0, +)$ and $\varphi_a : A^* \to a^*$ are surjective morphisms.

2 Automata

Introduction

The idea of an automaton has its origins in the scientific determinism of the seventeenth century enlightenment. Increasingly exact understanding of the physical world seemed to give the universe the characteristics of a machine, and as early as Descartes (1596–1650) there was speculation that the behaviour of animals, and even of man himself, could be described in mathematical terms.

Arguments about determinism and freedom of the will in respect of human behaviour have raged for many centuries and will probably rage for many more. But as far as I know there has never been a claim that computers possess freedom of the will, and so while there may be some doubt as to the legitimacy of mathematical machine theories in describing the complex subtleties of human brain functions, there is general agreement that these theories help greatly not only in the design and operation of computers but also in understanding the limitations of a computational approach.

One of the most important theoretical machines is the *Turing machine,* so named in honour of Alan Turing (1912–1954). To give an adequate treatment of Turing machines would require not a chapter but a whole book — for a highly readable account see M. Davis (1958). In Chapter 6 we do give a brief account of some aspects of Turing machine theory, but for the next two chapters we shall be exploring a much more primitive type of machine called a 'finite state automaton', or more simply just an 'automaton', using a definition that has its origins in a paper by Kleene (1956). The title 'Representation of events in nerve sets' of Kleene's paper gives a hint of its motivation, but perhaps not surprisingly the ideas in the paper have thus far proved more useful in computer science than in biology or human psychology. Indeed the concept of an automaton, though very simple, has led to results of significance both in mathematics itself and in theoretical computer science.

Automata are in fact very familiar objects, in the shape of coin machines. To keep the details as simple as possible, let us envisage a coin machine accepting 5p, 10p and 20p coins, and dispensing a cup of coffee for 20p. (It is of course easy to translate this example into other currencies.) The coins are the *inputs* to the machine, and the machine *accepts,* or *recognizes,* any sequence of inputs that adds up to precisely 20p. The 'language' accepted by the machine is the set of input sequences

$$(20),\ (10, 10),\ (10, 5, 5),\ (5, 10, 5),\ (5, 5, 10),\ (5, 5, 5, 5).$$

The machine has five *states* q_0, q_5, q_{10}, q_{15}, q_{20}, (the subscript indicating the amount of money received). The state q_0, when the user begins to insert coins, is the *initial* state; the state q_{20}, when the machine delivers the cup of coffee, is the *terminal* state. The behaviour of the machine can be described pictorially as

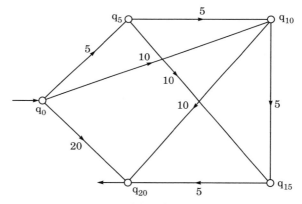

Each input sequence recognized by the machine corresponds to a path from the initial state q_0 to the terminal state q_{20}.

This chapter is devoted to an account of the basic properties of automata, culminating in Kleene's Theorem, which establishes an equivalence between the set of so-called 'rational' languages and the set of languages recognized by automata.

2.1 Finite state automata

Abstractly, and in the simplest possible terms, a *finite state automaton* (or more simply just an *automaton*) $\mathcal{A}$ is an ordered pair (Q, A) of finite sets, with an 'action' of A on Q. The elements of Q are called the *states* of the automaton, A is called the *alphabet* of the automaton, and the elements of A are called *inputs*. The action of A on Q associates to every

(q, a) in $Q \times A$ a state $q.a$ (usually written just as qa). Thus the input a has the effect of changing the state of the automaton from q to a new state denoted by qa.

If we allow ourselves to consider successive actions of inputs from A on states q then in effect we extend the input set to A^*. It is clear that, for q in Q and a_1, a_2 in A, we would want to define $q(a_1 a_2)$ as $(qa_1)a_2$ and that this would extend, in an obvious notation, to

$$q(a_1 a_2 \ldots a_n) = ((\cdots (qa_1)a_2)\cdots)a_n.$$

We can define this more formally in a recursive way by stipulating that

$$
\begin{aligned}
q1 &= q \ (q \in Q) \\
q(wa) &= (qw)a \ (q \in Q, w \in A^*, a \in A).
\end{aligned}
\tag{2.1.1}
$$

Observe now that the action of A^* on Q has the property that

$$q(wz) = (qw)z \tag{2.1.2}$$

for all q in Q and all w, z in A^*. We prove this by induction on $|z|$, the case $|z| = 1$ being clear from (2.1.1). Next, if $z = z'a$ with $|z| = n$ and $|z'| = n - 1$ we have

$$
\begin{aligned}
q(wz) = q\big(w(z'a)\big) &= q\big((wz')a\big) = \big(q(wz')\big)a \ \text{by (2.1.1)} \\
&= \big((qw)z'\big)a \ \text{by the induction hypothesis} \\
&= (qw)(z'a) \ \text{by (2.1.1)} \\
&= (qw)z
\end{aligned}
$$

as required.

We often find it convenient to summarize the action of A on Q by means of a table. For example, if $Q = \{0, 1, 2, 3\}$ and $A = \{a, b\}$ we might use the table

$$
\begin{array}{c|cc}
 & a & b \\
\hline
0 & 0 & 0 \\
1 & 2 & 1 \\
2 & 2 & 3 \\
3 & 0 & 1
\end{array}
\tag{2.1.3}
$$

to summarize the information that

$$0a = 0b = 0; \ 1a = 2, \ 1b = 1; 2a = 2, \ 2b = 3; \ 3a = 0, \ 3b = 1.$$

More conveniently still, we can describe an automaton via its *state diagram*. Technically, this is a 'labelled digraph' — see Harary (1969) —whose vertices are the elements of Q and in which $q \longrightarrow q'$ is an edge (labelled by a) if and only if $qa = q'$. Just how easy and useful this is can be seen by drawing the state diagram of the automaton described by the Table (2.1.3):

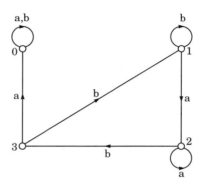

Figure 2.1.4

To find, for example, the state $1(aba)$ we simply examine the path beginning at 1 and following, in succession, the arrows labelled a, b and a. In this way it is then very easy to see (for example) that

$$1(aba) = 2(aba) = 3(aba) = 0(aba) = 0. \tag{2.1.5}$$

This observation will turn out to be quite important in understanding the operation of this automaton, and we shall return to it.

Note 2.1.6. The automaton A depends of course not only on the state set Q and the input set A but also on the nature of the action of A on Q. It is sometimes useful to have a more precise notation for this action and to give a name, such as φ, to the underlying function from $Q \times A$ into Q. We then write $\varphi(q, a)$ rather than qa. While this can be a useful clarification, it has the disadvantage that the formulae (2.1.1) and (2.1.2) take the less memorable forms

$$\varphi(q, 1) = q \ (q \in Q),$$
$$\varphi(q, wa) = \varphi(\varphi(q, w), a) \ (q \in Q, w \in A^*, a \in A) \tag{2.1.7}$$

and

$$\varphi(q, wz) = \varphi(\varphi(q, w), z) \ (q \in Q, w, z \in A^*). \tag{2.1.8}$$

So usually we shall prefer the simpler notation, but if for any reason we need to emphasize the precise way in which A acts on Q we shall denote the automaton A by (Q, A, φ).

Note 2.1.9. What we have defined in this section is often called a *complete deterministic automaton*; *complete* because qa is defined for every q and every a, and *deterministic* because the effect of the input a on the state q is a completely determined state qa. We shall see later (Section 2.3) that it can be convenient to relax both of these conditions.

2.2 Initial and final states; the behaviour of an automaton

Clearly an automaton $\mathcal{A} = (Q, A)$, as we have defined it, is some kind of rudimentary machine, whose internal state can be altered by various inputs. What we can actually do with such a machine is much less clear. The key lies in enhancing the basic definition with the idea of 'initial' and 'terminal' states.

We designate a certain state i in Q to be the *initial* state, and a non-empty subset T of Q (not necessarily a proper subset) to be the set of *terminal* states. Then we say that the automaton *recognizes* (or *accepts*) a word w in A^* if $iw \in T$. The *language $L(\mathcal{A})$ recognized by the automaton $\mathcal{A}$* (sometimes called the *behaviour* of $\mathcal{A}$) is then defined as the set of words in A^* that are recognized by $\mathcal{A}$.

In the state diagram it is traditional to draw attention to the initial state by means of an inward arrow and to the terminal states by means of outward arrows. A state that is both initial and terminal is decked with a double arrow $\longleftrightarrow$. Thus if in the example (2.1.4) we take $i = 1$ and $T = \{1, 2, 3\}$ then we get the following revised state diagram for $\mathcal{A} = \big(\{0, 1, 2, 3\}, \{a, b\}\big)$:

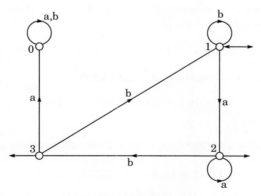

Figure 2.2.1

It is then easy to see that the word $b^2 a^3 b^2 a^4$ is recognized by $\mathcal{A}$, since
$$1(b^2 a^3 b^2 a^4) = 2.$$

By contrast, since $1(b^2a^3ba^4) = 0$, the word $b^2a^3ba^4$ is not recognized by $\mathcal{A}$. From the formula (2.1.5) we readily see that $1w = 0$ for every word w containing aba, and so such a word is *not* recognized by $\mathcal{A}$. In fact, as we shall now show, the converse is also true: the set $L(\mathcal{A})$ consists precisely of those words in $\{a, b\}^*$ not containing aba as a segment.

To see this let us first observe that if $1z = 3$ (where $z \in A^*$ and $|z| \geq 2$) then z must end in ab. For z ends in one or other of the two-letter words a^2, ab, ba, b^2; that is,

$$z = z'a^2 \text{ or } z = z'ab \text{ or } z = z'ba \text{ or } z = z'b^2,$$

where $z' \in \{a, b\}^*$ and where $1z' \in \{0, 1, 2, 3\}$. Now

$$
\begin{array}{ll}
1a^2 = 2a^2 = 2, & 3a^2 = 0a^2 = 0, \\
1ab = 2ab = 3, & 3ab = 0ab = 0, \\
1ba = 3ba = 2, & 2ba = 0ba = 0, \\
1b^2 = 2b^2 = 3b^2 = 1, & 0b^2 = 0;
\end{array}
$$

hence no matter what value we give to $1z'$ it is not possible to have $1z = 3$ unless z takes the form $z'ab$.

Now we prove, by induction on $|w|$, that every word w not containing aba is recognized by $\mathcal{A}$. Certainly the word 1 of length 0 and the words a, b of length 1 are recognized by $\mathcal{A}$, since $11 = 1 \in T$, $1a = 2 \in T$ and $1b = 1 \in T$. Now suppose that $|w| = n$, that w does not contain aba, and that all words of length less than n and not containing aba are recognized by $\mathcal{A}$. Then either (i) $w = za$ or (ii) $w = zb$, where $|z| = n - 1$ and z does not contain aba. By the induction hypothesis $z \in L(\mathcal{A})$ and so $1z \in \{1, 2, 3\}$. In case (i) we must in fact have $1z \in \{1, 2\}$, since $1z = 3$ would imply by the argument of the last paragraph that z ends in ab and hence that w ends in aba. So in case (i) we have that

$$1w = 1za \in \{1a, 2a\} = \{2\}$$

and thus $w \in L(\mathcal{A})$. In case (ii) we have that

$$1w = 1zb \in \{1b, 2b, 3b\} = \{1, 3\}$$

and so again $w \in L(\mathcal{A})$.

The conclusion is therefore that for the automaton $\mathcal{A}$ described by Figure 2.2.1,

$$L(\mathcal{A}) = \{a, b\}^* \backslash \{a, b\}^* aba \{a, b\}^*. \tag{2.2.2}$$

If we wish to draw attention not only to the basic structure of the automaton $\mathcal{A}$ but also to its initial state i and its set T of terminal states — and we usually do — then we write it as a quadruple $\mathcal{A} = (Q, A, i, T)$ or (see Note 2.1.6) as a quintuple $\mathcal{A} = (Q, A, \varphi, i, T)$.

Example 2.2.3. The automaton whose state diagram is

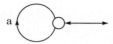

recognizes a^*.

Example 2.2.4. The automaton whose state diagram is

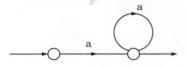

recognizes a^+.

Example 2.2.5. The automaton whose state diagram is

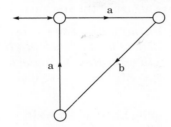

recognizes $(aba)^*$.

In describing elements recognized by an automaton $\mathcal{A} = (Q, A, i, T)$ it is frequently helpful to use a form of words deriving from the state diagram. If $q(a_1 a_2 \ldots a_n) = q'$, then the states q, q' in the automaton are connected by a *path*

$$q \xrightarrow{a_1} q_2 \xrightarrow{a_2} q_3 \longrightarrow \cdots \longrightarrow q_n \xrightarrow{a_n} q'$$

in the state diagram. The word $a_1 a_2 \ldots a_n$ is called the *label* of the path. A path will be called *successful* if it begins with the initial state i and ends with a terminal state t in T. Thus $w \in L(\mathcal{A})$ *if and only if there exists a successful path with label w*. For example, in the automaton described by the state diagram (2.2.1) the path

$$1 \xrightarrow{b} 1 \xrightarrow{a} 2 \xrightarrow{a} 2 \xrightarrow{b} 3$$

is successful, with label $ba^2 b$.

A subset L of A^* is called *recognizable* if there exists an automaton $\mathcal{A}$ such that $L(\mathcal{A}) = L$. Thus, for example, formula (2.2.2) tells us that when $A = \{a, b\}$ the set $A^* \backslash A^* aba A^*$ is recognizable. The major aim of this chapter is to give a useful alternative description of the recognizable subsets of a finitely generated free monoid A^*.

It may be interesting at this stage to give an example of a subset of $\{a, b\}^*$ that is *not* recognizable.

Theorem 2.2.6. *The subset*

$$L = \{a^n b^n : n \geq 1\}$$

of $\{a, b\}^*$ is not recognizable.

Proof. Suppose, by way of contradiction, that there exists an automaton $\mathcal{A} = (Q, A, i, T)$ recognizing L, and denote ia^n by q_n for $n = 1, 2, \ldots$. Then $q_n b^n \in T$ for every n. Since Q is finite there must exist m, n with $m \neq n$ and $q_m = q_n$. But then we have

$$ia^m b^n = q_m b^n = q_n b^n \in T$$

and so $a^m b^n \in L$. This is a contradiction. ∎

The problem of proving that a given set is not recognizable is frequently made easier by the use of the so-called 'Pumping Lemma' :

Theorem 2.2.7 (The Pumping Lemma). *Let L be an infinite recognizable language in A^*. Then there exists a positive integer N such that every word z in L of length exceeding N can be factorized as $z = uvw$ in such a way that* (i) $u, w \in A^*$, $v \in A^+$, (ii) $|uv| \leq N$, (iii) $uv^m w \in L$ *for every $m \geq 0$.*

Proof. The name 'Pumping Lemma' arises from the notion that we can 'inflate' the word uvw with as many copies of v as we wish.

We have $L = L(\mathcal{A})$, where $\mathcal{A} = (Q, A, i, T)$ is a finite state automaton. Let $|Q| = N$, and suppose that $z = a_1 a_2 \ldots a_k \in L$, where $|z| = k > N$. In the successful path

$$i = q_0 \xrightarrow{a_1} q_1 \xrightarrow{a_2} \cdots \xrightarrow{a_k} q_k = t(\in T)$$

with label $z = a_1 a_2 \ldots a_k$ there must be at least one repetition among the $N + 1$ states $q_0, q_1, \ldots, q_N$. To be definite, let q_r (with $r \geq 0$) be the first state in the sequence $q_1, q_2, \ldots, q_N$ that repeats and let q_{r+s} be its first repetition. Notice that $r \geq 0$, $s > 0$, $r + s \leq N$. We thus have $z = uvw$, where $u \ (\in A^*)$, $v \ (\in A^+)$, $w \ (\in A^*)$ label the paths

$$q_0 \longrightarrow \cdots \longrightarrow q_r, \quad q_r \longrightarrow \cdots \longrightarrow q_{r+s},$$

$$q_{r+s} \longrightarrow \cdots \longrightarrow q_N \longrightarrow \cdots \longrightarrow q_k$$

respectively. The successful path from q_0 to q_k thus includes a loop

$$q_r \longrightarrow \cdots \longrightarrow q_{r+s} = q_r$$

with label v. We obtain another successful path if we choose to go round this loop a number of times, or indeed if we leave it out altogether.

So $uv^mw \in L$ for all $m \geq 0$. We complete the proof by remarking that $|uv| = r + s \leq N$. ∎

As was indicated in the sentence introducing the Pumping Lemma, the main use of the lemma is negative. For example, it gives us an alternative proof that $L = \{a^nb^n : n \geq 1\}$ is not recognizable. For if it is recognized by a finite state automaton we consider the integer N that the Pumping Lemma gives us and then look at a word $z = a^nb^n$ for which $n > N$. The lemma then implies that z can be written as uvw, with $|v| \geq 1$ and $|uv| \leq N < n$, in such a way that $uv^mw \in L$ for all $m \geq 0$. The condition $|uv| \leq N < n$ means that both u and v must come from the initial block of a's in z :

$$u = a^p, \ v = a^q, \ z = a^rb^n$$

(with $p, r \geq 0, q > 0, p + q + r = n$). But then

$$uv^0w = a^{n-q}b^n \in L$$

and so we have a contradiction. Hence L is not recognizable.

Example 2.2.8. Let $A = \{a, b, \ldots\}$ be a finite alphabet containing at least two letters, and let P be the set of *palindromes* in A^*, i.e., the set $\{w \in A^* : w^R = w\}$. (Recall that if $w = a_1a_2\ldots a_n \in A^*$, then w^R is the *reversed* word $a_na_{n-1}\ldots a_1$.) We show that P is not recognizable. For otherwise consider the integer N given by the Pumping Lemma and let us look at a palindrome $z = a^nba^n$, where $n > N$. Then z can be expressed as uvw, with $|v| \geq 1$, $|uv| \leq N < n$, in such a way that $uv^mw \in P$ for all $m \geq 0$. Now the restriction $|uv| \leq N < n$ means that both u and v must come from the first block of a's in z :

$$u = a^p, \ v = a^q, \ w = a^rba^n$$

(with $p \geq 0, r \geq 0, q > 0, p + q + r = n$). But then the lemma implies that $uv^0w = a^{n-q}ba^n \in P$ — a contradiction, since $a^{n-q}ba^n$ is certainly not a palindrome.

Note 2.2.9. We have arrived gradually at the full notion of a finite state automaton. For easy reference we now record a set of formal mathematical definitions. Let Q, A be finite, non-empty sets, let φ : $Q \times A \to Q$ be a function, let $i \in Q$ and let T be a non-empty subset of Q. A *finite state automaton* is a quintuple $\mathcal{A} = (Q, A, \varphi, i, T)$. The elements of Q are called the *states* of $\mathcal{A}$; the set A is called the set of *inputs* or the *alphabet* of $\mathcal{A}$; i is called the *initial* state and the elements of T are called the *terminal* states. An element w of A^* is said to be *recognized* by $\mathcal{A}$ if $\varphi(i, w) \in T$. The *language* $L(\mathcal{A})$ recognized by $\mathcal{A}$ is the set of all elements w in A^* that are recognized by $\mathcal{A}$.

Note 2.2.10. It is legitimate to ask why an automaton is defined to have a unique initial state and a multiplicity of terminal states. The answer is of course that ultimately a definition is a matter of choice. It turns out to be convenient to do it this way, but in fact the crucial concept of recognizability is not changed if we allow a multiplicity of initial states. (See Exercise 2.13.) If, however, we insist on a single terminal state then some subsets of A^* that are recognizable in the usual sense cease to be recognizable. (See Exercise 2.14.)

2.3 Incomplete and non-deterministic automata

It is frequently useful to work with a more general definition of an automaton. As before we have a finite non-empty set Q of states and a finite non-empty alphabet A. But now the action of A on Q is described by means of a function φ from $Q \times A$ into $\mathcal{P}(Q)$, the set of all subsets of Q. Informally, we have that qa may be *undefined* (which corresponds to the case where $\varphi(q, a) = \emptyset$); or qa may be *ambiguous* (which corresponds to the case where $\varphi(q, a)$ contains more than one element of Q). We can still present the properties of such an automaton in tabular form: for example, if $Q = \{1, 2, 3\}$ and $A = \{a, b\}$ we might have

	a	b
1	$\{1, 2\}$	$\{2, 3\}$
2	1	$\emptyset$
3	3	$\{2, 3\}$

(2.3.1)

If we designate 1 as the initial state and $\{2\}$ as the set of terminal states then the corresponding state diagram is

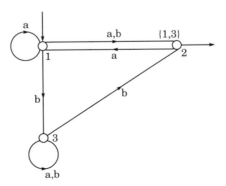

Figure 2.3.2

As in the earlier case we can (by analogy with formula (2.1.1)) extend the action of A on Q to an action of A^* on Q by the rules that

$$q1 = \{q\} \ (q \in Q)$$

$$q(wa) = \bigcup_{p \in qw} pa \ (q \in Q, w \in A^*, a \in A). \tag{2.3.3}$$

If we make the reasonable convention that for a subset P of Q the notation Pa will mean

$$\bigcup_{p \in P} pa$$

then this second definition becomes

$$q(wa) = (qw)a, \tag{2.3.4}$$

exactly as in formula (2.1.1), though now the equality is between subsets rather than between elements of Q. A proof formally identical to the proof of (2.1.2) gives

$$q(wz) = (qw)z \ (q \in Q, w, z \in A^*). \tag{2.3.5}$$

Within an automaton of this more general kind we can still designate an initial state i and a non-empty set T of terminal states, and we say that a word w in A^* is *recognized by* A, and write $w \in L(A)$, if $iw \cap T \neq \emptyset$.

We now appear to have a new and more general concept of recognizability: a subset L of A^* is *recognizable* if there exists an automaton A (not necessarily complete and deterministic) for which $L = L(A)$. Perhaps surprisingly, this concept is no different from the previous one.

Theorem 2.3.6. *Let* $A = (A, Q, \varphi, i, T)$ *be an automaton (not necessarily complete and deterministic) and let* $L(A)$ *be the language recognized by* A. *Then there exists a complete deterministic automaton* A' *such that* $L(A') = L(A)$.

Proof. We define A' to have the same alphabet A as A, but the set of states of A' is taken as $\mathcal{P}(Q)$, the set of all subsets of Q. The action of A on $\mathcal{P}(Q)$ is a function $\psi : \mathcal{P}(Q) \times A \to \mathcal{P}(Q)$ defined in the obvious way by

$$\psi(P, a) = \bigcup_{p \in P} \varphi(p, a) \ (P \in \mathcal{P}(Q), a \in A).$$

In more compact notation this becomes

$$Pa = \bigcup_{p \in P} pa \ (P \in \mathcal{P}(Q), a \in A). \tag{2.3.7}$$

Leaving aside for the moment the question of initial and terminal states, we note that, as we have defined it thus far, $\mathcal{A}' = (\mathcal{P}(Q), A, \psi)$ is a complete deterministic automaton. The action of A on $\mathcal{P}(Q)$ extends in the usual way to an action of A^* on $\mathcal{P}(Q)$ by the rule that

$$P1 = P \quad (P \in \mathcal{P}(Q)),$$
$$P(wa) = (Pw)a \quad (P \in \mathcal{P}(Q), w \in A^*, a \in A). \tag{2.3.8}$$

In fact we now show, by induction on $|w|$, that for all w in A^* and all P in $\mathcal{P}(Q)$

$$Pw = \bigcup_{p \in P} pw. \tag{2.3.9}$$

For $|w| = 1$ this is simply formula (2.3.7). So assume now that $|w| = n \geq 1$, and let us write $w = za$, with $z \in A^*$, $a \in A$. Then

$$Pw = P(za) = (Pz)a \text{ by (2.3.8)}$$

$$= \left(\bigcup_{p \in P} pz \right) a \text{ by the induction hypothesis}$$

$$= \bigcup_{p \in P} (pz)a = \bigcup_{p \in P} p(za) \text{ by (2.3.4)}$$

$$= \bigcup_{p \in P} pw$$

as required.

Strictly speaking, formula (2.3.9) should read

$$\psi(P, w) = \bigcup_{p \in P} \varphi(p, w), \tag{2.3.10}$$

but we shall use this version only when the simpler version would cause confusion. Notice in particular that

$$\psi(\{q\}, w) = \varphi(q, w) \quad (q \in Q, w \in A^*).$$

We now complete the definition of the automaton $\mathcal{A}'$ by specifying its initial state as the singleton set $\{i\}$ and its terminal state as

$$T' = \{P \in \mathcal{P}(Q) : P \cap T \neq \emptyset\}. \tag{2.3.11}$$

We complete the proof by noting that

$$w \in L(\mathcal{A}) \text{ if and only if } \varphi(i, w) \cap T \neq \emptyset,$$
$$\text{i.e., if and only if } \psi(\{i\}, w) \cap T \neq \emptyset,$$
$$\text{i.e., if and only if } \psi(\{i\}, w) \in T',$$
$$\text{i.e., if and only if } w \in L(\mathcal{A}'). \qquad \blacksquare$$

Returning now to the automaton $\mathcal{A}$ whose state diagram is given by Figure 2.3.2, we see that the corresponding complete deterministic automaton $\mathcal{A}'$ has $2^3 = 8$ states, and the tabular representation is

	$\emptyset$	$\{1\}$	$\{2\}$	$\{3\}$	$\{2,3\}$	$\{1,3\}$	$\{1,2\}$	$\{1,2,3\}$
a	$\emptyset$	$\{1,2\}$	$\{1\}$	$\{3\}$	$\{1,3\}$	$\{1,2,3\}$	$\{1,2\}$	$\{1,2,3\}$
b	$\emptyset$	$\{2,3\}$	$\emptyset$	$\{2,3\}$	$\{2,3\}$	$\{2,3\}$	$\{2,3\}$	$\{2,3\}$

The initial state is $\{1\}$ and the terminal states are

$$\{2\}, \{2,3\}, \{1,2\} \text{ and } \{1,2,3\}.$$

The state diagram is

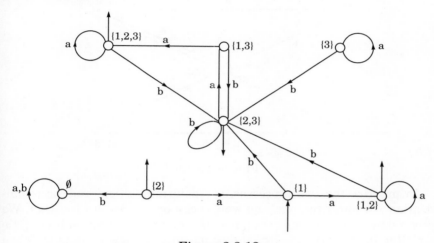

Figure 2.3.12

This latest automaton, while it has the attractive feature of being complete and deterministic, has some unattractive aspects as well. To explain these we require some more definitions.

In a deterministic automaton $\mathcal{A} = (Q, A, i, T)$ we say that a state q is *accessible* if there exists w in A^* such that $iw = q$. The automaton $\mathcal{A}$ is called *accessible* if every state is accessible. Notice that the automaton $\mathcal{A}'$ described by the state diagram (2.3.12) is not accessible. Specifically, the states $\emptyset$, $\{2\}$ and $\{3\}$ are not accessible.

Now the behaviour of an automaton $\mathcal{A}$ is described in terms of successful paths, i.e., paths from i into T, and only states which are accessible can ever be visited by successful paths. All other states are irrelevant to the behaviour of $\mathcal{A}$ and so it seems that it ought to be possible to remove them, leaving an accessible automaton with the same behaviour as $\mathcal{A}$.

This can in fact be done in a very simple-minded way. Given a complete deterministic automaton $\mathcal{A} = (Q, A, i, T)$ we define $\mathcal{A}^b = (Q', A, i, T)$, where Q' is obtained from Q by removing all states that are not accessible, and where the action of A on Q is simply the restriction to Q' of the action of A on Q. Since no successfiul path is hindered (or indeed helped) by this change, we have $L(\mathcal{A}^b) = L(\mathcal{A})$, and $\mathcal{A}^b$ is accessible.

If we apply this to the automaton given by Figure 2.3.12 we obtain the accessible automaton whose state diagram is given in Figure 2.3.13.This has the same behaviour as the original automaton (2.3.2) but is complete, deterministic and accessible.

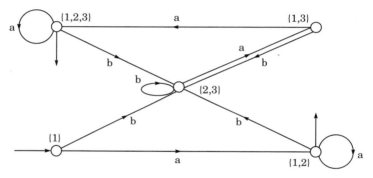

Figure 2.3.13

In practice the two processes, of 'determinization' and 'accessibilization' (to coin two appalling words) can be carried out together. We proceed by listing the subsets $\{i\}a$ of Q for a in A, then the subsets $\{i\}w$ for all w in A^* with $|w| = 2$, and so on until no new subsets of Q are obtained. The next example illustrates this process.

Example 2.3.14. Consider the (incomplete non-deterministic) automaton $\mathcal{A}$ whose state diagram is given by Figure 2.3.15. It is clear that $L(\mathcal{A}) = \{ab^2, a^2b\}$. Now notice that

$$\{1\}a = \{2,5\}, \quad \{1\}b = \emptyset;$$

$$\{2,5\}a = \{6\}, \quad \{2,5\}b = \{3\}, \emptyset a = \emptyset b = \emptyset;$$

$$\{6\}a = \{3\}a = \emptyset, \quad \{6\}b = \{7\}, \quad \{3\}b = \{4\};$$

$$\{7\}a = \{7\}b = \{4\}a = \{4\}b = \emptyset.$$

So a complete, deterministic, accessible automaton $\mathcal{A}'$ recognizing $L(\mathcal{A})$ is described by Figure 2.3.16.

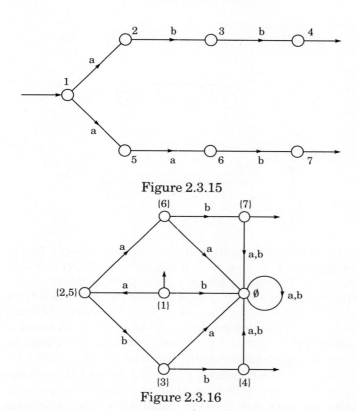

Figure 2.3.15

Figure 2.3.16

Example 2.3.17. It is easy to write down an incomplete, non-deterministic automaton recognizing $L = \{a, b\}^* ab \{a, b\}^*$. Its state diagram is

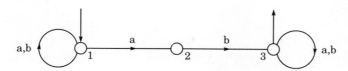

To find a complete, deterministic, accessible automaton $\mathcal{A}'$ recognizing L, we calculate

$$\{1\}a = \{1, 2\}, \qquad\qquad \{1\}b = \{1\},$$
$$\{1, 2\}a = \{1, 2\}, \qquad\qquad \{1, 2\}b = \{1, 3\},$$
$$\{1, 3\}a = \{1, 2, 3\}, \qquad\qquad \{1, 3\}b = \{1, 3\},$$
$$\{1, 2, 3\}a = \{1, 2, 3\}, \qquad \{1, 2, 3\}b = \{1, 3\}.$$

We obtain a state diagram

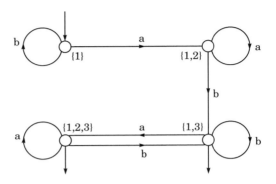

for $\mathcal{A}'$.

The idea of accessibility has an obvious dual. We say that a state q in an automaton $\mathcal{A} = (Q, A, i, T)$ is *coaccessible* if there exists w in A^* such that $qw \in T$, and the automaton $\mathcal{A}$ is itself called *coaccessible* if all its states are coaccessible. Again, since $L(\mathcal{A})$ is described in terms of successful paths, i.e., paths from i into T, and since only states that are coaccessible can be visited by successful paths, the behaviour of $\mathcal{A}$ will not be affected if we purge it of all states that are not coaccessible. An automaton which is both accessible and coaccessible is called *trim*.

The automaton described by Figure 2.3.13 and the automaton $\mathcal{A}'$ of Example 2.3.17 are both trim. However, there is no guarantee that the routine we have developed to produce a complete, deterministic, accessible automaton will also achieve coaccessibility. For example, the automaton $\mathcal{A}'$ of Example 2.3.14 has a state $\emptyset$ which is not coaccessible. If we purge it we obtain an automaton $\mathcal{A}''$ with state diagram

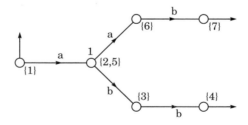

This new automaton is still deterministic and is also trim, but it is not complete, since (for example) $\{7\}a$ and $\{7\}b$ are now undefined.

This example illustrates the dilemma in which we find ourselves. Given an automaton $\mathcal{A}$ of the most general sort, we can guarantee to produce a deterministic and accessible automaton $\mathcal{A}'$ such that $L(\mathcal{A}') =$

$L(\mathcal{A})$, but beyond that we may have to choose between completeness and coaccessibility. Anyway, the details of this section are less important than the main message, which is that for the most part we do not need to be particulary precise about what sort of automaton $\mathcal{A}$ we are dealing with. Usually we are interested only in $L(\mathcal{A})$; so if $\mathcal{A}$ is incomplete we can complete it, if $\mathcal{A}$ is non-deterministic we can 'determinize' it, if $\mathcal{A}$ is not trim we can trim it; and we can do all of these (though not quite all simultaneously) without changing $L(\mathcal{A})$.

2.4 Finite sets

We begin with a very simple result:

Theorem 2.4.1. *Let A be a finite alphabet. All singleton sets in A^* are recognizable.*

Proof. Let $w = a_1 a_2 \ldots a_n \in A^+$. If $\mathcal{A}$ is the (incomplete) automaton with state diagram

then $L(\mathcal{A}) = \{w\}$. If $w = 1$, let

$$\mathcal{A} = \big(\{i, 0\}, A, i, \{i\}\big),$$

with $ia = 0a = 0$ for all a in A. The state diagram is

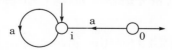

and clearly $L(\mathcal{A}) = \{1\}$. ∎

Next, we have an even simpler result:

Theorem 2.4.2. *The empty set is recognizable.*

Proof. Consider the automaton

$$\mathcal{A} = \big(\{i, 0\}, A, i, \{0\}\big),$$

with $ia = 0a = i$ for all a in A. The state diagram is

and evidently $L(\mathcal{A}) = \emptyset$. ∎

Next, we have

Theorem 2.4.3. *Let A be a finite alphabet and let L_1, L_2 be recognizable subsets of A^*. Then $L_1 \cup L_2$ is a recognizable subset of A^*.*

Proof. Let $L_1 = L(\mathcal{A}_1)$, $L_2 = L(\mathcal{A}_2)$, where

$$\mathcal{A}_1 = (Q_1, A, \varphi_1, i_1, T_1), \quad \mathcal{A}_2 = (Q_2, A, \varphi_2, i_2, T_2)$$

are complete deterministic automata. Let $\mathcal{A} = (Q, A, \varphi, i, T)$, where $Q = Q_1 \times Q_2$,

$$\varphi\big((q_1, q_2), a\big) = \big(\varphi_1(q_1, a), \varphi_2(q_2, a)\big) \ \big((q_1, q_2) \in Q, a \in A\big), \qquad (2.4.4)$$

$i = (i_1, i_2)$ and $T = (T_1 \times Q_2) \cup (Q_1 \times T_2)$. In our simpler notation formula (2.4.4) becomes $(q_1, q_2)a = (q_1 a, q_2 a)$, and indeed we have the more general result that

$$(q_1, q_2)w = (q_1 w, q_2 w) \ \big((q_1, q_2) \in Q, w \in A^*\big).$$

If $w \in L_1$ then $i_1 w \in T_1$ and so

$$(i_1, i_2)w \in T_1 \times Q_2 \subseteq T.$$

Equally, $w \in L_2$ gives $(i_1, i_2)w \in T$, and so we conclude that

$$L_1 \cup L_2 \subseteq L(\mathcal{A}).$$

Conversely, if $w \in L(\mathcal{A})$ then either $i_1 w \in T_1$ or $i_1 w \in T_2$ and so $w \in L_1 \cup L_2$. We conclude that

$$L(\mathcal{A}) = L_1 \cup L_2. \qquad \blacksquare$$

From these three theorems we now immediately deduce our first important result concerning recognizable sets:

Theorem 2.4.5. *Let A be a finite alphabet. Every finite subset of A^* is recognizable.*

Theorems 2.4.1 and 2.4.3 together give a routine for constructing an automaton recognizing a given finite set. The next example shows that there is a more direct way and provides the idea for an alternative proof of Theorem 2.4.5.

Example 2.4.6. Let $A = \{a, b\}$ and let $L = \{ab, ab^2, b^2, bab\}$. Then the automaton with state diagram

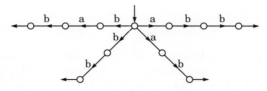

Figure 2.4.7

recognizes L. It is of course *not* the automaton obtained by using the proofs of Theorems 2.4.1 and 2.4.3. which would have $3 \times 4 \times 3 \times 4 = 144$ states instead of 11. A still smaller automaton recognizing L is given by

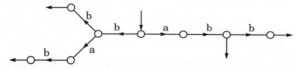

and this has the advantage, which (2.4.7) does not, of being deterministic.

2.5 Rational sets and Kleene's Theorem

Let A be a finite non-empty set. We have seen that all finite subsets of A^* are recognizable but that not every subset of A^* is recognizable. To give a set-theoretic description of the recognizable subsets of A^* we need some new concepts.

Given $L_1, L_2 \subseteq A^*$ we define their *product* $L_1 L_2$ in the obvious way:

$$L_1 L_2 = \{uv : u \in L_1, v \in L_2\}.$$

Given a subset L of A^* we define

$$L^* = \{u_1 u_2 \ldots u_n : n \geq 0, u_1, u_2, \ldots, u_n \in L\}. \tag{2.5.1}$$

That is, L^* is the submonoid of A^* generated by L. This is the standard notation for what is often called *Kleene's star operation*, but it does open the door to a possible misunderstanding. It is important to note that L^* does not necessarily coincide with the free monoid generated by the set L. It *may* coincide with the free monoid (e.g., when $L = A$), but more commonly it does not. (See Exercise 1.8.) Fortunately the context will always make it clear whether X^* is to be interpreted as the free monoid on X or as the submonoid of A^* generated by a subset X of A^*. Notice that for every $L \subseteq A^*$

$$L^* = \{1\} \cup L \cup L^2 \cup \ldots = \bigcup_{n=0}^{\infty} L^n.$$

A subset of A^* is called *rational* if it can be obtained from finite subsets of A^* by finitely many applications of $\cup$ (union), . (multiplication) and * (Kleene's star operation). For example, if $A = \{a, b\}$ then

$$(ab)^*, \; \{ab, ba\}^*, \; (a^* \cup b^*)^*.\{a, b\}^*$$

are all rational subsets of A^*. Less transparently, $\{a, b\}^+$ is rational, since

$$\{a, b\}^+ = a\{a, b\}^* \cup b\{a, b\}^*.$$

Finding examples of sets that are *not* rational is harder, but we shall soon learn how to do this.

The set of all rational subsets of A^* is denoted by $\operatorname{Rat} A^*$. The main result of this chapter is

Theorem 2.5.2 (Kleene's Theorem). *Let A be a finite alphabet and let L be a subset of A^*. Then L is a recognizable set if and only if L is rational.*

Proof. Let us denote the set of all recognizable subsets of A^* by $\operatorname{Rec} A^*$. From Theorem 2.4.5 we know that all finite subsets of A^* are in $\operatorname{Rec} A^*$. If we can show that

$$L_1, L_2 \in \operatorname{Rec} A^* \;\Rightarrow\; L_1 \cup L_2 \in \operatorname{Rec} A^*, \tag{2.5.3}$$

$$L_1, L_2 \in \operatorname{Rec} A^* \;\Rightarrow\; L_1.L_2 \in \operatorname{Rec} A^*, \tag{2.5.4}$$

and

$$L \in \operatorname{Rec} A^* \;\Rightarrow\; L^* \in \operatorname{Rec} A^*, \tag{2.5.5}$$

then it will follow that every rational set is recognizable, i.e., that $\operatorname{Rat} A^* \subseteq \operatorname{Rec} A^*$.

The first of these implications has already been proved in Theorem 2.4.3. To prove the implication (2.5.4), suppose that L_1, L_2 are recognized respectively by complete deterministic automata

$$\mathcal{A}_1 = (Q_1, A, \varphi_1, i_1, T_1), \; \mathcal{A}_2 = (Q_2, A, \varphi_2, i_2, T_2).$$

We may certainly assume that $Q_1 \cap Q_2 = \emptyset$. We now produce a new (non-deterministic) automaton by a procedure that can be thought of as coupling $\mathcal{A}_1$ and $\mathcal{A}_2$ 'in series'. Precisely, we define

$$\mathcal{A} = (Q_1 \cup Q_2, A, \psi, i_1, T_2),$$

where, for $q_1 \in Q_1$, $q_2 \in Q_2$ and $a \in A$, we have

$$\psi(q_1, a) = \begin{cases} \{\varphi_1(q_1, a)\} & \text{if } \varphi_1(q_1, a) \notin T_1, \\ \{\varphi_1(q_1, a), i_2\} & \text{if } \varphi_1(q_1, a) \in T_1, \end{cases}$$
$$\psi(q_2, a) = \{\varphi_2(q_2, a)\}.$$

Suppose now that $w = z_1 z_2 \in L_1 L_2$, with $z_1 \in L_1$, $z_2 \in L_2$. Then $\varphi_1(i_1, z_1) \in T_1$ and so $\psi(i_1, z_1)$ contains i_2. (Notice that $\psi(i_1, z_1)$ may be larger than $\{\varphi_1(i_1, z_1), i_2\}$ since the path from i_1 to $\varphi_1(i_1, z_1)$ may have visited T_1 several times previously.) It now follows that $\psi(i_1, z_1 z_2)$ contains $\varphi_2(i_2, z_2)$ and so has a non-empty intersection with T_2. Hence $w \in L(\mathcal{A})$. We have shown that $L_1 L_2 \subseteq L(\mathcal{A})$.

Conversely, suppose that $w \in L(\mathcal{A})$, so that $\psi(i_1, w)$ contains an element of T_2. The path in $\mathcal{A}$ from i_1 to $\psi(i_1, w)$ can enter Q_2 only if $\varphi_1(i_1, u) \in T_1$ for some left factor u of w. This may in fact happen for several left factors, and it may be helpful to complicate the notation slightly by supposing that it happens for left factors $u_1, u_2, \ldots, u_k$, where

$$|u_1| < |u_2| < \cdots < |u_k|$$

and where $u_j v_j = w$ for $j = 1, 2, \ldots, k$. Thus $\varphi_1(i_1, u_j) \in T_1$ and hence $u_j \in L_1$ for each j.

At each stage we have $i_2 \in \psi(i_1, u_j)$ and so $\psi(i_1, w)$ includes $\varphi_2(i_2, v_j)$. In fact,

$$\psi(i_1, w) = \{\varphi_1(i_1, w), \varphi_2(i_2, v_1), \ldots, \varphi_2(i_2, v_k)\}.$$

By assumption, at least one of $\varphi_2(i_2, v_1), \ldots, \varphi_2(i_2, v_k)$ belongs to T_2. So there exists j in $\{1, 2, \ldots, k\}$ such that $v_j \in L_2$. Thus $w = u_j v_j \in L_1 L_2$ as required. This establishes the implication (2.5.4).

As regards the implication (2.5.5), it will be enough to prove it in the case where $1 \notin L$. That this is so follows from the two statements

$$\left(L \backslash \{1\}\right)^* = L^*, \tag{2.5.6}$$

$$L \in \operatorname{Rec} A^* \ \Rightarrow \ L \backslash \{1\} \in \operatorname{Rec} A^*; \tag{2.5.7}$$

for in the case where $1 \in L$ we then have the chain of implications

$$L \in \operatorname{Rec} A^* \ \Rightarrow L \backslash \{1\} \in \operatorname{Rec} A^*$$
$$\Rightarrow \left(L \backslash \{1\}\right)^* \in \operatorname{Rec} A^*$$
$$\Rightarrow L^* \in \operatorname{Rec} A^*.$$

Of the two statements (2.5.6) and (2.5.7) the first is obvious. To prove the second, suppose that $L = L(\mathcal{A})$, where

$$\mathcal{A} = (Q, A, \varphi, i, T).$$

If

$$\mathcal{A}' = \big(Q \times \{0,1\}, A, \varphi', (i,0), T \times \{1\}\big),$$

with

$$\varphi'\big((q,0), a\big) = \varphi'\big((q,1), a\big) = \big(\varphi(q,a), 1\big)$$

for all q in Q and all a in A, then

$$\varphi'\big((i,0), w\big) = \begin{cases} \big(\varphi(i,w), 1\big) & \text{if } w \neq 1, \\ \big(\varphi(i,w), 0\big) & \text{if } w = 1. \end{cases}$$

Hence $\varphi'\big((i,0), w\big) \in T \times \{1\}$ if and only if $w \in L\backslash\{1\}$. Thus

$$L(\mathcal{A}') = L\backslash\{1\}.$$

So now suppose that $1 \notin L$ and that $L = L(\mathcal{A})$, where

$$\mathcal{A} = (Q, A, \varphi, i, T).$$

This time we create a new automaton $\mathcal{A}^+$ by 'looping' $\mathcal{A}$ onto itself. Precisely, we define

$$\mathcal{A}^+ = (Q, A, \varphi^+, i, T),$$

where

$$\varphi^+(q,a) = \begin{cases} \{\varphi(q,a)\} & \text{if } \varphi(q,a) \notin T, \\ \{\varphi(q,a), i\} & \text{if } \varphi(q,a) \in T. \end{cases}$$

Suppose now that $w = z_1 z_2 \ldots z_n \in L^+ \ (= L^*\backslash\{1\})$, with $n \geq 1$ and $z_1, z_2, \ldots, z_n \in L$. Since $\varphi(i, z_1) \in T$ we see that $\varphi^+(1, z_1)$ contains both i and an element of T. If we now suppose inductively that $\varphi^+(i, z_1 z_2 \ldots z_{k-1})$ contains both i and an element of T, we see that $\varphi^+(i, z_1 z_2 \ldots z_k)$ contains

$$\varphi^+(i, z_k) = \{\varphi(i, z_k), i\}$$

(for $\varphi(i, z_k) \in T$ by assumption). Thus $\varphi^+(i, z_1 z_2 \ldots z_k)$ contains both i and an element of T, as required. By induction it now follows that $\varphi^+(i, z_1 z_2 \ldots z_n)$ contains both i and an element of T, and so $w \in L(\mathcal{A}^+)$. We have shown that $L^+ \subseteq L(\mathcal{A}^+)$.

Conversely, suppose that $w \in L(\mathcal{A}^+)$, so that $\varphi^+(i, w)$ contains an element of T. Let z_1 be the shortest non-trivial left factor of w such that $\varphi^+(i, z_1)$ contains an element of T. (There must *be* such a factor, even if it has to be w itself.) Then $\varphi^+(i, z_1) = \{\varphi(i, z_1), i\}$, and $z_1 \in L$. If $z_1 = w$ then $w \in L \subseteq L^+$ and we are finished. Otherwise $w = z_1 w_1$, and we let z_2 be the shortest non-trivial left factor of w_1 such that $\varphi^+(i, z_1 z_2)$ contains an element of T. Then

$$\varphi^+(i, z_1 z_2) = \{\varphi(i, z_1 z_2), \varphi(i, z_2), i\},$$

and either $z_1 z_2 \in L$ or $z_2 \in L$. In either case $z_1 z_2 \in L^+$. If $w = z_1 z_2$ then we are finished. Otherwise $w = z_1 z_2 w_2$ and we let z_3 be the shortest non-trivial left factor of w_2 such that $\varphi^+(i, z_1 z_2 z_3)$ contains an element of T. Then

$$\varphi^+(i, z_1 z_2 z_3) = \{\varphi(i, z_1 z_2 z_3), \varphi(i, z_2 z_3), \varphi(i, z_3), i\},$$

and either $z_1 z_2 z_3 \in L$ or $z_2 z_3 \in L$ or $z_3 \in L$. In any case $z_1 z_2 z_3 \in L^+$. Since w has finite length this process must eventually terminate, and we conclude that $w \in L^+$.

We have shown that

$$L \in \operatorname{Rec} A^* \Rightarrow L^+ \in \operatorname{Rec} A^*.$$

To show the implication (2.5.5) we now need only remark that if L^+ is recognizable then so is $L^* = L \cup \{1\}$, by virtue of Theorems 2.4.1 and 2.4.3.

Taking stock at this stage in the proof of Kleene's Theorem, we note that we have now shown that every rational set is recognizable. Symbolically, we have

$$\operatorname{Rat} A^* \subseteq \operatorname{Rec} A^*.$$

To show the opposite inclusion, consider now a recognizable set L. Thus for some complete deterministic automaton $\mathcal{A} = (Q, A, i, T)$ we have that

$$L = L(\mathcal{A}) = \{w \in A^* : iw \in T\}.$$

Clearly L is a finite union

$$L = \bigcup_{t \in T} L(i, t),$$

where $L(i, t) = \{w \in A^* : iw = t\}$, and it will clearly be enough if we show that each $L(i, t)$ is rational. In fact it will be convenient to show the somewhat more general result that $L(p, q)$ is rational for every p, q in Q, where

$$L(p, q) = \{w \in A^* : pw = q\},$$

the set of labels of all the paths in $\mathcal{A}$ from p to q.

For each path

$$p \longrightarrow \cdots \longrightarrow q$$

from p to q we can define the *territory* as the set of vertices in Q visited by the path. Notice that the territory includes p and q. For p, q in Q and for each subset R of Q containing p and q we define $L(p, R, q)$ as the set of words labelling paths from p to q whose territory lies within R. Formally, we define

$$L(p, R, q) = \{w \in A^* : pw = q, pz \in R \text{ for all left factors } z \text{ of } w\}. \quad (2.5.8)$$

A *strict* path from p to q is one that does not visit either p or q on the way. If V is a subset of Q such that $p \notin V$, $q \notin V$ then we can define $Z(p, V, q)$ as the set of words in A^* labelling strict paths from p to q whose territory lies in $V \cup \{p, q\}$. Formally, we define

$$Z(p, V, q) = \{w \in A^* : pw = q, pz \in V \text{ for all proper left factors } z \text{ of } w\}. \quad (2.5.9)$$

Our aim is to show that $L(p, q) \left(= L(p, Q, q)\right)$ is a rational set, and we do this by proving the still more general result that $L(p, R, q)$ and $Z(p, V, q)$ are both rational for every subset R of Q containing p and q and for every subset V of Q not containing p and q. We do this by induction on the number of elements in R and V.

Before we begin the induction, it is useful to introduce one final piece of new notation: for p and q in Q, let

$$A(p, q) = \{a \in A : pa = q\}. \quad (2.5.10)$$

The set $A(p, q)$, being finite, is certainly rational. Of course, it may well be empty.

To anchor the induction, notice that $|R| = 1$ is possible only if $q = p$, and that

$$L(p, \{p\}, p) = \left(A(p, p)\right)^*$$

is rational (being obtained from a finite set by one application of Kleene's star operation). The case $|V| = 0$ is possible, and

$$Z(p, \emptyset, q) = A(p, q)$$

is rational. If $|V| = 1$ then $V = \{r\}$ for some $r \neq p, q$, and

$$Z(p, \{r\}, q) = A(p, r)\left(A(r, r)\right)^* A(r, q)$$

is rational.

Let $n \geq 2$ and suppose inductively that $L(p, R, q)$ (with $p, q \in R$) and $Z(p, V, q)$ (with $p, q \notin V$) are rational whenever $|R|$ and $|V|$ are less than n. Now consider $L(p, R, q)$, where $|R| = n$, and let $w \in L(p, R, q)$. We may suppose that the path from p to q having w as label visits p several times. To be more precise, we may write

$$w = z_1 z_2 \ldots z_k w'$$

for some $k \geq 0$, where

$$z_1, z_2, \ldots, z_k \in Z(p, R \setminus \{p\}, p).$$

If $q = p$ then $w' = 1$ and we have

$$L(p, R, p) = \left[Z(p, R \setminus \{p\}, p) \right]^*.$$

By the induction hypothesis $Z(p, R \setminus \{p\}, p)$ is rational, and hence $L(p, R, p)$ is rational.

So let us suppose that $q \neq p$, and let w_1 be the shortest left factor of w' for which $pw_1 = q$. Thus

$$w_1 \in Z(p, R \setminus \{p, q\}, q)$$

and $w = z_1 z_2 \ldots z_k w_1 w_2$, with

$$w_2 \in L(q, R \setminus \{p\}, q).$$

In fact we have

$$L(p, R, q) = \left[Z(p, R \setminus \{p\}, p) \right]^* Z(p, R \setminus \{p, q\}, q) L(q, R \setminus \{p\}, q) \quad (2.5.11)$$

and so, by the inductive hypothesis, $L(p, R, q)$ is rational.

Now consider w in $Z(p, V, q)$, where $|V| = n$ and $p, q \notin V$. Then there is a path

$$p \xrightarrow{a} r \longrightarrow \cdots \longrightarrow s \xrightarrow{b} q$$

with label w, where $w = aw'b$ and $r, s \in V$. Evidently we have $w' \in L(r, V, s)$. Indeed we have

$$Z(p, V, q) = \bigcup_{r, s \in V} \left[A(p, r) L(r, V, s) A(s, q) \right]. \quad (2.5.12)$$

By the argument of the previous paragraph $L(r, V, s)$ is rational. Hence $Z(p, V, q)$ is rational also. This completes the proof of Theorem 2.5.2. ∎

Exercises 2

2.1. Describe the languages recognized by the following automata. In all cases the alphabet is $\{a, b\}$.

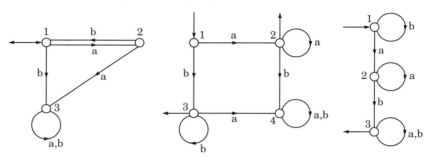

2.2. Consider the (incomplete, non-deterministic) automaton $\mathcal{A}$ given by the diagram

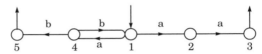

(i) Describe the language $L(\mathcal{A})$ recognized by $\mathcal{A}$.

(ii) Use the routine described in Examples 2.3.14 and 2.3.17 to obtain (a) a complete deterministic accessible automaton recognizing $L(\mathcal{A})$; (b) a trim deterministic automaton recognizing $L(\mathcal{A})$.

2.3. Consider the (incomplete non-deterministic) automaton $\mathcal{A}$ given by the diagram

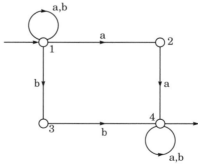

(i) Describe the language $L(\mathcal{A})$ recognized by $\mathcal{A}$.

(ii) Use the routine described in Examples 2.3.14 and 2.3.15 to obtain a complete deterministic trim automaton recognizing $L(\mathcal{A})$.

2.4. Let $A = \{a\}$, an alphabet consisting of one letter only, and let L be an infinite subset of a^*. Show that L is recognizable if and only if there exist $k, l \geq 1$ such that, for all $n \geq l$,

$$a^n \in L \text{ if and only if } a^{n+k} \in L.$$

2.5. Show that the subsets $\{a^{n^2} : n = 1, 2, \ldots\}$ and $\{a^p : p \text{ is prime}\}$ of a^* are not recognizable.

2.6. Use the Pumping Lemma to show that the following languages are not recognizable:

$$\text{(i) } \{a^n b^{2n} : n \geq 1\}; \quad \text{(ii) } \{a^n b^n a^n : n \geq 1\};$$
$$\text{(iii) } \{a^{n!} b^{n!} : n \geq 1\}; \quad \text{(iv) } \{z^2 : z \in \{a, b\}^*\}.$$

2.7. (i) Suppose that L $(\subseteq A^*)$ is recognized by an automaton $\mathcal{A} = (Q, A, i, T)$. Describe an automaton recognizing the complement $A^* \backslash L$ of L in A^*.
(ii) Suppose that L_1, L_2 $(\subseteq A^*)$ are recognized respectively by automata

$$\mathcal{A}_1 = (Q_1, A, \varphi_1, i_1, T_1), \quad \mathcal{A}_2 = (Q_2, A, \varphi_2, i_2, T_2).$$

Describe an automaton recognizing the intersection $L_1 \cap L_2$ of L_1 and L_2.

2.8. Show that Rat A^* is closed under 'relative complementation'. That is, show that if $L, M \in$ Rat A^* then so does $L \backslash M$ (defined as $L \cap (A^* \backslash M)$).

2.9. For a subset L of A^*, define

$$L^R = \{w^R : w \in L\}.$$

Show that
$$(L_1 \cup L_2)^R = L_1^R \cup L_2^R,$$
$$(L_1 L_2)^R = L_2^R L_1^R,$$
$$(L^*)^R = (L^R)^*,$$

and deduce that $L \in$ Rat A^* if and only if $L^R \in$ Rat A^*.

2.10. Show that $L = \{a^m b^n : m \geq n\}$ is not recognizable.

2.11. (i) Show that $\{ww^R : w \in \{a, b\}^*\}$ (the set of palindromes of even length) is not recognizable.

(ii) Let $A = \{a, b, c\}$. Show that the language $\{wcw^R : w \in \{a, b\}^*\}$ in A^* is not recognizable.

2.12. Show that $\{w \in \{a, b\}^* : |w|_a = |w|_b\}$ is not recognizable. What about $\{w \in \{a, b\}^* : |w|_a \geq |w|_b\}$?

2.13. Define a 'polyautomaton' to be just like an automaton, but with possibly more than one initial state. Let Polyrec A^* be the set of subsets of A^* that are recognized by a polyautomaton. Show that

$$\text{Polyrec } A^* = \text{Rec } A^*;$$

that is, show that no advantage is gained by generalizing an automaton in this way.

2.14. Define a 'monoautomaton' to be just like an automaton, but with a single terminal state. Let Monorec A^* be the set of subsets of A^* that are recognizable by a monoautomaton. Let $\mathcal{A} = (Q, A, i, t)$ be a monoautomaton. Show (i) that if $1 \in L(\mathcal{A})$ then $t = i$; (ii) that if $t = i$ then (in the notation of formula (2.5.7))

$$L(\mathcal{A}) = \left(Z(i, Q\backslash\{i\}, i)\right)^*.$$

Deduce that, for example, a^*b^* belongs to Rat A^* but does not belong to Monorec A^*.

3 The syntactic monoid

Introduction

In this chapter we obtain another characterization of rational languages in terms of a monoid $\mathrm{Syn}(L)$, called the *syntactic monoid* of L, associated with every $L \subseteq A^*$. The computation of $\mathrm{Syn}(L)$ is then discussed, via a notion important in its own right, namely the *minimal automaton* of L.

3.1 The syntactic congruence

Let A be a finite alphabet and let $L \subseteq A^*$. The *syntactic congruence* σ_L on A^* is defined by

$$\sigma_L = \{(w, z) \in A^* \times A^* : (\forall u, v \in A^*)$$
$$uwv \in L \text{ if and only if } uzv \in L\}. \tag{3.1.1}$$

The use of the word 'syntactic' perhaps requires some justification. We can think of the elements u and v as providing a *context* for w and z. We may then think of w and z as *syntactically equivalent* if in every context either both uwv and uzv are in the language or both uwv and uzv are outside the language. We shall return to this idea in Chapter 4.

More urgently, we must of course justify the word 'congruence'. In fact we shall show the more general result that for every monoid M and every subset P of M the relation

$$\sigma_P = \{(x, y) \in M \times M : (\forall u, v \in M)$$
$$uxv \in P \text{ if and only if } uyv \in P\}. \tag{3.1.2}$$

is a congruence on M. To see this, notice first that σ_P is obviously an equivalence. Suppose now that $(x, y) \in \sigma_P$ and $a \in M$. Then (for all u, v in M) $u(ax)v \in P$ if and only if $(ua)xv \in P$, i.e., if and only if $(ua)yv \in P$, i.e., if and only if $u(ay)v \in P$. Hence $(ax, ay) \in \sigma_P$, and similarly $(xa, ya) \in \sigma_P$. Thus σ_P is a congruence.

Let us explore this general situation a little further before returning to the case where $M = A^*$. Suppose that $x \ \sigma_P \ y$ and that $x \in P$. Then $1x1 \in P$ and so by the definition (3.1.2) $y = 1y1 \in P$. In effect $x \ \sigma_P \ y$ implies that either both x and y are in P or both x and y are outside P. In the terminology of Section 1.5, the congruence σ_P *respects* P. In fact we have

Theorem 3.1.3. *Let M be a monoid and let P be a subset of M. The syntactic congruence σ_P is the unique maximum congruence on M respecting P.*

Proof. Suppose that ρ is a congruence on M respecting P, and let $(x, y) \in \rho$. Then $(uxv, uyv) \in \rho$ for all u, v in M, and so for each choice of u, v either uxv, $uyv \in P$ or uxv, $uyv \notin P$. That is, for all u, v in M,

$$uxv \in P \text{ if and only if } uyv \in P.$$

Thus $(x, y) \in \sigma_P$. ∎

Returning now to the case where $M = A^*$ and using Definition 3.1.1, we now define $\mathrm{Syn}(L)$, the *syntactic monoid* of L, to be A^*/σ_L. Before stating the result that demonstrates the importance of this notion, we introduce another idea. We say that a language L in A^* is *recognized by a monoid* M if there exists a morphism $\varphi : A^* \to M$ and a subset P of M such that

$$\varphi^{-1}(P) = L.$$

A language L is certainly recognized by its syntactic monoid $\mathrm{Syn}(L)$, for if $P = \sigma_L^\natural(L)$ then

$$L = (\sigma_L^\natural)^{-1}(P).$$

We now have

Theorem 3.1.4. *Let A be a finite alphabet and let $L \subseteq A^*$. The following statements are equivalent:*
  (i) *L is a rational subset of A^*;*
  (ii) *$\mathrm{Syn}(L)$ is finite;*
  (iii) *L is recognized by a finite monoid M.*

Proof. (i)$\Rightarrow$(ii). Suppose that L is a rational subset of A^*. Then by Kleene's Theorem (2.6.2) $L = L(\mathcal{A})$ for some deterministic automaton $\mathcal{A} = (Q, A, \varphi, i, T)$. Define a relation τ on A^* by the rule that

$$w \ \tau \ z \text{ if and only if } (\forall q \in Q) \ qw = qz. \tag{3.1.5}$$

It is clear that τ is an equivalence, and we now show that it is a congruence. Let $w \tau z$ and let $u \in A^*$. Then from the fact that $qw = qz$ for all q in Q we easily deduce that

$$q(wu) = (qw)u = (qz)u = q(zu),$$

and hence that $wu \tau zu$. Also, since $qu \in Q$, we can deduce from (3.1.5) that $(qu)w = (qu)z$ for all q in Q, and hence that $uw \tau uz$.

The crucial point now is that the monoid A^*/τ is finite, and this follows from the finiteness of Q. For suppose that $Q = \{q_1, \ldots, q_n\}$. Then a τ-class $w\tau$ ($w \in A^*$) is determined by the values from Q taken by $q_1 w, \ldots, q_n w$. It follows that the total number of τ-classes cannot exceed n^n.

Suppose now that $w, z \in A^*$ and that $w \tau z$. Then $uvw \tau uzw$ for all u, v in A^* and so, for every q in Q,

$$quwv = quzv.$$

In particular
$$iuwv \in T \text{ if and only if } iuzv \in T;$$

that is,
$$uwv \in L \text{ if and only if } uzv \in L.$$

Thus $(w, z) \in \sigma_L$.

We have shown that $\tau \subseteq \sigma_L$. By Theorem 1.5.8 we thus have

$$\mathrm{Syn}(L) = A^*/\sigma_L \simeq (A^*/\tau)/(\sigma_L/\tau),$$

a quotient monoid of the finite monoid A^*/τ. Thus $\mathrm{Syn}(L)$ is finite.

It is clear that (ii)$\Rightarrow$(iii).

(iii)$\Rightarrow$(i). Suppose that we have a finite monoid M, a morphism $\varphi : A^* \to M$ and a subset P of M such that $L = \varphi^{-1}(P)$. We wish to show that L is rational. By Kleene's Theorem (2.5.2) it is enough to show that $L = L(\mathcal{A})$ for some finite state automaton $\mathcal{A} = (Q, A, i, T)$. Let $Q = M$. (This is certainly a finite set, which is all that we require for Q.) Define the action of A on Q by the rule that

$$q.a = q\varphi(a) \quad (q \in Q, \ a \in A).$$

This action extends in the usual way to an action of A^* on Q, and we have
$$q.w = q\varphi(w) \quad (q \in Q, \ w \in A^*).$$

Let $i = 1$ (the identity element of M) and let $T = P$. Then $w \in L(\mathcal{A})$ if and only if $i.w \in T$, i.e., if and only if $1\varphi(w) \in P$, i.e., if and only if $w \in \varphi^{-1}(P) = L$. ∎

There is a sense in which the syntactic monoid $\mathrm{Syn}(L)$ is the 'best' monoid recognizing L. This is important enough to be worth recording precisely and in more detail. To do so we need the concept first introduced in Section 1.5: given two monoids S and T, we say that S *divides* T, and write $S\,|\,T$, if there is a submonoid U of T and a surjective morphism from U to S. In other words, S is a *morphic image of a submonoid* of T. Then we have

Theorem 3.1.6. *Let A be a finite alphabet, let $L \subseteq A^*$ and let M be a monoid. The following statements are equivalent:*
  (i) *M recognizes L;*
  (ii) $\mathrm{Syn}(L)$ *divides M.*

Proof. (i)$\Rightarrow$(ii). By assumption there is a morphism $\varphi : A^* \to M$ and a subset P of M such that $\varphi^{-1}(P) = L$. We may certainly suppose that P is contained in the submonoid $S = \mathrm{im}\,\varphi$ of M, and so in effect we have a diagram

$$A^* \quad \xrightarrow{\;\varphi\;} \quad S$$
$$\Big\downarrow {\scriptstyle \sigma_L^\natural}$$
$$\mathrm{Syn}(L)$$

in which both the morphisms are surjective.

Now the congruence $\ker \varphi$ respects L, for

$$x \in L \text{ and } (x,y) \in \ker \varphi$$
$$\Rightarrow \varphi(y) = \varphi(x) \in P \;\Rightarrow\; y \in L.$$

Hence, by Theorem 3.1.3, $\ker \phi \subseteq \sigma_L$. By Theorem 1.5.7 there is a morphism θ from S onto $\mathrm{Syn}(L)$ such that $\theta \circ \varphi = \sigma_L^\natural$. We have thus shown that $\mathrm{Syn}(L)$ is a homomorphic image of a submonoid S of M, and so $\mathrm{Syn}(L)$ divides M as required.

(ii)$\Rightarrow$(i). Suppose that M has a submonoid S and that there exists a surjective morphism $\psi : S \to \mathrm{Syn}(L)$:

$$M \;\supseteq\; S \;\xrightarrow{\;\psi\;}\; \mathrm{Syn}(L)$$
$$\Big\uparrow {\scriptstyle \sigma_L^\natural}$$
$$A^*$$

For each a in A we can choose $\theta(a)$ to be some element in $\psi^{-1}\big(\sigma_L^\natural(a)\big)$; then

$$\psi(\theta(a)) = \sigma_L^\natural(a).$$

(Since ψ is surjective the choice is always possible.) We thus have a map $\theta : A \to S$, and by Theorem 1.7.3 this can be extended uniquely to a morphism $\theta : A^* \to S$. Since the maps $\psi \circ \theta$ and $\sigma_L^\natural$ coincide on A, their unique extensions to A^* also coincide. That is, $\psi \circ \theta = \sigma_L^\natural$, as morphisms from A^* into $\mathrm{Syn}(L)$. Now let $P = \theta(L)$. Then certainly $\theta^{-1}(P) \supseteq L$. To show that $\theta^{-1}(P) = L$, suppose that $w \in \theta^{-1}(P)$. Then $\theta(w) \in P$ and so there exists z in L such that $\theta(z) = \theta(w)$. It follows that $\psi(\theta(z)) = \psi(\theta(w))$ and hence that $\sigma_L^\natural(z) = \sigma_L^\natural(w)$. Hence $z \, \sigma_L \, w$. Since $z = 1z1 \in L$ it now follows by the definition of σ_L that $w \in L$. Thus $L = \theta^{-1}(P)$ and so M recognizes L. ∎

3.2 The minimal automaton

Before introducing the idea of the *minimal automaton* associated with a language L in $\mathrm{Rat}\, A^*$, we need a little more algebraic background and notation. Let M be a monoid or semigroup, let X be an arbitrary subset of M, and let $u \in M$. We define

$$u^{-1}X = \{s \in M : us \in X\}. \tag{3.2.1}$$

Notice carefully that this notation does *not* imply the existence of an inverse u^{-1} for u in M. The only meaning to be attached to $u^{-1}X$ is the meaning given by the definition (3.2.1) — though, as it happens, when M is a group the set $u^{-1}X$ coincides with the product of the element u^{-1} and the set X.

The other point to note at the outset is that $u^{-1}X$ may well be empty. For future use we record the obvious result that

$$1^{-1}X = X \tag{3.2.2}$$

and the only slightly less obvious result that

$$(uv)^{-1}X = v^{-1}(u^{-1}X) \quad (u, v \in M) \tag{3.2.3}$$

which we verify as follows:

$$(uv)^{-1}X = \{s \in M : uvs \in X\}$$
$$= \{s \in M : vs \in u^{-1}X\} = v^{-1}(u^{-1}X).$$

Recall now from Section 1.1 that $\mathcal{P}(A^*)$ denotes the set of all subsets of A^*. We define an action of A^* on $\mathcal{P}(A^*)$ by the rule that, for $X \subseteq A^*$ and $u \in A^*$,

$$X.u = u^{-1}X. \tag{3.2.4}$$

Notice now that

$$X.1 = X, \ X.(uv) = (X.u).v, \quad (u, v \in A^*),$$

and so we have gone some way towards defining an automaton. Of course $\mathcal{P}(A^*)$, the obvious candidate for the set of states, is infinite, but we shall soon see how to get round this point.

Let L be an arbitrary subset of A^*. Define

$$Q_L = \{u^{-1}L : u \in A^*\}, \tag{3.2.5}$$

$$i_L = L \,(= 1^{-1}L \in Q_L), \ T_L = \{u^{-1}L : u \in L\}. \tag{3.2.6}$$

An alternative characterization of T_L is given by

$$T_L = \{X \in Q_L : 1 \in X\};$$

for $1 \in u^{-1}L$ if and only if $u.1 \in L$, i.e., if and only if $u \in L$.

We do not yet know whether Q_L is finite, but at this stage we can define a (possibly infinite) automaton

$$\mathcal{A}_L = (Q_L, A, i_L, T_L), \tag{3.2.7}$$

and we notice right away that

$$i_L.w \in T_L \text{ if and only if } w^{-1}L \in T_L,$$
$$\text{i.e., if and only if } w \in L.$$

So if Q_L is finite then the automaton $\mathcal{A}_L$ recognizes L.

In fact, we have

Theorem 3.2.8. *The set Q_L defined by (3.2.5) is finite if and only if L is a rational set.*

Proof. Suppose first that Q_L is finite. Then, as we have seen, the automaton $\mathcal{A}_L$ defined by (3.2.7) recognizes L, and so L is rational by Kleene's Theorem (2.5.2).

Conversely, suppose that L is rational, so that $L = L(\mathcal{A})$ for some accessible deterministic automaton $\mathcal{A} = (Q, A, i, T)$. Define a map $\zeta : Q \to \mathcal{P}(A^*)$ by the rule that

$$\zeta(q) = \{w \in A^* : qw \in T\}.$$

In fact $\zeta(q) \in Q_L$ for all q in Q; for $q = iu$ for some u in A^* by the accessibility property, and so

$$w \in \zeta(q) \text{ if and only if } iuw \in T$$
$$\text{i.e., if and only if } uw \in L$$
$$\text{i.e., if and only if } w \in u^{-1}L.$$

That is, $\zeta(q) = u^{-1}L \in Q_L$, and we may thus regard ζ as a map from Q into Q_L. In fact the map $\zeta : Q \to Q_L$ is surjective, since for all $u^{-1}L$ in Q_L we can consider the state $q = iu$ of $\mathcal{A}$ and then observe that

$$\zeta(q) = \{w \in A^* : iuw \in T\}$$
$$= \{w \in A^* : uw \in L\} = u^{-1}L.$$

Our assumption is that Q is finite. Since ζ maps Q onto Q_L it follows that Q_L is finite also.   ∎

It is a consequence of our method of proof that if L is a rational subset of A^* and $\mathcal{A} = (Q, A, i, T)$ is a deterministic accessible automaton recognizing L, then $|Q| \geq |Q_L|$. For this reason it is natural to call $\mathcal{A}_L$, defined by (3.2.7), the *minimal automaton* recognizing L. We shall return to this point later.

The construction of $\mathcal{A}_L$ is elegant algebraically, but it would be wrong to claim too much for it. If L is specified in a way that is at all complicated then $\mathcal{A}_L$ can be hard to compute. In simple cases, however, the routine is easy enough.

Example 3.2.9. Let $A = \{a, b\}$, and let

$$L = A^*abA^* = \{w \in A^* : ab \text{ is a segment of } w\}.$$

Then $i_L = L$, and

$$L.a = \{w \in A^* : aw \in L\} = L \cup ba^* = L_1 \text{ (say)},$$
$$L.b = \{w \in A^* : bw \in L\} = L,$$
$$L_1.a = \{w \in A^* : a^2w \in L\} = L \cup ba^* = L_1,$$
$$L_1.b = \{w \in A^* : abw \in L\} = A^* = L_2 \text{ (say)},$$
$$L_2.a = L_2.b = L_2.$$

The set of terminal states is given by

$$T_L = \{X \in Q_L : 1 \in X\} = \{L_2\},$$

and the state diagram of $\mathcal{A}_L$ is

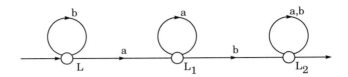

Example 3.2.10. Let $A = \{a, b, c\}$ and let

$$L = a^+ b^+ c^+ = \{a^m b^n c^p : m, n, p \geq 1\}.$$

Here $i_L = L$ and

$$L.a = \{w \in A^* : aw \in L\} = L \cup b^+ c^+ = L_1,$$
$$L.b = \{w \in A^* : bw \in L\} = \emptyset,$$
$$L.c = \{w \in A^* : cw \in L\} = \emptyset,$$
$$L_1.a = \{w \in A^* : a^2 w \in L\} = L \cup b^+ c^+ = L_1,$$
$$L_1.b = \{w \in A^* : abw \in L\} = b^* c^+ = L_2,$$
$$L_1.c = \{w \in A^* : acw \in L\} = \emptyset,$$
$$\emptyset.a = \emptyset.b = \emptyset.c = \emptyset,$$
$$L_2.a = \{w \in A^* : abaw \in L\} = \emptyset,$$
$$L_2.b = \{w \in A^* : ab^2 w \in L\} = b^* c^+ = L_2,$$
$$L_2.c = \{w \in A^* : abcw \in L\} = c^* = L_3,$$
$$L_3.a = \{w \in A^* : abcaw \in L\} = \emptyset,$$
$$L_3.b = \{w \in A^* : abcbw \in L\} = \emptyset,$$
$$L_3.c = \{w \in A^* : abc^2 w \in L\} = c^* = L_3.$$

Here $T_L = \{L_3\}$, and the state diagram of $\mathcal{A}_L$ is

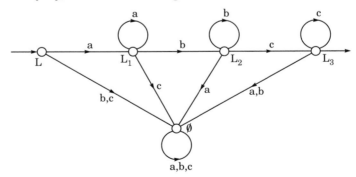

Notice that in each case the finiteness of the minimal automaton provides a verification that L is rational. It is instructive to apply the routine to an L that is known not to be rational.

Example 3.2.11.   Let $A = \{a, b\}$ and let

$$L = \{a^n b^n : n \geq 1\}.$$

By Theorem 2.2.3 the set L is not rational. We have

$$i_L = L,$$

$$L.a = \{w \in A^* : aw \in L\} = \{a^{n-1} b^n : n \geq 1\} = L_1 \text{ (say)},$$

$$L.b = \{w \in A^* : bw \in L\} = \emptyset,$$

$$\emptyset.a = \emptyset.b = \emptyset.$$

More generally,

$$L.a^r = \{a^{n-r} b^n : n \geq r\} = L_r \text{ (say)},$$

and

$$L_r.b^s = \{w \in A^* : b^s w \in L_r\} = \begin{cases} \emptyset & \text{if } s > r, \\ \{b^{r-s}\} & \text{if } s \leq r. \end{cases}$$

(Notice that L_r contains b^r.) We obtain a state diagram

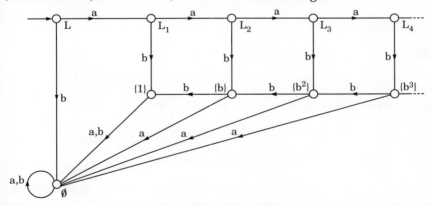

The fact that this is infinite constitutes a demonstation that L is not rational — but of course we have already seen a simpler proof of this fact.

3.3   Reduced automata

In Section 2.3 we looked at various desirable properties that an automaton might have, such as completeness, determinacy, accessibility and coaccessibility. In order to keep the confusion of that section to a tolerable level we have until now delayed consideration of a final desirable feature of an automaton.

Let $\mathcal{A} = (Q, A, i, T)$ be a complete, deterministic, accessible automaton. With an obvious extension of the notation (3.2.1), for each q in Q let us denote the subset $\{w \in A^* : qw \in T\}$ of A^* by $q^{-1}T$, and let us define an equivalence relation ρ on Q by

$$\rho = \{(q_1, q_2) \in Q \times Q : q_1^{-1}T = q_2^{-1}T\}. \tag{3.3.1}$$

We say that the automaton $\mathcal{A}$ is *reduced* if ρ is the identical relation on Q , i.e., if

$$q_1^{-1}T = q_2^{-1}T \;\Rightarrow\; q_1 = q_2.$$

Theorem 3.3.2. *For every rational language L the minimal automaton $\mathcal{A}_L$ given by (3.2.7) is reduced.*

Proof. Let $q_1 = u_1^{-1}L$ and $q_2 = u_2^{-1}L$ belong to Q_L and suppose that $q_1 \,\rho\, q_2$, so that

$$\{w \in A^* : q_1.w \in T_L\} = \{w \in A^* : q_2.w \in T_L\}.$$

That is,

$$\{w \in A^* : w^{-1}u_1^{-1}L \in T_L\} = \{w \in A^* : w^{-1}u_2^{-1}L \in T_L\}.$$

Thus

$$\{w \in A^* : (u_1 w)^{-1}L \in T_L\} = \{w \in A^* : (u_2 w)^{-1}L \in T_L\}.$$

Hence, from the definition (3.2.6) of T_L,

$$\{w \in A^* : u_1 w \in L\} = \{w \in A^* : u_2 w \in L\}.$$

That is, $u_1^{-1}L = u_2^{-1}L$. Thus $q_1 = q_2$ as required. ∎

Let $\mathcal{A} = (Q, A, \varphi, i, T)$ be a complete, deterministic, accessible automaton, and let ρ be the equivalence on Q defined by (3.3.1). We define the *quotient automaton* $\mathcal{A}/\rho$ to be

$$(Q/\rho, A, \psi, i\rho, T'), \tag{3.3.3}$$

where

$$T' = \{t\rho : t \in T\} \tag{3.3.4}$$

and (for all q in Q and a in A)

$$\psi(q\rho, a) = \big(\varphi(q, a)\big)\rho. \tag{3.3.5}$$

We shall usually want to write this more simply as

$$(q\rho)a = (qa)\rho. \tag{3.3.6}$$

We must of course convince ourselves that this action of A on Q/ρ is well-defined, i.e., that $q_1\rho = q_2\rho$ implies $\psi(q_1\rho, a) = \psi(q_2\rho, a)$. So suppose that $q_1\rho = q_2\rho$. Then, for all w in A^*, $q_1w \in T$ if and only if $q_2w \in T$. This holds in particular for all w in aA^*. Thus, for all z in A^*, $q_1az \in T$ if and only if $q_2az \in T$. From this we deduce that $(q_1a)\rho = (q_2a)\rho$ — which is exactly what we require. Thus A/ρ is indeed an automaton. We show also that

$$T' = \{q\rho \in Q/\rho : q\rho \cap T \neq \emptyset\} = \{q\rho \in Q/\rho : q\rho \subseteq T\}.$$

The first of these equalities is clear from (3.3.4). As for the second equality, it is clear that $q\rho \subseteq T$ implies that $q\rho \cap T \neq \emptyset$. To show the opposite implication, notice that if $q\rho \cap T \neq \emptyset$ then there exists t in $q\rho$ such that $t \in T$. But then $q' \, \rho \, t$ for *every* q' in $q\rho$ and so $(q')^{-1}T \; (= t^{-1}T)$ contains 1. It follows that $q' \in T$, as required.

Theorem 3.3.7. *Let* $A = (Q, A, \varphi, i, T)$ *be a complete, deterministic, accessible automaton. Then the automaton* A/ρ *defined by (3.3.3) is reduced, and* $L(A/\rho) = L(A)$.

Proof. The formula (3.3.6) extends in an obvious way: for all q in Q and w in A^*

$$(q\rho)w = (qw)\rho.$$

Let $q_1\rho$, $q_2\rho \in Q/\rho$ and suppose that

$$(q_1\rho)^{-1}T' = (q_2\rho)^{-1}T'.$$

Then

$$\{w : (q_1\rho)w \in T'\} = \{w : (q_2\rho)w \in T'\}.$$

That is,

$$\{w : (q_1w)\rho \in T'\} = \{w : (q_2w)\rho \in T'\}.$$

That is, by the definition of T',

$$\{w : q_1w \in T\} = \{w : q_2w \in T\}.$$

Thus $q_1^{-1}T = q_2^{-1}T$, and so $q_1\rho = q_2\rho$. Thus A/ρ is reduced.

Next, $w \in L(A/\rho)$ if and only if $(i\rho)w \in T'$, i.e., if and only if $iw \in T$, i.e., if and only if $w \in L(A)$. ∎

This theorem tells us that if we have a complete, deterministic, accessible automaton recognizing a given rational language — and we saw in Section 2.3 how this can be achieved — then we can arrange for it to be reduced. (It is not hard to see that our routine for producing a reduced automaton does not destroy the completeness, the determinism, or the accessibility.)

Example 3.3.8.   Let $L = A^*abA^* \cup A^*baA^*$, where $A = \{a, b\}$. An obvious (incomplete and nondeterministic) automaton recognizing L is the one with state diagram

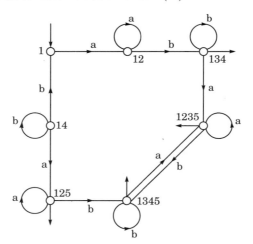

If we copy the routine of Section 2.3 we obtain a complete, deterministic accessible automaton $\mathcal{A}$ such that $L(\mathcal{A}) = L$. Its state diagram is

The automaton $\mathcal{A}$ is not reduced:

$$T = \{\{1, 2, 5\}, \{1, 3, 4\}, \{1, 2, 3, 5\}, \{1, 3, 4, 5\}\},$$

and

$$\{1\}^{-1}T = A^*abA^* \cup A^*baA^* = L,$$

$$\{1, 2\}^{-1}T = L \cup bA^*, \quad \{1, 4\}^{-1}T = L \cup aA^*,$$

$$\{1, 2, 5\}^{-1}T = \{1, 3, 4\}^{-1}T = \{1, 2, 3, 5\}^{-1}T = \{1, 3, 4, 5\}^{-1}T = A^*.$$

Thus ρ identifies the four terminal states of $\mathcal{A}$, and $\mathcal{A}/\rho$ has the much simpler state diagram

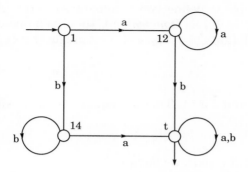

This new automaton is reduced. If we apply to this same language
L the routine for finding the minimal automaton, we find

$$L.a = L \cup bA^* = L_1, \; L.b = L \cup aA^* = L_2,$$

$$L_1.a = L_1, \; L_1.b = A^*,$$

$$L_2.a = A^*, \; L_2.b = L_2,$$

$$A^*.a = A^*.b = A^*.$$

The state diagram is

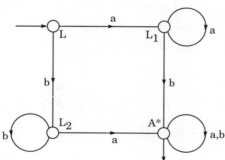

The fact that the state diagrams of $\mathcal{A}/\rho$ and $\mathcal{A}_L$ are in effect identical
is not an accident, as we shall see.

The proof of Theorem 3.2.8 involved a map $\zeta : Q \to Q_L$ given in our
new notation by

$$\zeta(q) = q^{-1}T \; (q \in Q), \tag{3.3.9}$$

and we showed that ζ maps Q onto Q_L. Notice that

$$\zeta(q) = \zeta(q') \;\Rightarrow\; q \, \rho \, q';$$

thus if the automaton $\mathcal{A} = (Q, A, i, T)$ is reduced then ζ is both injective
and surjective.

It is reasonable to define two automata $\mathcal{A}_1 = (Q_1, A, \varphi_1, i_1, T_1)$ and $\mathcal{A}_2 = (Q_2, A, \varphi_2, i_2, T_2)$ as *isomorphic* if there is a one-one map ζ from Q_1 onto Q_2 such that $\zeta(i_1) = i_2$, $q_1 \in T_1$ if and only if $\zeta(q_1) \in T_2$, and

$$\zeta\big(\varphi_1(q_1, a)\big) = \varphi_2\big(\zeta(q_1), a\big)$$

for all q_1 in Q_1 and a in A. This last equality is more simply expressed as

$$\zeta(q_1 a) = \zeta(q_1)a.$$

With these definitions, we obtain

Theorem 3.3.10. *Let L be a rational language in A^*, let $\mathcal{A}_L$ be the minimal automaton defined by (3.2.7), and let $\mathcal{A} = (Q, A, i, T)$ be any complete, deterministic, accessible, reduced automaton recognizing L. Then $\mathcal{A}$ is isomorphic to $\mathcal{A}_L$.*

Proof. The one-one map ζ from Q onto Q_L is given by (3.3.9). We have

$$\zeta(i) = i^{-1}T = \{w \in A^* : iw \in T\} = L = i_L,$$

and also

$$q \in T \text{ if and only if } 1 \in q^{-1}T,$$
$$\text{i.e., if and only if } q^{-1}T \in T_L,$$
$$\text{i.e., if and only if } \zeta(q) \in T_L.$$

Finally,

$$\zeta(qa) = (qa)^{-1}T = a^{-1}(q^{-1}T) = (q^{-1}T).a \quad \textbf{(by definition)}$$
$$= \zeta(q).a. \qquad \blacksquare$$

There are thus two routes towards the minimal automaton for a rational language L. We may either take the direct algebraic route and compute $\mathcal{A}_L$, or we may guess at some kind of an automaton recognizing L and then proceed by stages to refine L until we arrive at a complete, deterministic, accessible, reduced automaton. Theorem 3.3.10 assures us that the two procedures lead in the end to the same automaton. Example 3.3.8 illustrates both approaches, and we have already noted that the two automata are identical in this case.

The second of these procedures looks on the face of it as though it would be much the longer, but of course not all of the stages in the refinement process will always be necessary. Another example may be helpful here:

Example 3.3.11. Let $A = \{a, b\}$. The automaton with state diagram

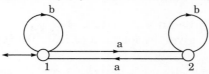

recognizes the language $L_1 = \{w \in A^* : |w|_a \text{ is even}\}$ and the automaton

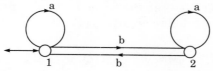

recognizes the language $L_2 = \{w \in A^* : |w|_b \text{ is even}\}$. The routine developed in Exercise 2.6.7 gives an automaton with state diagram

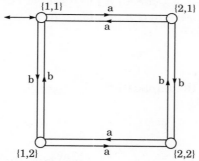

recognizing $L_1 \cap L_2$. Notice that

$$T = \{(1,1)\}, \quad (1,1)^{-1}T = L_1 \cap L_2,$$

$$(2,1)^{-1}T = \{w \in A^* : |w|_a \text{ is odd}, |w|_b \text{ is even}\},$$
$$(1,2)^{-1}T = \{w \in A^* : |w|_a \text{ is even}, |w|_b \text{ is odd}\},$$
$$(2,2)^{-1}T = \{w \in A^* : |w|_a \text{ and } |w|_b \text{ are odd}\}.$$

Thus the automaton is already reduced.

If we use the other routine, we have

$$L = \{w \in A^* : |w|_a \text{ and } |w|_b \text{ are even}\},$$
$$L.a = \{w \in A^* : |w|_a \text{ is odd}, |w|_b \text{ is even}\} = L_1,$$
$$L.b = \{w \in A^* : |w|_a \text{ is even}, |w|_b \text{ is odd}\} = L_2,$$
$$L_1.b = \{w \in A^* : |w|_a \text{ is even}, |w|_b \text{ is odd}\} = L_3,$$

$$L_1.a = L, \ L_2.b = L, \ L_2.a = L_3, \ L_3.a = L_2, \ L_3.b = L_1,$$

and we get an effectively identical diagram.

3.4 The transformation monoid of an automaton

Let $\mathcal{A} = (Q, A, i, T)$ be an automaton. Let us now look again at the congruence τ on A^* given by formula (3.1.5). We reproduce it here for convenience:

$$w \ \tau \ z \text{ if and only if } (\forall q \in Q) \ qw = qz. \qquad (3.4.1)$$

We saw in Section 3.1 that the monoid A^*/τ is finite. Denoting the τ-class of w by $\bar{w}$, we define an action of A^*/τ on Q by the rule that

$$q\bar{w} = qw \quad (q \in Q, \ w \in A^*).$$

This definition makes sense precisely because of the definition of τ : if $\bar{w}_1 = \bar{w}_2$ then $qw_1 = qw_2$ for all q in Q and so $q\bar{w}_1 = q\bar{w}_2$ for every q. Notice now that for all q in Q and all $\bar{w}, \bar{z}$ in A^*/τ we have

$$q(\bar{w}\bar{z}) = (q\bar{w})\bar{z}.$$

Moreover, we have the implication

$$(\forall q \in Q)q\bar{w} = q\bar{z} \quad \Rightarrow \quad \bar{w} = \bar{z},$$

and so — see formula (1.6.7) — $(Q, A^*/\tau)$ is a *transformation monoid*. We shall call $(Q, A^*/\tau)$ *the transformation monoid of the automaton $\mathcal{A}$* and denote it by $\mathrm{TM}(\mathcal{A})$.

To summarize, we have shown that there is a transformation monoid $\mathrm{TM}(\mathcal{A})$ associated in a natural way with every automaton $\mathcal{A}$. Conversely, if we are given a finite transformation monoid $X = (Q, M)$ then we can define an automaton $(Q, M\backslash\{1\}, \varphi)$ (in the primitive sense of the word, where no initial or terminal states are specified) by specifying that

$$\varphi(q, s) = qs \quad (q \in Q, \ s \in M\backslash\{1\}).$$

If we regard $M\backslash\{1\}$ as merely a set, notice that this does define an action of $(M\backslash\{1\})^*$ on Q in the usual way. But of course if the words $s_1 s_2 \ldots s_m$ and $t_1 t_2 \ldots t_n$ in $(M\backslash\{1\})^*$ are equal when regarded as products in $\mathbf{M}$ then their effect on each q in Q is the same.

We refer to this automaton as *the automaton of the transformation monoid $X = (Q, M)$* and denote it by $\mathrm{Aut}(X)$.

Example 3.4.2. Consider the automaton $\mathcal{A}$ given by the state diagram

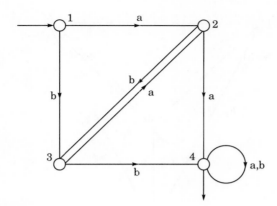

The actions of $\bar{a}$, $\bar{b}$ on Q are given by

$$\bar{a} = \begin{pmatrix} 1\,2\,3\,4 \\ 2\,4\,2\,4 \end{pmatrix}, \; \bar{b} = \begin{pmatrix} 1\,2\,3\,4 \\ 3\,3\,4\,4 \end{pmatrix}.$$

Then

$$\bar{a}^2 = \bar{b}^2 = \begin{pmatrix} 1\,2\,3\,4 \\ 4\,4\,4\,4 \end{pmatrix} = z \text{ (say)},$$

$$\bar{a}\bar{b} = \begin{pmatrix} 1\,2\,3\,4 \\ 3\,4\,3\,4 \end{pmatrix} = e \text{ (say)},$$

$$\bar{b}\bar{a} = \begin{pmatrix} 1\,2\,3\,4 \\ 2\,2\,4\,4 \end{pmatrix} = f \text{ (say)},$$

$$\bar{a}\bar{b}\bar{a} = \begin{pmatrix} 1\,2\,3\,4 \\ 2\,4\,2\,4 \end{pmatrix} = \bar{a}, \; \bar{b}\bar{a}\bar{b} = \begin{pmatrix} 1\,2\,3\,4 \\ 3\,3\,4\,4 \end{pmatrix} = \bar{b}.$$

So TM($\mathcal{A}$) consists of exactly 6 elements, namely 1, $\bar{a}$, $\bar{b}$, z, e and f. All are transformations of $\{1, 2, 3, 4\}$, and the multiplication table is easy to compute.

If we start with this transformation monoid

$$X = (Q, M) = \left(\{1, 2, 3, 4\}, \; \{1, \bar{a}, \bar{b}, e, f, z\}\right)$$

and construct $\mathrm{Aut}(X)$ according to the recipe we have described then we do not get back to the original automaton. We obtain a new automaton whose alphabet has 5 letters $\bar{a}$, $\bar{b}$, e, f, z, and whose state diagram is

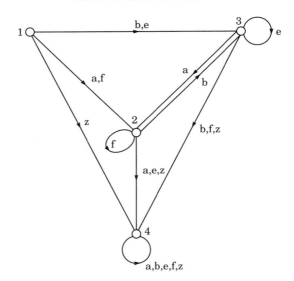

(For convenience we are dropping the bar on the letters a and b.)
However, if we now compute the transformation monoid of this new
automaton we get precisely the same 6-element monoid as before.

This example illustrates the imperfect nature of the correspondence
between automata and transformation monoids. We may well have
$\mathrm{Aut}\big(\mathrm{TM}(\mathcal{A})\big) \neq \mathcal{A}$ for an automaton $\mathcal{A}$. On the other hand we do have

Theorem 3.4.3. *For every finite transformation monoid $X = (Q, M)$,*

$$\mathrm{TM}\big(\mathrm{Aut}(X)\big) \simeq X.$$

Proof. Let $X = (Q, M)$, where Q is finite. Then

$$\mathrm{Aut}(X) = (Q, M\backslash\{1\}, \varphi),$$

where

$$\varphi(q, s) = qs \quad (q \in Q, s \in M),$$

the action of s on Q being determined by the transformation monoid.
We then define $\mathrm{TM}\big(\mathrm{Aut}(X)\big)$ as $\big(Q, (M\backslash\{1\})^*/\tau\big)$, where

$$(s_1 s_2 \ldots s_m)\tau = (t_1 t_2 \ldots t_n)\tau$$

if and only if $q(s_1 s_2 \ldots s_m) = q(t_1 t_2 \ldots t_n)$ for all q in Q.
    Define a map θ from $(M\backslash\{1\})^*$ onto M by the rule that $\theta(s_1 s_2 \ldots s_m)$
is the product $s_1 s_2 \ldots s_m$ of the elements $s_1, s_2, \ldots, s_m$ in M, and θ maps
the empty word 1 in $(M\backslash\{1\})^*$ to the identity 1 of M. (The abuse

of language involved in using the same expression $s_1 s_2 \ldots s_m$ for two distinct entities will not lead to any long-lasting confusion.) For any two elements $s_1 s_2 \ldots s_m$ and $t_1 t_2 \ldots t_n$ of $\left(M\backslash\{1\}\right)^*$,

$$\theta(s_1 s_2 \ldots s_m) = \theta(t_1 t_2 \ldots t_n)$$

if and only if $s_1 s_2 \ldots s_m = t_1 t_2 \ldots t_n$ as elements of M,

i.e., if and only if $(\forall q \in Q) \, qs_1 s_2 \ldots s_m = qt_1 t_2 \ldots t_n$,

i.e., if and only if $(s_1 s_2 \ldots s_m) \, \tau \, (t_1 t_2 \ldots t_n)$.

It thus follows by Theorem 1.5.6 that there is an isomorphism $\alpha :$ $\left(M\backslash\{1\}\right)^*/\tau \to M$ such that the diagram

$$
\begin{array}{ccc}
\left(M\backslash\{1\}\right)^* & \xrightarrow{\;\theta\;} & M \\[2pt]
\Big\downarrow{\scriptstyle \tau^{\natural}} & \nearrow{\scriptstyle \alpha} & \\[2pt]
\left(M\backslash\{1\}\right)^*/\tau & &
\end{array}
$$

commutes. Moreover, for each $s_1 s_2 \ldots s_m$ in $\left(M\backslash\{1\}\right)^*$ and every q in Q,

$$q\Big(\alpha\big((s_1 s_2 \ldots s_m)\tau\big)\Big) = q\big(\theta(s_1 s_2 \ldots s_m)\big) = q(s_1 s_2 \ldots s_m);$$

hence, recalling the definition in Section 1.6, we see that

$$(1_Q, \alpha)$$

is a TM-isomorphism from $(Q, (M\backslash\{1\})^*/\tau)$ onto (Q, M). ∎

Remark 3.4.4. Let $X = (Q, M)$ be a transformation monoid and let G be a set of generators for M. It is always possible to take G as $M\backslash\{1\}$, but there may well be a smaller generating set. Then, just as for the case $G = M\backslash\{1\}$, the automaton $\mathcal{A}' = (Q, G, \varphi)$, with

$$\varphi(q, g) = qg \; (q \in Q, g \in G),$$

has the property that $\mathrm{TM}(\mathcal{A}') = X$. If in the example (3.4.2), where

$$M = \{1, \bar{a}, \bar{b}, z, e, f\},$$

we choose $G = \{\bar{a}, \bar{b}\}$, then the automaton $\mathcal{A}' = (Q, G, \varphi)$ is in effect the automaton $\mathcal{A}$ from which we started.

Remark 3.4.5.   The state diagram of the automaton (Q, G, φ) is essentially the generalization to monoids of the Cayley diagrams of group theory. (See, for example, Coxeter and Moser 1964.)

3.5 The calculation of the syntactic monoid

Let A be a finite alphabet, let L be a rational subset of A^* and let $\mathcal{A}_L = (Q_L, A, i_L, T_L)$ be the minimal automaton of L, as described in Section 3.2. Recall that the states of $\mathcal{A}_L$ are subsets of the form $u^{-1}L$, where $u \in A^*$. The transformation monoid of $\mathcal{A}_L$ is $(Q_L, A^*/\tau)$, where $w \, \tau \, z$ if and only if

$$(u^{-1}L).w = (u^{-1}L).z$$

for all u in A^*, i.e.,— see (3.2.4) — if and only if

$$w^{-1}(u^{-1}L) = z^{-1}(u^{-1}L)$$

for all u in A^*. We can rewrite this condition in more elementary terms as

$$\{v : wv \in u^{-1}L\} = \{v : zv \in u^{-1}L\},$$

or as

$$\{v : uvw \in L\} = \{v : uzv \in L\}.$$

Equivalently we have, for all u, v in A^*,

$$uwv \in L \text{ if and only if } uzv \in L.$$

Thus τ is precisely the syntactic congruence σ_L defined in formula (3.1.2). We may express the conclusion formally as follows:

Theorem 3.5.1. *Let A be a finite alphabet and let L be a rational subset of A^*. The syntactic monoid of L coincides with the transformation monoid of the minimal automaton of L.* ∎

Example 3.5.2.   Let $A = \{a, b\}$ and let $L = A^* aba A^*$. The minimal automaton $\mathcal{A}_L$ of L is calculated as follows:

$$i_L = L,$$
$$L.a = \{w : aw \in L\} = L \cup baA^* = L_1,$$
$$L.b = \{w : bw \in L\} = L,$$
$$L_1.a = \{w : a^2 w \in L\} = L_1,$$
$$L_1.b = \{w : abw \in L\} = L \cup aA^* = L_2,$$
$$L_2.a = \{w : abaw \in L\} = A^* = L_3,$$
$$L_2.b = \{w : ab^2 w \in L\},$$
$$L_3.a = L_3.b = L_3.$$

The state diagram is

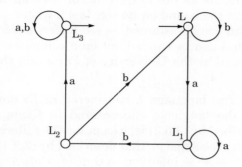

and (simplifying the notation by replacing L, L_1, L_2, L_3 by 1, 2, 3, 4 respectively) we have elements of the transformation monoid of $\mathcal{A}_L$ as follows. We write $ijkl$ for the map

$$\begin{pmatrix} 1\,2\,3\,4 \\ i\,j\,k\,l \end{pmatrix}$$

and denote the map 4444 by 0.

$a = 2244$	$b = 1314$	$a^2 = 2244 = a$	$ab = 3344$
$ba = 2424$	$b^2 = 1114$	$a^3 = a$	$a^2b = ab$
$aba = 4444 = 0$	$ab^2 = 1144$	$ba^2 = ba$	$bab = 3434$
$b^2a = 2224$	$b^3 = b^2$	$a^4 = a$	$a^3b = ab$
$a^2ba = 0$	$a^2b^2 = ab^2$	$aba^2 = 0$	$abab = 0$
$ab^2a = a$	$ab^3 = ab^2$	$ba^3 = ba$	$ba^2b = bab$
$baba = 0$	$bab^2 = 1414$	$b^2a^2 = b^2a$	$b^2ab = 3334$
$b^3a = b^2a$	$b^4 = b^2$	$a^5 = a$	$a^4b = ab$
$a^3ba = 0$	$a^3b^2 = ab^2$	$a^2ba^2 = 0$	$a^2bab = 0$
$a^2b^2a = a$	$a^2b^3 = ab^2$	$aba^3 = 0$	$aba^2b = 0$
$ababa = 0$	$abab^2 = 0$	$ab^2a^2 = a$	$ab^2ab = ab$
$ab^3a = a$	$ab^4 = ab^2$	$ba^4 = ba$	$ba^3b = bab$
$ba^2ba = 0$	$ba^2b^2 = bab^2$	$baba^2 = 0$	$babab = 0$
$bab^2a = ba$	$bab^3 = bab^2$	$b^2a^3 = b^2a$	$b^2a^2b = b^2ab$
$b^2aba = 0$	$b^2ab^2 = b^2$	$b^3a^2 = b^2a$	$b^3ab = b^2ab$
$b^4a = b^2a$	$b^5 = b^2.$		

Since the 32 words of length 5 give rise to no new elements, no longer word will give a new element either. So we conclude that the syntactic monoid consists of the 12 elements 1, a, b, ab, ba, b^2, 0, ab^2, bab, b^2a, bab^2 and b^2ab. It is easy to draw up a Cayley table if this is required.

Remark 3.5.4. The procedure adopted is to consider the two words of length 1, then the four words of length 2, then the eight words of length

3, and so on until no new elements appear. This can of course be a fairly lengthy procedure. As we saw in the proof of Theorem 3.1.4 the monoid A^*/τ is finite, and the bound on its size is n^n, where $n = |Q|$. But this fairly large bound can be attained — see Exercise 3.8. Evidently the process is one that can be carried out on a computer, and a program has been developed at the University of Paris and the University of Nebraska.

Remark 3.5.5. The language L considered in Example 3.5.2 is the complement of the language encountered in Example 2.2.1. The automaton described by (3.5.3) is, except for the different specification of terminal states, the same as that described by (2.2.1). The syntactic monoid of $A^*\backslash A^* aba A^*$ is identical to that of $A^* aba A^*$ — see Exercise 3.1.

Exercises 3

3.1. Show that $\sigma_P = \sigma_{M\backslash P}$ for every subset P of a monoid M.

3.2. Let M be a monoid and let P be a subset of M. Show that

$$(x, y) \in \sigma_P \text{ if and only if } (\forall u \in M)\, x^{-1} u^{-1} P = y^{-1} u^{-1} P.$$

3.3. Show that if L is a rational subset of A^* then the set

$$\{u^{-1} L v^{-1} : u, v \in A^*\}$$

is finite.

3.4. For a language L and a word w in A^* define

$$\begin{aligned} K(w) &= \{(u, v) \in A^* \times A^* : uwv \in L\} \\ &= \{(u, v) \in A^* \times A^* : w \in u^{-1} L v^{-1}\}. \end{aligned}$$

Notice that $w\, \sigma_L\, w'$ if and only if $K(w) = K(w')$, and deduce that

$$w\sigma_L = \left[\bigcap_{(u,v)\in K(w)} u^{-1} L v^{-1} \right] \Big\backslash \left[\bigcup_{(u,v)\notin K(w)} u^{-1} L v^{-1} \right].$$

3.5. Consider a singleton set $L = \{w\}$ in A^*, with $|w| = l \geq 1$. Show that (for x, y in A^*) $x\, \sigma_{\{w\}}\, y$ if and only if either (i) $x = y$ and both

x, y are segments of w, or (ii) neither x nor y is a segment of w. Show that the $\sigma_{\{w\}}$-class

$$\{x : x \text{ is not a segment of } w\}$$

is a zero element in the syntactic monoid $\mathrm{Syn}(\{w\})$. Denote it by 0. Show that for segments x, y of w the multiplication in $\mathrm{Syn}(\{w\})$ is given by

$$\{x\}\{y\} = \begin{cases} \{xy\} & \text{if } xy \text{ is a segment of } w, \\ 0 & \text{otherwise.} \end{cases}$$

Show that
$$\big(\mathrm{Syn}(L)\backslash\{1\}\big)^{l+1} = 0.$$

3.6. More generally, consider a finite set $L = \{w_1, w_2, \ldots, w_n\}$, and let

$$l = \max\{|w_1|, |w_2|, \ldots, |w_n|\}.$$

In the notation of Exercise 3.4, show that $K(x) = \emptyset$ whenever $|x| \geq l+1$, and show that in this case $x\sigma_L$ is a zero element for $\mathrm{Syn}(L)$. Show that

$$\big(\mathrm{Syn}(L)\backslash\{1\}\big)^{l+1} = 0.$$

3.7. Illustrate the ideas in the last two exercises by computing the Cayley table for the syntactic monoid M (of order 8) of $\{aba, bab\}$. Notice that $\big(M\backslash\{1\}\big)^4 = 0$.

3.8. Deduce from Exercise 1.7 that the transformation monoid of the automaton whose state diagram is

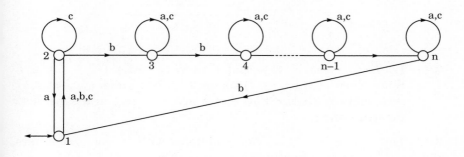

is T_n, of order n^n.

3.9. Let $A = \{a, b\}$. Find the minimal automaton of $L = A^*a^2A^* \cup A^*b^2A^*$, first by 'refining' the automaton

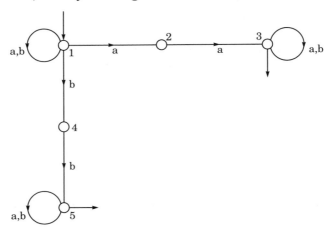

and then by direct algebraic calculation. Find the syntactic monoid of L.

3.10. Repeat this routine for

    (i) $L = a^+b^+$ and the automaton

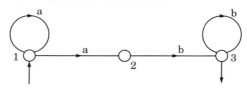

    (ii) $L = (ab)^+$ and the automaton

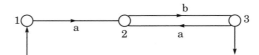

3.11. Let L be an *ideal* in A^* — that is, suppose that $A^*L \subseteq L$, $LA^* \subseteq L$. Show that L (which by Theorem 3.1.3 is a union of σ_L-classes) is a single σ_L-class. Show that L acts as a zero in the syntactic monoid $\mathrm{Syn}(L)$.

3.12. Let $A = \{a, b, \ldots\}$ be a finite alphabet with at least two letters, and let P be the set of *palindromes* in A^*; that is, let $P = \{w \in A^* : w^R = w\}$. Show that in the 'minimal automaton' of P the states $P, P.a, P.a^2, \ldots$ are all distinct. Deduce that P is not rational.

3.13. Let $\mathcal{A} = (Q, A, i, T)$ be a finite state automaton. Recall the equivalence

$$\tau = \{(q_1, q_2) \in Q \times Q : q_1^{-1}T = q_2^{-1}T\},$$

defined by (3.3.1). On the face of it, the verification that $q_1^{-1}T = q_2^{-1}T$ involves the computation of two infinite sets $\{w \in A^* : q_1 w \in T\}$ and $\{w \in A^* : q_2 w \in T\}$. Show that the calculations are in fact finite, by carrying out the following steps.

  (i) Let $A_m = \{w \in A^* : |w| = m\}$ $(m = 0, 1, 2, \ldots)$, and let

$$\tau_m = \{(q_1, q_2) \in Q \times Q : (\forall l \leq m)\, q_1^{-1}T \cap A_l = q_2^{-1}T \cap A_l\}.$$

Show that $\tau_0 \supseteq \tau_1 \supseteq \tau_2 \supseteq \cdots$ and that

$$\tau = \bigcap_{m \geq 0} \tau_m.$$

  (ii) Show that $q_1\, \tau_{m+1}\, q_2$ if and only if $q_1\, \tau_0\, q_2$ and $(\forall a \in A)\, q_1 a\, \tau_m q_2 a$.
  (iii) Show that if $\tau_m = \tau_{m+1}$ then $\tau_{m+1} = \tau_{m+2}$. Deduce that $\tau = \tau_m$ in this case.
  (iv) From the finiteness of Q and from

$$|Q/\tau_0| \leq |Q/\tau_1| \leq |Q/\tau_2| \leq \cdots$$

deduce that $\tau = \tau_m$ for some $m \leq |Q| - 1$.

3.14. Let A be a finite alphabet and let $L \subseteq A^*$. We obtain an infinite automaton recognizing L by defining $Q = A^*$, $i = 1$, $T = L$ and

$$\varphi(w, a) = wa \quad (w \in A^*,\ a \in A).$$

Show that the equivalence

$$\rho_L = \{(w_1, w_2) \in A^* \times A^* : w_1^{-1}L = w_2^{-1}L\}$$

(see (3.3.1)) is the maximum right congruence on A^* that respects L. Deduce that L is rational if and only if the largest right congruence on A^* respecting L has a finite number of classes.

4 Languages

Introduction

Up to this point we have freely used the word 'language' for any subset whatever of A^* (where A is as usual our finite alphabet). Clearly this is for most purposes much too general a notion: any language spoken and understood by people or accepted by computers must have a degree of form and structure, and a proper theory of formal languages ought to mirror this fact.

The key contribution in this area was made by Chomsky (1959), whose theory of formal grammars has had a major influence in the development of the subject. In this chapter we shall examine the so-called 'phrase structure grammars' and then go on to study two special cases. Languages generated by 'regular' grammars turn out to be the familiar rational languages first encountered in Chapter 2 and hence are associated with finite state automata in the way that we have seen (Theorem 2.5.2). 'Context-free' grammars, which are studied in Sections 4.3 and 4.4, turn out to be connected with a more powerful kind of machine called a 'pushdown automaton', but this connection is a large enough topic to merit a new chapter, and so much of what we have to say regarding context-free languages must wait until Chapter 5. Equally, phrase structure grammars correspond to the still more powerful Turing machines, which will be studied in Chapter 6.

4.1 Phrase structure grammars

A couple of generations ago the 'parsing' of sentences was a standard exercise in primary schools. Let us take as an example not 'The cat sat on the mat' but a rather more complicated sentence (one that corresponds more closely to the behaviour of real cats).

Example 4.1.1. The sentence 'The black cat lay silently on a soft chair' has a structure which can be illustrated as in Figure 4.1.2.

We now make some very formal definitions, which we shall quickly relate to our example. A *formal grammar*, or *phrase structure grammar*, is a quadruple $\Gamma = (V, A, \pi, \sigma)$, where V is a finite set of symbols called the *vocabulary* of Γ, A is a non-empty subset of V called the *terminal alphabet* of Γ, π is a finite subset of $(V\backslash A)^+ \times V^*$, and $\sigma \in V\backslash A$. The elements (u, v) of π are called the *productions* of Γ, and we write $u \overset{*}{\to}\Gamma v$ (or, if no confusion will result, simply $u \to v$) whenever $(u, v) \in \pi$.

To return to our example, we have

$$A = \{\text{the, black, cat, lay, silently, on, a, soft, chair}\},$$

$$V = A \cup \{\text{sentence, nounphrase, verbphrase, preposition,}$$
$$\text{adjective, article, noun, verb, adverb}\},$$

$$\sigma = \text{sentence},$$

and the productions are

$$\text{sentence} \to (\text{nounphrase})(\text{verbphrase})(\text{preposition})(\text{nounphrase}),$$
$$\text{nounphrase} \to (\text{article})(\text{adjective})(\text{noun}),$$
$$\text{verbphrase} \to (\text{verb})(\text{adverb}),$$
$$\text{article} \to \text{the,}\quad \text{article} \to \text{a,}$$
$$\text{adjective} \to \text{black,}\quad \text{adjective} \to \text{soft,}$$
$$\text{noun} \to \text{cat,}\quad \text{noun} \to \text{chair,}$$
$$\text{verb} \to \text{lay,}\quad \text{adverb} \to \text{silently.}$$

We then envisage the grammar as 'generating' sentences. In our example we have

$$\text{sentence} \Rightarrow (\text{nounphrase})(\text{verbphrase})(\text{preposition})(\text{nounphrase})$$
$$\Rightarrow \cdots \Rightarrow (\text{article})(\text{adjective})(\text{noun})(\text{verb})(\text{adverb})(\text{preposition})$$
$$(\text{article})(\text{adjective})(\text{noun})$$
$$\Rightarrow \cdots \Rightarrow (\text{the})(\text{black})(\text{cat})(\text{lay})(\text{silently})(\text{on})(\text{a})(\text{soft})(\text{chair}).$$

Formally, and in the general case, for w, w' in V^* we write $w \Rightarrow w'$ if there exist x, y in V^* and a production $u \to v$ in π such that

$$w = xuy, \quad w' = xvy.$$

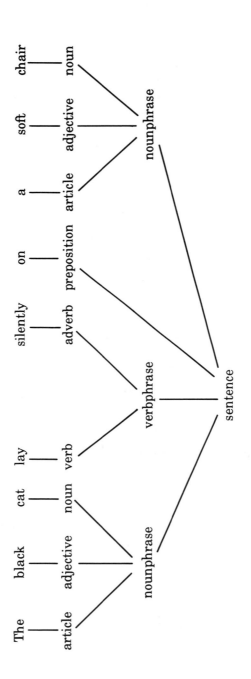

Figure 4.1.2

We say that w' *derives from* w. We write $w \overset{*}{\Rightarrow} z$ if either $w = z$ or there exist $w_1, w_2, \ldots, w_n$ (with $n \geq 2$) in V^* such that

$$w = w_1, z = w_n \text{ and } w_i \Rightarrow w_{i+1} \ (i = 1, 2, \ldots, n - 1).$$

We refer to this chain of transformations as a *derivation* in Γ, and there is no great harm in extending the meaning of the words 'derives from' to cover this more general situation: z *derives from* w.

The *language* $L(\Gamma)$ *generated by* Γ is the set of words *in the terminal alphabet* A that can be derived in this way from σ:

$$L(\Gamma) = \{w \in A^* : \sigma \overset{*}{\Rightarrow} w\}. \tag{4.1.3}$$

So in our example we have

$$L(\Gamma) = \left\{ \begin{array}{c} a \\ the \end{array} \right\} \left\{ \begin{array}{c} black \\ soft \end{array} \right\} \left\{ \begin{array}{c} cat \\ chair \end{array} \right\} \text{ lay silently on}$$
$$\left\{ \begin{array}{c} a \\ the \end{array} \right\} \left\{ \begin{array}{c} black \\ soft \end{array} \right\} \left\{ \begin{array}{c} cat \\ chair \end{array} \right\}, \tag{4.1.4}$$

a total of 2^6 elements of A^*. The sentences we obtain are of varying degrees of reasonableness in terms of the meaning they carry, but this is not the point here. They are all grammatical sentences, *syntactically* acceptable, though not all *semantically* acceptable. The sentence 'The black chair lay silently on a soft cat' is ludicrous precisely because it is syntactically correct; the sentence 'Black soft the lay a chair silently cat on' is merely meaningless.

The simplest type of grammar is a *regular* (or *right linear*) grammar, where every production in π is either of the form

$$\alpha \rightarrow x\beta \quad (x \in A^+, \ \alpha, \beta \in V \backslash A) \tag{4.1.5}$$

or of the form

$$\alpha \rightarrow y \quad (\alpha \in V \backslash A, \ y \in A^*). \tag{4.1.6}$$

Slightly less restrictive is a *context-free* grammar, where every production is of the form

$$\alpha \rightarrow z \quad (\alpha \in V \backslash A, \ z \in V^*). \tag{4.1.7}$$

These will be the subject of Sections 4.3 and 4.4 and will be further examined in Chapter 5. At this stage we make the definitions that a language is *regular* [*context-free*] if it can be generated by a regular [context-free] grammar. We immediately have

Theorem 4.1.8. *Every regular language is context-free.* ∎

Remark 4.1.9. The one-sidedness in the definition of a regular grammar is only apparent. If we define a right linear language to be a language generated by a right linear grammar and a left linear language to be a language generated by a grammar whose productions are all of the form (4.1.6) or of the form

$$\alpha \to \beta x \; (x \in A^+, \alpha, \beta \in V \backslash A),$$

then it turns out that the class of right linear languages is identical to the class of left linear languages. (See Exercise 4.3.)

Remark 4.1.10. The name 'context-free' is quite descriptive. Consider a typical production (4.1.7). If $w = u\alpha v$ then we may think of u and v as providing a *context* for the symbol α, but the transformation $u\alpha v \Rightarrow uzv$ can take place irrespective of the context, i.e., for arbitrarily chosen u, v in V^*. A *context-sensitive* language has productions of the form

$$x\alpha y \to xzy \; (x, y, z \in V^*, \alpha \in V \backslash A),$$

so that α can change to z only when flanked by x and y, i.e., only in an appropriate context.

4.2 Regular and rational languages

The grammar specified in Example 4.1.1 is easily seen to be context-free, and so the language consisting of the 2^6 sentences given by (4.1.4) is context-free. In fact the language is regular, because of the following result.

Theorem 4.2.1. *Every finite language is regular.*

Proof. Let $L = \{w_1, w_2, \dots w_m\}$ be a finite subset of A^*, where A is an arbitrary finite alphabet. Let

$$\Gamma = (A \cup \{\sigma, \gamma_1, \gamma_2, \dots, \gamma_m\}, A, \pi, \sigma),$$

where π consists of the productions

$$\sigma \to w_1, \; \sigma \to w_2, \; \dots, \; \sigma \to w_m.$$

Then Γ is a regular grammar, and

$$L(\Gamma) = \{w_1, w_2, \dots, w_m\} = L.$$ ∎

The fact that the language given by (4.1.4) is regular, though generated by a grammar that is merely context-free, illustrates an important point and a potential source of misunderstanding. To show that a language L is not regular we need to show that *there cannot be* a regular grammar generating L.

Example 4.2.2. Let $A = \{a, b\}$, let $V \backslash A = \{\sigma, \lambda, \mu\}$ and let $\Gamma = (V, A, \pi, \sigma)$ have productions

$$\sigma \to a\sigma, \ \sigma \to b\sigma, \ \sigma \to a\lambda, \ \lambda \to b.$$

This is a regular grammar. A typical derivation is

$$\sigma \Rightarrow a\sigma \Rightarrow a^2\sigma \Rightarrow a^2 b\sigma \Rightarrow a^2 ba\sigma$$
$$\Rightarrow a^2 ba^2 \lambda \Rightarrow a^2 ba^2 b\mu \Rightarrow a^2 ba^2 b.$$

In fact $L(\Gamma) = A^* ab$. To see this, notice that for all w in A^* there is a derivation

$$\sigma \overset{*}{\Rightarrow} w\sigma$$

and a derivation

$$w\sigma \Rightarrow wa\lambda \Rightarrow wab\mu \Rightarrow wab.$$

So certainly $A^* ab \subseteq L(\Gamma)$. Conversely, suppose that $w \in L(\Gamma)$, so that there is a derivation $\sigma \overset{*}{\Rightarrow} w$ in Γ. Since the only production involving no non-terminal symbol on the right is $\mu \to 1$, the penultimate stage in the derivation of w must be $w\mu$. From the nature of the productions of Γ it is clear that no word occurring in a derivation can ever have more than one non-terminal symbol, and moreover that the single non-terminal symbol must be on the right-hand edge of the word. Since $\lambda \to b\mu$ is the only production in which μ appears on the right and since $\sigma \to a\lambda$ is the only production in which λ appears on the right, the final three steps of the derivation must be

$$z\sigma \Rightarrow za\lambda \Rightarrow zab\mu \Rightarrow zab.$$

Thus $L(\Gamma) \subseteq A^* ab$.

Notice that the language $L(\Gamma)$ of this example is rational. This, together with Theorem 4.2.1, gives a hint that there might be a connection between regular and rational languages. But before stating and proving the theorem that clarifies that connection it is helpful to examine the notion of a regular grammar a little more closely. Let us decide to call a grammar *hyper-regular* if all productions have the form

$$\alpha \to a\beta \quad (\alpha, \beta \in V \backslash A, \ a \in A)$$

or

$$\alpha \to 1 \quad (\alpha \in V\backslash A).$$

This certainly seems more special than a regular grammar, since in (4.1.5) and (4.1.6) x may be any word in A^+ and y any word in A^*. Calling a language *hyper-regular* if it can be generated by a hyper-regular grammar, we see right away that every hyper-regular language is regular. In fact the somewhat unimaginative term 'hyper-regular' is going to be of temporary interest only, since we have

Lemma 4.2.3. *Every regular language is hyper-regular.*

Proof. Let $L = L(\Gamma)$, where $\Gamma = (V, A, \pi, \sigma)$ is a regular grammar. We shall define a hyper-regular grammar $\Gamma' = (V', A, \pi', \sigma)$ such that $L(\Gamma') = L(\Gamma)$. For each production

$$\alpha \to a_1 a_2 \ldots a_m \beta \tag{4.2.4}$$

in π, where $m \geq 2$, $a_1, a_2, \ldots, a_m \in A$, and $\beta \in V\backslash A$, we define non-terminal symbols $\zeta_1, \zeta_2, \ldots, \zeta_{m-1}$ in V' and within π' mimic the production (4.2.4) by means of productions

$$\alpha \to a_1 \zeta_1, \ \zeta_1 \to a_2 \zeta_2, \ldots, \zeta_{m-1} \to a_m \beta. \tag{4.2.5}$$

For each production

$$\alpha \to b_1 b_2 \ldots b_n \tag{4.2.6}$$

in π, where $n \geq 1$ and $b_1, b_2, \ldots b_n \in A$, we define non-terminal symbols $\eta_1, \eta_2, \ldots, \eta_n$ in V' and within π' mimic the production (4.2.6) by productions

$$\alpha \to b_1 \eta_1, \eta_1 \to b_2 \eta_2, \ldots, \eta_{n-1} \to b_n \eta_n, \eta_n \to 1. \tag{4.2.7}$$

Certainly Γ' is a hyper-regular grammar. Moreover, it is clear that the productions (4.2.5) give that $\alpha \overset{*}{\Rightarrow} a_1 a_2 \ldots a_m \beta$ in Γ'. We similarly obtain that $\alpha \overset{*}{\Rightarrow} b_1 b_2 \ldots b_n$ in Γ' from the productions (4.2.7). Thus every derivation $\sigma \overset{*}{\Rightarrow} w$ in Γ can be simulated by a (longer) derivation $\sigma \overset{*}{\Rightarrow} w$ in Γ'. We conclude that $L(\Gamma) \subseteq L(\Gamma')$.

To prove the reverse inclusion, suppose that $w \in L(\Gamma')$. Then certainly $w \in L(\Gamma'')$, where $\Gamma'' = (V', A, \pi \cup \pi', \sigma)$ has all the productions in Γ together with all the productions in Γ'. Certainly there is a derivation

$$\sigma \overset{*}{\Rightarrow} w \tag{4.2.8}$$

of w in Γ''. We shall show that there is a derivation of w in Γ, by induction on the number of symbols from $V'\backslash V$ appearing in the derivation (4.2.8). If no such symbols appear then (4.2.8) is already a derivation in Γ.

Otherwise the first appearance of a symbol of $V'\backslash V$ is based either on a production

$$\alpha \to a_1\zeta_1$$

(where $\alpha \to a_1 a_2 \ldots a_m\beta$ is a production in Γ) or on a production

$$\alpha \to b_1\eta_1$$

(where $\alpha \to b_1 b_2 \ldots b_n$ is a production in Γ). Consider the first of these cases. Since the final word w in the derivation (4.2.8) has no non-terminal symbols, and since the grammar Γ'' has no productions of the type $\zeta_i \to y$ $(y \in A^*)$ for any of the symbols of $V'\backslash V$, the only way in which ζ_1, once introduced, can subsequently disappear must involve changes from ζ_1 to $a_2\zeta_2$, ..., ζ_{m-1} to $a_m\beta$. But then the sequence of transitions

$$\alpha \to a_1\zeta_1, \zeta_1 \to a_2\zeta_2, \ldots, \zeta_{m-1} \to a_m\beta$$

can be replaced by a single transition

$$\alpha \to a_1 a_2 \ldots a_m\beta$$

in Γ'', and the number of symbols from $V'\backslash V$ has been reduced.

Equally, in the second case the derivation must involve subsequent changes from η_1 to $b_2\eta_2$, ..., η_{n-1} to $b_n\eta_n$, and these n transitions can be replaced by a single transition in Γ'' from α to $b_1 b_2 \ldots b_n$.

In both cases the derivation (4.2.8) is replaced by one with fewer occurrences of symbols from $V'\backslash V$. By induction it now follows that $L(\Gamma'') \subseteq L(\Gamma)$. Hence certainly $L(\Gamma') \subseteq L(\Gamma)$, as required. ∎

We now have

Theorem 4.2.9. *The set of regular languages coincides with the set of rational languages.*

Proof. Suppose that $L \in \mathrm{Rat}\, A^*$. Then by Kleene's Theorem (2.5.2) there exists a complete deterministic automaton

$$\mathcal{A} = (Q, A, \varphi, i, T)$$

such that $L = L(\mathcal{A})$. We define a regular grammar Γ as follows: $\Gamma = (Q \cup A, A, \pi, i)$, where π consists of the productions

$$q \to a\varphi(q, a) \;\; (= a(qa))$$

for all $q \in Q$, $a \in A$, and

$$t \to 1$$

for all t in T. We show that $L(\Gamma) = L$.

Suppose first that $w = a_1 a_2 \ldots a_n \in L(\Gamma)$, so that there is a derivation

$$i \overset{*}{\Rightarrow} a_1 a_2 \ldots a_n$$

of w in Γ. Writing $\varphi(q, a)$ as qa in the usual way, we see that this derivation must be of the form

$$1 \Rightarrow a_1 q_1 \Rightarrow a_1 a_2 q_2 \Rightarrow \cdots \Rightarrow a_1 a_2 \ldots a_n q_n$$
$$\Rightarrow a_1 a_2 \ldots a_n,$$

where $i \to a_1 q_1$, $q_1 \to a_2 q_2$, $\ldots, q_{n-1} \to a_n q_n$ and $q_n \to 1$ are productions in Γ. Thus

$$i a_1 = q_1, \; q_1 a_2 = q_2, \ldots q_{n-1} a_n = q_n$$

in the automaton $\mathcal{A}$, and $q_n \in T$. Hence $i a_1 a_2 \ldots a_n \in T$ and so $a_1 a_2 \ldots a_n \in L(\mathcal{A}) = L$.

Conversely, suppose that $w = a_1 a_2 \ldots a_n \in L = L(\mathcal{A})$, so that there is a successful path

$$i \to q_1 \to q_2 \to \cdots \to q_n$$

in $\mathcal{A}$ with $i a_1 = q_1$, $q_1 a_2 = q_2$, $\ldots$, $q_{n-1} a_n = q_n$ and $q_n \in T$. Then there are productions

$$i \to a_1 q_1, \; q_1 \to a_2 q_2, \; \ldots, \; q_{n-1} \to a_n q_n, \; q_n \to 1$$

in Γ and so there is a derivation

$$1 \Rightarrow a_1 q_1 \Rightarrow a_1 a_2 q_2 \Rightarrow \cdots \Rightarrow a_1 a_2 \ldots a_n q_n$$
$$\Rightarrow a_1 a_2 \ldots a_n,$$

in Γ, giving $a_1 a_2 \ldots a_n \in L(\Gamma)$. Thus $L(\Gamma) = L$.

We have shown that every rational language is regular. Conversely, suppose that $L = L(\Gamma)$, where $\Gamma = (V, A, \pi, \sigma)$ is a regular grammar. By Lemma 4.2.3 we can assume that the grammar Γ is hyper-regular, i.e., that all productions are of the form $\alpha \to a\beta$ (with $a \in A$, $\alpha, \beta \in V \backslash A$) or $\alpha \to 1$ (with $\alpha \in V \backslash A$). Let $\mathcal{A}$ be the (non-deterministic) automaton $(V \backslash A, A, \varphi, \sigma, T)$, where

$$T = \{\alpha \in V \backslash A : \alpha \to 1 \text{ is in } \pi\}$$

and where $\beta \in \varphi(\alpha, a)$ whenever $\alpha \to a\beta$ is in π.

Suppose first that $w = a_1 a_2 \ldots a_n \in L = L(\Gamma)$. The derivation of w must be of the form

$$\sigma \Rightarrow a_1 \beta_1 \Rightarrow a_1 a_2 \beta_2 \Rightarrow \cdots a_1 a_2 \ldots a_n \beta_n$$
$$\Rightarrow a_1 a_2 \ldots a_n,$$

where

$$\sigma \to a_1\beta_1, \ \beta_1 \to a_2\beta_2, \ \dots, \ \beta_{n-1} \to a_n\beta_n, \ \beta_n \to 1$$

are productions in π. Thus $\beta_n \in T$ and

$$\sigma \xrightarrow{a_1} \beta_1 \xrightarrow{a_2} \beta_2 \xrightarrow{a_3} \cdots \xrightarrow{a_n} \beta_n$$

is a successful path in $\mathcal{A}$ with label w. Thus $w \in L(\mathcal{A})$.

We have shown that $L \subseteq L(\mathcal{A})$. To show the reverse inclusion, suppose that $w = a_1 a_2 \dots a_n \in L(\mathcal{A})$, so that there is a successful path

$$\sigma \xrightarrow{a_1} \beta_1 \xrightarrow{a_2} \beta_2 \xrightarrow{a_3} \cdots \xrightarrow{a_n} \beta_n \ (\in T)$$

with label w. Thus

$$\beta_1 \in \varphi(\sigma, a_1), \ \beta_2 \in \varphi(\beta_1, a_2), \ \dots, \ \beta_n \in \varphi(\beta_{n-1}, a_n)$$

and so

$$\sigma \to a_1\beta_1, \ \beta_1 \to a_2\beta_2, \ \dots, \ \beta_{n-1} \to a_n\beta_n, \ \beta_n \to 1$$

are productions in π. Hence there is a derivation

$$\sigma \Rightarrow a_1\beta_1 \Rightarrow a_2 a_2 \beta_2 \Rightarrow \cdots a_1 a_2 \dots a_n \beta_n$$
$$\Rightarrow a_1 a_2 \dots a_n$$

in Γ, and so $w \in L(\Gamma) = L$. ∎

Example 4.2.10. Consider the rational language $L = aA^*bA^*a$, where $A = \{a, b\}$. With a bit of experience it is possible to guess at a regular grammar that will generate L. Otherwise we can use the routine of Section 3.2 to find the minimal automaton of L and then use the method of Theorem 4.2.9 to determine the grammar. The calculation of the minimal automaton gives $i_L = L$ and

$$L.a = A^*bA^*a = L_1, \ L.b = \emptyset,$$
$$L_1.a = L_1, \ L_1.b = L_1 \cup A^*a = L_2,$$
$$L_2.a = L_2 \cup \{1\} = L_3, \ L_2.b = L_2,$$
$$L_3.a = L_3, \ L_3.b = L_2$$

and so (leaving out the 'empty' state $\emptyset$) we get state diagram

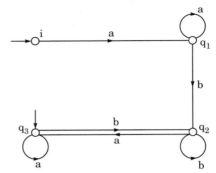

Hence finally we get $L = L(\Gamma)$, where

$$\Gamma = \big(\{i, q_1, q_2, q_3, a, b\}, \{a, b\}, \pi, i\big)$$

and the productions in π are

$$i \to aq_1, \quad q_1 \to aq_1, \quad q_1 \to bq_2, \quad q_2 \to aq_3,$$

$$q_2 \to bq_2, \quad q_3 \to aq_3, \quad q_3 \to bq_2, \quad q_3 \to 1.$$

It may be helpful to end this section by drawing together the various characterizations we have found for rational sets, in the form of an extended version of Kleene's Theorem. (See Theorems 2.5.2, 3.1.4 and 4.2.9.)

Theorem 4.2.11. *Let A be a finite alphabet and let L be a subset of A^*. Then the following statements are equivalent:*

(i) *L is rational;*

(ii) *$L = L(\mathcal{A})$ for some finite state automaton $\mathcal{A}$;*

(iii) *the syntactic monoid $\mathrm{Syn}(L)$ is finite;*

(iv) *L is recognized by a finite monoid;*

(v) *$L = L(\Gamma)$ for some regular grammar Γ.*   ∎

From now on we can regard the words 'rational', 'regular' and 'recognizable' (by a finite state automaton or by a finite monoid) as interchangeable when applied to a language $L \subseteq A^*$.

4.3 Context-free languages

Recall that a grammar $\Gamma = (V, A, \pi, \sigma)$ is *context-free* if all productions are of the form

$$\alpha \to z$$

where $\alpha \in V \backslash A$, $z \in V^*$.

Example 4.3.1. Let $\Gamma = (V, A, \pi, \sigma)$, where $A = \{a, b\}$, $V \backslash A = \{\sigma, \lambda\}$ and π consists of the productions

$$\sigma \to a\lambda b, \ \lambda \to a\lambda b, \ \lambda \to 1.$$

Then $L(\Gamma) = \{a^n b^n : n \geq 1\}$, and so (recalling Theorem 2.2.6) we now have an example of a language, namely $\{a^n b^n : n \geq 1\}$, which is context-free but not regular.

Example 4.3.2. Let $\Gamma = (V, A, \pi, \sigma)$, where $A = \{a, b\}$, $V \backslash A = \{\sigma\}$, and π consists of the productions

$$\sigma \to a\sigma a, \ \sigma \to b\sigma b, \ \sigma \to a, \ \sigma \to b, \ \sigma \to 1.$$

Here we have that

$$L(\Gamma) = \{w \in A^* : w = w^R\},$$

the set of all *palindromes* in the alphabet A. Again, recalling Example 2.2.8, we see that this language is context-free but not regular.

Example 4.3.3. Consider the language $L \subset \{a, b\}^*$ consisting of all words beginning and ending in different letters. The context-free grammar with productions

$$\sigma \to a\lambda b, \ \sigma \to b\lambda a, \ \lambda \to a\lambda, \ \lambda \to b\lambda, \ \lambda \to 1$$

generates L. In this case $L = aA^*b \cup bA^*a$ is rational. Its minimal automaton has state diagram

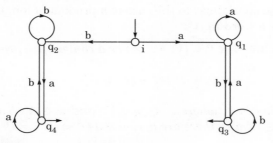

and a regular grammar generating L is

$$\Gamma = \left(\{1, q_1, q_2, q_3, q_4, a, b\}, \{a, b\}, \pi, i\right),$$

where π consists of the productions

$$i \rightarrow aq_1, \; i \rightarrow bq_2, \; q_1 \rightarrow aq_1, \; q_1 \rightarrow bq_3,$$

$$q_2 \rightarrow aq_4, \; q_2 \rightarrow bq_2, \; q_3 \rightarrow aq_1, \; q_3 \rightarrow bq_3,$$

$$q_4 \rightarrow aq_4, \; q_4 \rightarrow bq_2, \; q_3 \rightarrow 1, \; q_4 \rightarrow 1.$$

The crucial lemma that enables us to keep track of derivations in a context-free grammar is as follows:

Lemma 4.3.4. *Let $\Gamma = (V, A, \pi, \sigma)$ be a context-free grammar and let $\zeta_1 \zeta_2 \overset{*}{\Rightarrow} \eta$ be a derivation in Γ, with $\zeta_1, \zeta_2, \eta \in V^*$. Then η can be expressed as $\eta_1 \eta_2$ in such a way that there are derivations $\zeta_1 \overset{*}{\Rightarrow} \eta_1$, $\zeta_2 \overset{*}{\Rightarrow} \eta_2$ in Γ, each of which is no longer than the derivation $\zeta_1 \zeta_2 \overset{*}{\Rightarrow} \eta$.*

Proof. The proof is by induction on the number of steps (i.e., the length) of the derivation $\zeta_1 \zeta_2 \overset{*}{\Rightarrow} \eta$. If there is only one step, it must be based on a production $\alpha \rightarrow z$ (with $\alpha \in V \backslash A$, $z \in V^*$), and α must lie inside ζ_1 or inside ζ_2. If we suppose without loss of generality that $\zeta_1 = u\alpha v$ with $u, v \in V^*$, then the derivation is $\zeta_1 \zeta_2 = u\alpha v \zeta_2 \Rightarrow uzv\zeta_2 = \eta$; so $\eta = \eta_1 \eta_2$ with $\eta_1 = uzv$, $\eta_2 = \zeta_2$ and we clearly have derivations $\zeta_1 \overset{*}{\Rightarrow} \eta_1$, $\zeta_2 \rightarrow \eta_2$, of lengths 1 and 0 respectively.

In general, if the derivation $\zeta_1 \zeta_2 \overset{*}{\Rightarrow} \eta$ is of length n we may suppose without essential loss of generality that the first step is $\zeta_1 \zeta_2 \Rightarrow \zeta_1' \zeta_2$, where $\zeta_1 = u\alpha v$, $\zeta_1' = uzv$ and $\alpha \rightarrow z$ is a production in Γ. By induction we may then assume that $\eta = \eta_1 \eta_2$ and that there are derivations $\zeta_1' \overset{*}{\Rightarrow} \eta_1$, $\zeta_2 \overset{*}{\Rightarrow} \eta_2$, each of length at most $n - 1$. But then clearly there is a derivation

$$\zeta_1 \Rightarrow \zeta_1' \overset{*}{\Rightarrow} \eta_1$$

of length at most n, as required.   ∎

It is clear that this lemma extends to products of arbitrary finite length. It also specializes to the case of a product of length 3 where one of $\zeta_1, \zeta_2, \zeta_3$ is a word in A^*:

Corollary 4.3.5. *Let $\Gamma = (V, A, \pi, \sigma)$ be a context-free grammar and let*

$$\zeta_1 w \zeta_2 \overset{*}{\Rightarrow} \eta$$

be a derivation in Γ, where $\zeta_1, \zeta_2, \eta \in V^*$ and $w \in A^*$. Then $\eta = \zeta_1' w \zeta_2'$, where $\zeta_1', \zeta_2' \in V^*$ and there are derivations $\zeta_1 \overset{*}{\Rightarrow} \zeta_1'$, $\zeta_2 \overset{*}{\Rightarrow} \zeta_2'$ in Γ.   ∎

It is convenient at this stage to record for future use a somewhat more technical corollary of Lemma 4.3.4:

Corollary 4.3.6. *Let* $\Gamma = (V, A, \pi, \sigma)$ *be a context-free grammar. Let* $\lambda \in V \backslash A$, $\zeta_1, \zeta_2 \in V^*$ *and suppose that* $\zeta_1 \lambda \zeta_2 \overset{*}{\Rightarrow} \zeta_1' \lambda \zeta_2'$ *is a derivation in* Γ *making use only of productions* $\alpha \to z$ *in which* λ *does not appear in* z. *Then there are derivations* $\zeta_1 \overset{*}{\Rightarrow} \zeta_1'$ $\zeta_2 \overset{*}{\Rightarrow} \zeta_2'$ *in* Γ. ■

As an illustration of the use of the lemma and its corollary we have

Theorem 4.3.7. *Let* $A = \{a, b\}$. *The language* $\{w \in A^* : |w|_a = |w|_b\}$ *is context-free.*

Proof. Notice that by Exercise 2.11 this language is not regular.
　　Let
$$\Gamma = (\{\sigma, \lambda, \mu, a, b\}, \{a, b\}, \pi, \sigma),$$
where π consists of the productions
$$\sigma \to a\mu, \ \sigma \to b\lambda, \ \lambda \to a, \ \lambda \to a\sigma,$$
$$\lambda \to b\lambda^2, \ \mu \to b, \ \mu \to b\sigma, \ \mu \to a\mu^2.$$
This is a context-free grammar.
　　We prove by induction on the length of w that
　　(i) there is a derivation $\sigma \overset{*}{\Rightarrow} w$ if and only if $|w|_a = |w|_b$;
　　(ii) there is a derivation $\lambda \overset{*}{\Rightarrow} w$ if and only if $|w|_a - |w|_b = 1$;
　　(iii) there is a derivation $\mu \overset{*}{\Rightarrow} w$ if and only if $|w|_b - |w|_a = 1$.
Ultimately the only one of these that interests us is the first, but to make the induction work we have to consider all three.
　　Let us look first at (i) for words w of length 2 (the smallest possible). The only possible derivations are
$$\sigma \Rightarrow a\mu \Rightarrow ab, \ \sigma \Rightarrow b\lambda \Rightarrow ba.$$
There is no way of deriving a^2 or b^2, and all other derivations lead to words of length exceeding 2. So (i) holds when $|w| = 2$. As far as (ii) and (iii) are concerned, the shortest words that can arise are of length 1 and the only possible derivations are $\lambda \Rightarrow a$, $\mu \Rightarrow b$. Thus (ii) and (iii) hold when $|w| = 1$.
　　Suppose now that (i), (ii) and (iii) hold for all w such that $|w| < n$, and suppose that we have a derivation $\sigma \overset{*}{\Rightarrow} w$, where $|w| = n \geq 2$. This must begin either with $\sigma \Rightarrow a\mu$ or with $\sigma \Rightarrow b\lambda$. By Corollary 4.3.5 we have either $w = aw'$ and $\mu \overset{*}{\Rightarrow} w'$ or $w = bw''$ and $\lambda \overset{*}{\Rightarrow} w''$. In the first case we can use the induction hypothesis to deduce that $|w'|_b - |w'|_a = 1$, while in the second case we similarly deduce that $|w''|_a - |w''|_b = 1$. In either case we have $|w|_a = |w|_b$.
　　We have shown that $\sigma \overset{*}{\Rightarrow} w$ and $|w| = n$ together imply that $|w|_a = |w|_b$. Suppose conversely that $|w| = n$ and $|w|_a = |w|_b$. Then either $w = aw'$

with $|w'|_b - |w'|_a = 1$ and $|w'| = n - 1$, or $w = bw''$ with $|w''|_a - |w''|_b = 1$ and $|w''| = n - 1$. By induction we thus have either a derivation $\mu \overset{*}{\Rightarrow} w'$ or a derivation $\lambda \overset{*}{\Rightarrow} w''$. Hence we have either a derivation

$$\sigma \Rightarrow a\mu \overset{*}{\Rightarrow} aw' = w$$

or a derivation

$$\sigma \Rightarrow b\lambda \overset{*}{\Rightarrow} bw'' = w.$$

Thus (i) holds with $|w| = n$.

To show that (ii) holds with $|w| = n$, suppose first that $\lambda \overset{*}{\Rightarrow} w$. The derivation must begin either $\lambda \Rightarrow a\sigma$ or $\lambda \Rightarrow b\lambda^2$. In the first case Corollary 4.3.5 implies that $w = aw'$, where $\sigma \overset{*}{\Rightarrow} w'$. Since $|w'| = n - 1$ we deduce from the induction hypothesis that $|w'|_a = |w'|_b$, and so $|w|_a - |w|_b = 1$. In the second case Lemma 4.3.4 gives $w = bw'w''$, and there are derivations $\lambda \overset{*}{\Rightarrow} w'$, $\lambda \overset{*}{\Rightarrow} w''$ in Γ. Since $|w'|, |w''| < n$ we may deduce from the induction hypothesis that

$$|w'|_a - |w'|_b = |w''|_a - |w''|_b = 1.$$

Hence $|w|_a - |w|_b = 1$.

Conversely, suppose that w is such that $|w| = n$, $|w|_a - |w|_b = 1$. Then either $w = aw'$ with $|w'| = n - 1$ and $|w'|_a = |w'|_b$, or $w = bw''$ with $|w''| = n - 1$ and $|w''|_a - |w''|_b = 2$. In the first case we deduce by the induction hypothesis that there is a derivation $\sigma \overset{*}{\Rightarrow} w'$ and hence that there is a derivation

$$\lambda \Rightarrow a\sigma \overset{*}{\Rightarrow} aw' = w.$$

In the second case let w_1 be the shortest left factor of w'' for which $|w_1|_a - |w_1|_b = 1$. Then $w'' = w_1 w_2$, with

$$|w_1|_a - |w_1|_b = |w_2|_a - |w_2|_b = 1.$$

Since $|w_1|, |w_2| < n$ we can use the induction hypothesis to deduce that there are derivations $\lambda \overset{*}{\Rightarrow} w_1$, $\lambda \overset{*}{\Rightarrow} w_2$. Hence there is a derivation

$$\lambda \Rightarrow b\lambda^2 \overset{*}{\Rightarrow} bw_1 w_2 = bw'' = w.$$

Thus (ii) holds when $|w| = n$.

The proof that (iii) holds when $|w| = n$ is essentially identical to the proof concerning (ii). Hence Theorem 4.3.7 is proved. ∎

4.4 Chomsky Normal Form

When in Section 4.2 we established a link between regular grammars and automata, it proved helpful to show first that the grammar could be made 'hyper-regular', i.e., could be put into a relatively simple 'normal form'. By analogy, we shall show that every context-free language can be generated by a grammar in what is called 'Chomsky Normal Form'. It simplifies matters for the moment if we confine ourselves to languages in A^* not containing the empty word 1, i.e., to what we shall call A^+-*languages*.

Definition 4.4.1. A context-free grammar $\Gamma = (V, A, \pi, \sigma)$ generating an A^+-language L is said to be in *Chomsky Normal Form* if each of its productions is of one of the two types

$$\lambda \to \mu\nu \ (\lambda, \mu, \nu \in V \backslash A),$$

$$\lambda \to a \ (\lambda \in V \backslash A, \ a \in A).$$

Even if L is an A^+-language, the grammar generating it may contain '1-productions', i.e., productions of the form $\lambda \to 1$ for some λ in $V \backslash A$. For example, the grammar whose productions are

$$\sigma \to a\lambda b, \ \lambda \to a\lambda b, \ \lambda \to 1 \tag{4.4.2}$$

generates the $\{a, b\}^+$-language $\{a^n b^n : n \geq 1\}$. The first stage in our reduction routine is to eliminate all 1-productions.

Lemma 4.4.3. *Every context-free A^+-language L can be generated by a grammar containing no 1-productions.*

Proof. The language L is generated by a context-free grammar $\Gamma = (V, A, \pi, \sigma)$ which may contain productions of the forbidden type. Let us refer to a non-terminal symbol as 1-*accessible* if there is a derivation $\lambda \overset{*}{\Rightarrow} 1$.

We now introduce a simple piece of notation. Let

$$\zeta = \zeta_1 \lambda_1 \zeta_2 \lambda_2 \ldots \zeta_k \lambda_k \zeta_{k+1}$$

be a word in V^* containing precisely k 1-accessible symbols

$$\lambda_1, \lambda_2, \ldots, \lambda_k.$$

If S is a subset of $\{1, 2, \ldots, k\}$, let ζ_S denote the word obtained from ζ by deleting precisely those λ_i for which $i \in S$. Notice that

$$\zeta_\emptyset = \zeta, \ \zeta_{\{1,2,\ldots,k\}} = \zeta_1 \zeta_2 \ldots \zeta_{k+1}.$$

We shall define a context-free grammar $\Gamma' = (V, A, \pi', \sigma)$ such that π' contains no 1-productions and such that $L(\Gamma') = L$. The set π' is formed from π by the following rules:

(4.4.4) Delete all 1-productions.

(4.4.5) If the production $\alpha \to \zeta$ in π is such that ζ contains k (≥ 1) occurrences of 1-accessible symbols, replace $\alpha \to \zeta$ by up to 2^k productions $\alpha \to \zeta_S$. Here S ranges over all subsets of $\{1, 2, \ldots, k\}$, but if $\zeta_S = 1$ (which happens if ζ consists entirely of 1-accessible symbols) we do not include it. Notice that the production $\alpha \to \zeta$ itself does survive into π', since $\zeta_\emptyset = \zeta$.

Now observe that the effect of any production in π' can be achieved within the original grammar Γ: to get from α to ζ_S in Γ we use the production $\alpha \to \zeta$ in π and then use $\lambda_i \overset{*}{\Rightarrow} 1$ for the appropriate selection of 1-accessible symbols in ζ. Hence $L(\Gamma') \subseteq L(\Gamma)$.

In fact $L(\Gamma') = L(\Gamma)$. To see this, suppose that $w \in L(\Gamma)$, so that there is a derivation $\sigma \overset{*}{\Rightarrow} w$ in the grammar Γ. If this derivation makes no use of 1-productions then it is already a derivation in Γ' and so $w \in L(\Gamma')$. Suppose inductively that $z \in L(\Gamma')$ for every z in $L(\Gamma)$ whose derivation $\sigma \overset{*}{\Rightarrow} z$ in Γ involves fewer than n 1-productions; and suppose now that we have a derivation $\sigma \overset{*}{\Rightarrow} w$ involving n 1-productions. Within this derivation we turn our attention to the *last* step that is based on a 1-production, and we suppose that the 1-production in question is $\lambda \to 1$. Then we can trace the 'history' of this symbol λ from its introduction to the word by means of a production $\beta \to \omega_1 \lambda \omega_2$ (with $\omega_1, \omega_2 \in V^*$), and perhaps through various steps based on productions $\lambda \to \xi_1 \lambda \xi_2$, to its final disappearance. Now fix attention on the last step involving a production in which this λ appears on the right hand side, and suppose in fact that the production in question is $\alpha \to \zeta_1 \lambda \zeta_2$. The derivation $\sigma \overset{*}{\Rightarrow} w$ in Γ takes the form

$$\sigma \overset{*}{\Rightarrow} \tau_1 \alpha \tau_2 \Rightarrow \tau_1 \zeta_1 \lambda \zeta_2 \tau_2 \overset{*}{\Rightarrow} \tau_1' \zeta_1' \lambda \zeta_2' \tau_2'$$
$$\Rightarrow \tau_1' \zeta_1' \zeta_2' \tau_2' \overset{*}{\Rightarrow} t_1 z_1 z_2 t_2 = w \qquad (4.4.6)$$

(with $\tau_1, \tau_2, \tau_1', \tau_2', \zeta_1', \zeta_2' \in V^*$ and $t_1, t_2, z_1, z_2 \in A^*$) and by Corollary 4.3.6 there are derivations

$$\tau_1 \overset{*}{\Rightarrow} \tau_1', \quad \tau_1' \overset{*}{\Rightarrow} t_1, \quad \tau_2 \overset{*}{\Rightarrow} \tau_2', \quad \tau_2' \overset{*}{\Rightarrow} t_2,$$
$$\zeta_1 \overset{*}{\Rightarrow} \zeta_1', \quad \zeta_1' \overset{*}{\Rightarrow} z_1, \quad \zeta_2 \overset{*}{\Rightarrow} \zeta_2', \quad \zeta_2' \overset{*}{\Rightarrow} z_2$$

in Γ. These derivations, as well as the derivation $\sigma \overset{*}{\Rightarrow} \tau_1 \alpha \tau_2$, all involve fewer than n 1-productions and so by induction can be replaced by derivations in Γ'. Since by definition (4.4.5) Γ' contains a production $\alpha \to \zeta_1 \zeta_2$ we may now replace the derivation $\sigma \overset{*}{\Rightarrow} w$ in Γ by a derivation

$$\sigma \overset{*}{\Rightarrow} \tau_1 \alpha \tau_2 \Rightarrow \tau_1 \zeta_1 \zeta_2 \tau_2 \overset{*}{\Rightarrow} \tau_1' \zeta_1' \zeta_2' \tau_2'$$

$$\overset{*}{\Rightarrow} t_1 z_1 z_2 t_2 = w$$

in Γ'.

The careful reader will have noticed that this argument fails if $\zeta_1\zeta_2 = 1$, since the specification of π' in (4.4.5) explicitly excludes this case. Here we have a production $\alpha \to \lambda$ and α is 1-accessible. We backtrack to the last appearance of α on the right hand side of a production (say $\beta \to \zeta_1''\alpha\zeta_2''$), decompose the derivation as

$$\sigma \overset{*}{\Rightarrow} \tau_1''\beta\tau_2'' \Rightarrow \tau_1'\zeta_1''\alpha\zeta_2''\tau_2'' \overset{*}{\Rightarrow} \tau_1\zeta_1\alpha\zeta_2\tau_2$$
$$\Rightarrow \tau_1\zeta_1\lambda\zeta_2\tau_2 \Rightarrow \tau_1\zeta_1\zeta_2\tau_2 \overset{*}{\Rightarrow} w$$

and replace it by a derivation

$$\sigma \overset{*}{\Rightarrow} \tau_1''\beta\tau_2'' \Rightarrow \tau_1''\zeta_1''\zeta_2''\tau_2'' \overset{*}{\Rightarrow} w.$$

Of course we may now have $\zeta_1''\zeta_2'' = 1$, but in that case β is 1-accessible and we backtrack still further to the last appearance of β on the right hand side of a production. This backtracking cannot go on for ever, since $w \neq 1$. ∎

Example 4.4.7. Consider again the grammar with productions

$$\sigma \to a\lambda b, \ \lambda \to a\lambda b, \ \lambda \to 1.$$

The method of the lemma enables us to replace this by

$$\sigma \to a\lambda b, \ \sigma \to ab, \ \lambda \to a\lambda b, \ \lambda \to ab,$$

and a typical derivation

$$\sigma \Rightarrow a\lambda b \Rightarrow a^2\lambda b^2 \Rightarrow a^3\lambda b^3 \Rightarrow a^3b^3$$

becomes

$$\sigma \Rightarrow a\lambda b \Rightarrow a^2\lambda b^2 \Rightarrow a^2.ab.b^2 = a^3b^3.$$

Remark 4.4.8. It is instructive to examine the proof of Lemma 4.4.3 to see where we assumed that $1 \notin L$. Notice that if $1 \in L$ then σ itself is 1-accessible, but this in itself creates no difficulty. The proof that $L(\Gamma') \subseteq L(\Gamma)$ is unaffected by the change, but the proof that $w \in L(\Gamma)$ implies $w \in L(\Gamma')$ fails precisely when $w = 1$, for without 1-productions a derivation $\sigma \overset{*}{\Rightarrow} 1$ in Γ' is impossible.

In fact we have

Lemma 4.4.9. *Let L be a context-free language. Then $L\backslash\{1\}$ is context-free and can be generated by a grammar containing no 1-productions.* ∎

Example 4.4.10. The grammar

$$\sigma \to a\sigma b, \ \sigma \to 1$$

generates the language $L = \{a^n b^n : n \geq 0\}$ containing 1. If we apply the method of Lemma 4.4.3 to this grammar we obtain a new grammar

$$\sigma \to a\sigma b, \ \sigma \to ab$$

which generates $L\backslash\{1\} = \{a^n b^n : n \geq 1\}$.

The next stage in our reduction routine is to eliminate from our context-free grammar all *trivial* productions $\lambda \to \mu$ (with $\lambda, \mu \in V\backslash A$). Since productions of this kind merely amount to a renaming of symbols it certainly seems plausible that we should be able to circumvent them. We have

Lemma 4.4.11. *Every context-free language L can be generated by a grammar containing no 1-productions and no trivial productions.*

Proof. By virtue of Lemma 4.4.3 we may assume that $L = L(\Gamma)$, where $\Gamma = (V, A, \pi, \sigma)$ is a context-free grammar with no 1-productions. We define a new grammar $\Gamma' = (V, A, \pi', \sigma)$, where π' is specified as follows:
  (4.4.12) Remove all trivial productions from Γ.
  (4.4.13) Let $\mu \to z$ (with $z \in V^+$) be a non-trivial production in Γ. For every λ in $V\backslash A$ that can be connected to μ by a finite sequence

$$\lambda \to \nu_1 \to \cdots \to \nu_n \to \mu \quad (n \geq 0)$$

of trivial productions in Γ, let π' have the production $\lambda \to z$.

Suppose now that $w \in L(\Gamma)$, so that there is a derivation $\sigma \overset{*}{\Rightarrow} w$ in Γ. If this derivation makes no use of trivial productions then it is a derivation in Γ' and so $w \in L(\Gamma')$. Now let $k \geq 1$ and suppose inductively that if a derivation $\sigma \overset{*}{\Rightarrow} z$ in Γ makes fewer than k uses of trivial productions then $z \in L(\Gamma')$. Let $\sigma \overset{*}{\Rightarrow} w$ involve k uses of trivial productions and consider the *last* trivial production $\lambda \to \mu$ featuring in the derivation. We have

$$\sigma \overset{*}{\Rightarrow} \tau_1 \lambda \tau_2 \overset{*}{\Rightarrow} \tau_1 \mu \tau_2$$
$$\overset{*}{\Rightarrow} \tau_1' \mu \tau_2' \Rightarrow \tau_1' z \tau_2' \overset{*}{\Rightarrow} w \qquad (4.4.14)$$

and there are derivations $\tau_1 \overset{*}{\Rightarrow} \tau_1'$, $\tau_2 \overset{*}{\Rightarrow} \tau_2'$ in Γ. These involve no trivial productions and so in fact are derivations in Γ'. By definition, Γ' includes a production $\lambda \to z$, and so (4.4.14) can be replaced by a derivation

$$\sigma \overset{*}{\Rightarrow} \tau_1 \lambda \tau_2 \overset{*}{\Rightarrow} \tau_1' \lambda \tau_2' \overset{*}{\Rightarrow} \tau_1 z \tau_2' \overset{*}{\Rightarrow} w$$

involving only $k-1$ trivial productions. By the induction hypothesis we deduce that $w \in L(\Gamma')$. We have shown that $L(\Gamma) \subseteq L(\Gamma')$.

Conversely, suppose that $w \in L(\Gamma')$, so that there is a derivation $\sigma \overset{*}{\Rightarrow} w$ in Γ'. If this derivation makes no use of productions $\lambda \to z$ not in Γ then $w \in L(\Gamma)$. Otherwise we have

$$\sigma \overset{*}{\Rightarrow} \tau_1 \lambda \tau_2 \Rightarrow \tau_1 z \tau_2 \overset{*}{\Rightarrow} w.$$

By the definition (4.4.13) of Γ' there is a sequence

$$\lambda \to \nu_1 \to \cdots \to \nu_n \to \mu$$

of trivial productions in Γ such that $\mu \to z$ is a production in Γ. So

$$\sigma \overset{*}{\Rightarrow} \tau_1 \lambda \tau_2 \Rightarrow \tau_1 \nu_1 \tau_2 \Rightarrow \cdots \Rightarrow \tau_1 \nu_n \tau_2 \Rightarrow \tau_1 \mu \tau_2$$
$$\Rightarrow \tau_1 z \tau_2 \overset{*}{\Rightarrow} w$$

is a derivation in $\Gamma'' = (V, A, \pi \cup \pi', \sigma)$ involving fewer uses of productions outside π. Continuing in this way we eventually obtain a derivation $\sigma \overset{*}{\Rightarrow} w$ in Γ. Thus $L(\Gamma) = L(\Gamma')$, as required.   ∎

We illustrate the operation of this lemma by means of an example.

Example 4.4.15. Let $\Gamma = (V, A, \pi, \sigma)$, where

$$A = \{a, b\}, \ V \backslash A = \{\sigma, \lambda, \mu\}$$

and π consists of the productions

$$\sigma \to \lambda, \qquad \lambda \to \mu, \qquad \mu \to \sigma,$$
$$\sigma \to a^2, \qquad \lambda \to a, \qquad \mu \to b.$$

Then $L = \{a^2, a, b\}$. We have sequences

$$\sigma \to \lambda \to \mu, \qquad \sigma \to \lambda,$$
$$\mu \to \sigma \to \lambda, \qquad \mu \to \sigma,$$
$$\lambda \to \mu \to \sigma, \qquad \lambda \to \mu$$

of trivial productions and so Γ' has productions

$$\sigma \to a^2,\ \lambda \to a,\ \mu \to b,\ \sigma \to b,\ \sigma \to a,$$

$$\mu \to a,\ \mu \to a^2,\ \lambda \to a^2,\ \lambda \to b.$$

This example, while a bit artificial, illustrates why in specifying π' in (4.4.13) we have to allow for *sequences* of trivial productions. If we dealt only with trivial productions themselves then in the example we would fail to include $\sigma \to b$, $\mu \to a$ and $\lambda \to a^2$ and we would obtain $L(\Gamma') = \{a^2, a\} \subset L(\Gamma)$.

Moving on now to the next stage in our reduction, let us say that a context-free grammar $\Gamma = (V, A, \pi, \sigma)$ is *pure* if every production in π is of one of the two types

$$\lambda \to \zeta \quad (\lambda \in V \backslash A,\ \zeta \in (V \backslash A)^+,\ |\zeta| > 1),$$

$$\lambda \to a \quad (\lambda \in V \backslash A,\ a \in A).$$

Notice that a pure grammar contains neither 1-productions nor trivial productions.

Lemma 4.4.16. *Every context-free A^+-language L can be generated by a pure grammar.*

Proof. We may suppose that $L = L(\Gamma)$, where $\Gamma = (V, A, \pi, \sigma)$ has neither 1-productions nor trivial productions. We specify a pure grammar $\Gamma' = (V', A, \pi', \sigma)$ as follows. First, let

$$V' \backslash A = (V \backslash A) \cup \{\beta_a : a \in A\}.$$

That is to say, for each a in A we adjoin a new non-terminal symbol β_a to V. Productions in π are either of the type $\lambda \to a$, with $a \in A$, or of the type $\lambda \to z$, with $z \in V^+$, $z \notin V \backslash A$. Productions of the former type are included in π' without change, but for each production $\lambda \to z$ of the latter type we assign a production $\lambda \to z'$ to π', where z' is obtained from z by replacing each occurrence of a non-terminal symbol a by β_a. Then finally we assign a production $\beta_a \to a$ to π' for each a in A.

Certainly $\Gamma' = (V', A, \pi', \sigma)$ is a pure grammar. Moreover, every production $\lambda \to z$ in Γ (with $z \notin A$) can be simulated by a derivation

$$\lambda \Rightarrow z' \overset{*}{\Rightarrow} z$$

in Γ' and so $L(\Gamma) \subseteq L(\Gamma')$.

To show the reverse conclusion, suppose that $w \in L(\Gamma')$, so that there is a derivation $\sigma \overset{*}{\Rightarrow} w$ in Γ'. If this derivation involves just one step, it must be $\sigma \Rightarrow a$ and is already a derivation in Γ. Otherwise it must begin

$$\sigma \Rightarrow \tau_1 \beta_{a_1} \tau_2 \beta_{a_2} \ldots \tau_n \beta_{a_n} \tau_{n+1},$$

where $\tau_1, \tau_2, \ldots, \tau_{n+1} \in (V \backslash A)^*$ and where

$$\sigma \to \tau_1 a_1 \tau_2 a_2 \ldots \tau_n a_n \tau_{n+1}$$

is a production in Γ. By Lemma 4.3.4 we deduce that

$$w = t_1 z_1 t_2 z_2 \ldots t_n z_n t_{n+1},$$

where $t_1, t_2, \ldots, t_{n+1}, z_1, z_2, \ldots, z_n \in A^*$ and where $\tau_i \overset{*}{\Rightarrow} t_i (i = 1, 2, \ldots, n+1)$ and $\beta_{a_i} \overset{*}{\Rightarrow} z_i (i = 1, 2, \ldots, n)$ are derivations in Γ'. Since the only production involving the symbol β_{a_i} is $\beta_{a_i} \to a_i$ we must in fact have

$$w = t_1 a_1 t_2 a_2 \ldots t_n a_n t_{n+1}.$$

Since the derivations $\tau_i \overset{*}{\Rightarrow} t_i$ are shorter than the derivation $\sigma \overset{*}{\Rightarrow} w$ we may assume inductively that $\tau_i \overset{*}{\Rightarrow} t_i$ in Γ for $i = 1, 2, \ldots, n$. Hence we have a derivation

$$\sigma \Rightarrow t_1 a_1 t_2 a_2 \ldots t_n a_n t_{n+1} = w$$

in Γ. Thus $L(\Gamma) = L(\Gamma')$. ∎

Example 4.4.17. Following on Example 4.4.7, we can now replace the productions
$$\sigma \to a\lambda b, \ \sigma \to ab, \ \lambda \to a\lambda b, \ \lambda \to ab$$

by

$$\sigma \to \beta_a \lambda \beta_b, \ \sigma \to \beta_a \beta_b, \ \lambda \to \beta_a \lambda \beta_b, \ \lambda \to \beta_a \beta_b,$$

$$\beta_a \to a, \ \beta_b \to b$$

to obtain a pure grammar. The derivation

$$\sigma \Rightarrow a\lambda b \Rightarrow a^2 \lambda b^2 \Rightarrow a^2.ab.b^2 = a^3 b^3$$

becomes

$$\sigma \Rightarrow \beta_a \lambda \beta_b \Rightarrow \beta_a^2 \lambda \beta_b^2 \Rightarrow \beta_a^2.\beta_a \beta_b.\beta_b^2 \overset{*}{\Rightarrow} a^3 b^3.$$

Now at last we are ready to refer back to Definition 4.4.1 and prove the main result of this section.

Theorem 4.4.18. *Every context-free A^+-language can be generated by a grammar in Chomsky Normal Form.*

Proof. By Lemma 4.4.16 we may assume that $L = L(\Gamma)$ where $\Gamma = (V, A, \pi, \sigma)$ is pure. A pure grammar is quite close to being in Chomsky Normal Form, but of course Γ may well contain productions of type

$$\lambda \to \alpha_1 \alpha_2 \ldots \alpha_k \qquad (4.4.19)$$

where $k \geq 3$ and $\alpha_1, \alpha_2, \ldots, \alpha_k \in V \backslash A$. We create a new grammar $\Gamma' = (V', A, \pi', \sigma)$ in the following way.

For each production of the form (4.4.19) we adjoin $k - 2$ new nonterminal symbols $\gamma_1, \gamma_2, \ldots, \gamma_{k-2}$ and replace (4.4.19) by productions

$$\lambda \to \alpha_1 \gamma_1, \ \gamma_1 \to \alpha_2 \gamma_2, \ldots, \gamma_{k-2} \to \alpha_{k-1} \alpha_k. \qquad (4.4.20)$$

Otherwise V' is the same as V and π' the same as π.

Certainly the grammar $\Gamma' = (V', A, \pi', \sigma)$ is in Chomsky Normal Form. Notice also that each production of type (4.4.19) in Γ can be simulated by a derivation

$$\lambda \Rightarrow \alpha_1 \gamma_1 \Rightarrow \alpha_1 \alpha_2 \gamma_2 \Rightarrow \cdots \Rightarrow \alpha_1 \alpha_2 \ldots \alpha_k$$

in Γ'. Thus $L(\Gamma) \subseteq L(\Gamma')$.

To show the reverse inclusion, suppose that

$$\sigma \Rightarrow \omega_1 \Rightarrow \cdots \Rightarrow \omega_m \Rightarrow w \qquad (4.4.21)$$

is a derivation in Γ'. If it involves no occurrences of the symbols γ_i in $V' \backslash V$ then it is a derivation in Γ and so certainly $w \in L(\Gamma)$. Otherwise let ω_j be the last word in the derivation (4.4.21) to contain a symbol not in V. Then in the notation of (4.4.20) the step $\omega_j \Rightarrow \omega_{j+1}$ must be of the form

$$\omega_j = \tau_1 \gamma_{k-2} \tau_2 \Rightarrow \tau_2 = \omega_{j+1}.$$

Now the symbol γ_{k-2} can enter the derivation only by means of a sequence of productions of the type (4.4.20). So in fact the derivation $\sigma \overset{*}{\Rightarrow} \omega_{j+1}$ must take the form

$$\sigma \overset{*}{\Rightarrow} \tau_1 \lambda \tau_2 \Rightarrow \tau_1 \alpha_1 \gamma_1 \tau_2 \overset{*}{\Rightarrow} \tau_1' \alpha_1' \gamma_1 \tau_2'$$
$$\Rightarrow \tau_1' \alpha_1' \alpha_2 \gamma_2 \tau_2' \overset{*}{\Rightarrow} \tau_1'' \alpha_1'' \alpha_2'' \gamma_2 \tau_2''$$
$$\Rightarrow \tau_1'' \alpha_1'' \alpha_2'' \alpha_3 \gamma_3 \tau_2'' \overset{*}{\Rightarrow} \cdots$$
$$\Rightarrow \quad \cdots$$
$$\Rightarrow \tau_1^{(k-3)} \alpha_1^{(k-3)} \ldots \alpha_{k-3}^{(k-3)} \alpha_{k-2} \gamma_{k-2} \tau_2^{(k-3)}$$
$$\overset{*}{\Rightarrow} \tau_1^{(k-2)} \alpha_1^{(k-2)} \ldots \alpha_{k-2}^{(k-2)} \gamma_{k-2} \tau_2^{(k-2)} \quad (= \omega_j)$$
$$\Rightarrow \tau_1^{(k-2)} \alpha_1^{(k-2)} \ldots \alpha_{k-2}^{(k-2)} \alpha_{k-1} \alpha_k \tau_2^{(k-2)} \quad (= \omega_{j+1}).$$

Here

$$\tau_1 \overset{*}{\Rightarrow} \tau_1' \overset{*}{\Rightarrow} \tau_1'' \overset{*}{\Rightarrow} \cdots \overset{*}{\Rightarrow} \tau_1^{(k-2)},$$
$$\tau_2 \overset{*}{\Rightarrow} \tau_2' \overset{*}{\Rightarrow} \tau_2'' \overset{*}{\Rightarrow} \cdots \overset{*}{\Rightarrow} \tau_2^{(k-2)}$$

and

$$\alpha_i \overset{*}{\Rightarrow} \alpha_i^{(i)} \overset{*}{\Rightarrow} \cdots \overset{*}{\Rightarrow} \alpha_i^{(k-2)} \quad (i = 1, 2, \ldots, k - 2)$$

are derivations in Γ', and since all of them, and also the derivation $\sigma \overset{*}{\Rightarrow} \tau_1 \lambda \tau_2$, involve fewer symbols of $V' \backslash V$ than the derivation $\sigma \overset{*}{\Rightarrow} w$ we may assume that all are derivations in Γ. Then

$$\sigma \overset{*}{\Rightarrow} \tau_1 \lambda \tau_2 \Rightarrow \tau_1 \alpha_1 \alpha_2 \ldots \alpha_k \tau_2$$
$$\overset{*}{\Rightarrow} \tau_1^{(k-2)} \alpha_1^{(k-2)} \ldots \alpha_{k-2}^{(k-2)} \alpha_{k-1} \alpha_k \tau_2^{(k-2)}$$
$$\overset{*}{\Rightarrow} w$$

is a derivation in Γ.   ■

Example 4.4.22. The grammar which started life as

$$\sigma \to a\lambda b, \; \lambda \to a\lambda b, \; \lambda \to 1$$

had, by the end of Example 4.4.17, become

$$\sigma \to \beta_a \lambda \beta_b, \; \sigma \to \beta_a \beta_b, \; \lambda \to \beta_a \lambda \beta_b, \; \lambda \to \beta_a \beta_b,$$
$$\beta_a \to a, \; \beta_b \to b.$$

The final stage in our reduction to Chomsky Normal Form makes it into

$$\sigma \to \beta_a \gamma, \qquad \gamma \to \lambda \beta_b, \qquad \sigma \to \beta_a \beta_b, \qquad \lambda \to \beta_a \delta,$$

$$\delta \to \lambda \beta_b, \qquad \lambda \to \beta_a \beta_b, \qquad \beta_a \to a, \qquad \beta_b \to b,$$

and one possible derivation of $a^3 b^3$ is

$$\sigma \Rightarrow \beta_a \gamma \Rightarrow a\gamma \Rightarrow a\lambda \beta_b \Rightarrow a\beta_a \delta \beta_b$$
$$\Rightarrow a^2 \delta \beta_b \Rightarrow a^2 \lambda \beta_b^2 \Rightarrow a^2 \beta_a \beta_b^3 \Rightarrow a^3 \beta_b^3 \qquad (4.4.23)$$
$$\Rightarrow a^3 b \beta_b^2 \Rightarrow a^3 b^2 \beta_b \Rightarrow a^3 b^3.$$

That is, of course, only one of many possible derivations of $a^3 b^3$ using the grammar in Chomsky Normal Form, but it is of special interest because it is an example of a *leftmost derivation*, in which changes to the word are made as far to the left as possible. In general, given a grammar $\Gamma = (V, A, \pi, \sigma)$ in Chomsky Normal Form, we say that a derivation

$$\sigma \Rightarrow \zeta_1 \Rightarrow \cdots \Rightarrow \zeta_n \Rightarrow w$$

of w in A^* is a *leftmost derivation* if $n = 0$ or if (for $j = 1, 2, \ldots, n$) there exist $z_j \in A^*$, $\zeta_j' \in (V \backslash A)^+$, $w_j \in A^*$ such that $w = z_j w_j$, $\zeta_j = z_j \zeta_j'$, and $\zeta_j' \overset{*}{\Rightarrow} w_j$ is a derivation in Γ.

In fact we have

Theorem 4.4.24. *Let $w \in L(\Gamma)$, where $\Gamma = (V, A, \pi, \sigma)$ is a grammar in Chomsky Normal Form. Then there exists a leftmost derivation of w.*

Proof. We shall prove in fact that for every λ in $V \backslash A$ every derivation $\lambda \overset{*}{\Rightarrow} w$ for which w is in A^* can be replaced by a leftmost derivation. The proof is by induction on the number n of steps in the derivation $\lambda \overset{*}{\Rightarrow} w$, it being obvious that if $n = 1$ then the derivation $\lambda \Rightarrow a$ is already in leftmost form.

If $n \geq 1$ then the derivation $\lambda \overset{*}{\Rightarrow} w$ must begin with $\lambda \Rightarrow \alpha\beta$. By Lemma 4.3.4, $w = uv$ and $\alpha \overset{*}{\Rightarrow} u$, $\beta \overset{*}{\Rightarrow} v$ are derivations in Γ. Moreover, each is shorter than the derivation $\lambda \overset{*}{\Rightarrow} w$ and so by the induction hypothesis each may be replaced by a leftmost derivation. Then

$$\lambda \Rightarrow \alpha\beta \overset{*}{\Rightarrow} u\beta \overset{*}{\Rightarrow} uv = w$$

is a leftmost derivation of w.   ∎

Example 4.4.25. Consider again the grammar

$$\Gamma = \big(\{\sigma, \lambda, \mu, a, b\}, \{a, b\}, \pi, \sigma\big)$$

introduced in the proof of Theorem 4.3.7, where π consists of the productions

$$\sigma \to a\mu, \qquad \sigma \to b\lambda, \qquad \lambda \to a, \qquad \lambda \to a\sigma,$$

$$\lambda \to b\lambda^2, \qquad \mu \to b, \qquad \mu \to b\sigma, \qquad \mu \to a\mu^2.$$

We have seen that

$$L(\Gamma) = \{w \in \{a, b\}^+ : |w|_a = |w|_b\}.$$

If we reduce the grammar to Chomsky Normal Form the productions become

$$\sigma \to \beta_a\mu, \quad \sigma \to \beta_b\lambda, \quad \lambda \to a, \quad \lambda \to \beta_a\sigma, \quad \lambda \to \beta_b\gamma,$$

$$\gamma \to \lambda^2, \quad \mu \to b, \quad \mu \to \beta_b\sigma, \quad \mu \to \beta_b\delta, \quad \delta \to \mu^2,$$

and a leftmost derivation of $a^2 b^3 a$ is

$$\sigma \Rightarrow \beta_a\mu \Rightarrow a\mu \Rightarrow a\beta_a\delta \Rightarrow a^2\delta$$

$$\Rightarrow a^2\mu^2 \Rightarrow a^2 b\mu \Rightarrow a^2 b\beta_b\sigma \Rightarrow a^2 b^2 \sigma$$

$$\Rightarrow a^2 b^2 \beta_b\lambda \Rightarrow a^2 b^3 \lambda \Rightarrow a^2 b^3 a.$$

Remark 4.4.26. We certainly do not want to make any permanent exclusion of languages containing 1. In Lemma 4.4.9 we remarked that

if we apply the first stage in the reduction (namely, the removal of
1-productions) to a language L containing 1 the effect was to show that
$L\backslash\{1\}$ is generated by a grammar containing no 1-productions. Thus if
we continue the reduction routine we eventually obtain a grammar in
Chomsky Normal Form generating $L\backslash\{1\}$. At this point, if necessary,
we can restore 1 to the language simply by inserting new productions
$\sigma \to \mu\nu$ (where μ, ν are new symbols), $\mu \to 1$, $\nu \to 1$. (Simply inserting
$\sigma \to 1$ will not do, since it may bring about more changes than we
bargain for.) The new grammar is not strictly in Chomsky Normal
Form as we have defined it, but it comes close. Indeed we have a
theorem which slightly modifies Theorem 4.4.18:

Theorem 4.4.27. *Every context-free language in A^* can be generated
by a grammar $\Gamma = (V, A, \pi, \sigma)$ in which each production is either of the
form*

$$\lambda \to \alpha\beta \quad (\lambda, \alpha, \beta \in V\backslash A)$$

or of the form

$$\lambda \to a \quad (a \in A \cup \{1\}). \qquad \blacksquare$$

Exercises 4

4.1. Describe the languages in $\{a, b\}^*$ generated by the regular grammars with productions:
   (i) $\sigma \to a\lambda,\ \lambda \to b\lambda,\ \lambda \to 1$;
   (ii) $\sigma \to a\lambda,\ \lambda \to a\lambda,\ \lambda \to b\lambda,\ \lambda \to b\mu,\ \mu \to 1$;
   (iii) $\sigma \to a\lambda,\ \sigma \to a\sigma,\ \sigma \to b\sigma,\ \lambda \to b\mu$,
         $\mu \to a\nu,\ \nu \to a\nu,\ \nu \to b\nu,\ \nu \to 1$.

4.2. Let $A = \{a, b\}$. For each of the following regular languages in
A^*, find the minimal automaton — see Section 3.2 —and hence,
using the method of Theorem 4.2.9, find a regular grammar
generating the language:
   (i) a^+b^+;   (ii) $A^*a^2A^*$;
   (iii) $A^*\backslash A^*a^2A^*$;   (iv) $\{w \in A^* : |w|_a = 2\}$.

4.3. Say that a grammar $\Gamma = (V, A, \pi, \sigma)$ is *left regular* if every
production is either of the form $\alpha \to \beta x$ (with $x \in A^+$, $\alpha, \beta \in V\backslash A$)
or of the form $\alpha \to z$ (with $\alpha \in V\backslash A$, $z \in A^*$); and say that a
language is *left regular* if it can be generated by a left regular

grammar. Show that a language is left regular if and only if it is regular.

4.4. Describe the languages in $\{a, b\}^*$ generated by the context-free grammars with productions:

    (i) $\sigma \rightarrow a\sigma b^2$, $\sigma \rightarrow 1$;

    (ii) $\sigma \rightarrow a\lambda a$, $\sigma \rightarrow b\lambda b$, $\lambda \rightarrow a\lambda a$, $\lambda \rightarrow b\lambda b$, $\lambda \rightarrow 1$.

Is either of these languages regular?

4.5. Find a context-free grammar generating each of the following languages:

    (i) $\{a^n b^{2n} : n \geq 1\}$;

    (ii) $\{a^p b^{p+q} a^q : p, q \geq 1\}$;

    (iii) $\{a^m b^n : m, n \geq 1, m \neq n\}$;

    (iv) $\{w \in \{a, b\}^+ : |w|_a = |w|_b + 1\}$;

    (v) $\{wcw^R : w \in \{a, b\}^*\}$ $(\subseteq \{a, b, c\}^*)$.

4.6. Find a context-free grammar generating the set of palindromes of odd length in $\{a, b\}^*$. Put the grammar into Chomsky Normal Form and write down a leftmost derivation of $a^2 baba^2$.

4.7. (i) Describe the language generated by the context-free grammar

$$\Gamma = \left(\{a, b, \sigma, \lambda\}, \{a, b\}, \pi, \sigma\right),$$

where π consists of the productions

$$\sigma \rightarrow a\lambda a, \ \sigma \rightarrow b\lambda a, \ \lambda \rightarrow a\lambda a,$$

$$\lambda \rightarrow b\lambda a, \ \lambda \rightarrow 1.$$

(ii) Show that this language is not regular.

(iii) Put the grammar into Chomsky Normal Form and give a leftmost derivation of $ab^2 a^5$.

4.8. Find a context-free grammar generating

$$L = \{a^p b^{p+2q} a^q : p, q \geq 0\}.$$

Noticing that $1 \in L$, put this grammar into the modified Chomsky Normal Form described in Theorem 4.4.27.

5 Pushdown automata

Introduction

Before introducing the main ideas of this chapter we shall find it useful to make a slight change in our informal 'imagery' concerning finite state automata. We recall that for such an automaton $\mathcal{A} = (Q, A, i, T)$ we have an action $q \mapsto qw$ of A^* on Q, and $w \in L(\mathcal{A})$ if and only if $iw \in T$. If $w = a_1 a_2 \ldots a_n$ (with $a_1, a_2, \ldots, a_n \in A$) we may think of the word w as printed on a tape

$$\boxed{a_1} \boxed{a_2} \boxed{ \cdots } \boxed{a_n}$$

with the initial state i 'scanning' the leftmost square. If we have

$$ia_1 = q_1, \quad q_1 a_2 = q_2, \quad \ldots, \quad q_{n-1} a_n = q_n$$

where of course $i, q_1, \ldots, q_n$ need not all be distinct) we may think of a 'computation' proceeding as follows:

$$(i, \ a_1 a_2 \ldots a_n) \rightarrow (q_1, \ a_2 \ldots a_n) \rightarrow \cdots \rightarrow (q_{n-1}, a_n) \rightarrow (q_n, 1).$$

At a typical stage q_j scans the leftmost square containing a_j, deletes a_j, changes to q_{j+1} and moves to scan the new leftmost square. At the end of the process the tape has been completely erased, and the computation has been 'successful' if and only if $q_n \in T$. If the automaton is incomplete there may be computations that terminate while the tape still has symbols on it; and if the automaton is non-deterministic there may be branchings in the computation.

The most serious defect of a finite state automaton — and since we plan shortly to discuss an improved version it will avoid confusion if we now use 'FSA' rather than 'automaton' as an abbreviation — is its lack of any kind of memory. We can illustrate this by looking first at the finite language

$$L_5 = \{a^n b^n : 1 \le n \le 5\}$$

and then at the infinite language

$$L = \{a^n b^n : n \geq 1\}.$$

The first, being finite, is certainly recognizable, for example by the FSA

$$\mathcal{A}_5 = (\{i, q_1, \ldots, q_5, t\}, \{a, b\}, i, \{t\})$$

in which

$$ia = q_1, \quad q_i a = q_{i+1} \ (i = 1, 2, 3, 4),$$
$$q_j b = q_{j-1} \ (j = 2, 3, 4, 5), \quad q_1 b = t,$$

a typical successful computation being

$$(i, a^3 b^3) \rightarrow (q_1, a^2 b^3) \rightarrow (q_2, ab^3) \rightarrow (q_3, b^3)$$
$$\rightarrow (q_2, b^2) \rightarrow (q_1, b) \rightarrow (t, 1).$$

In effect the FSA $\mathcal{A}_5$ is using the states $q_1, q_2, \ldots q_5$ to count the number of a's in the first half of the word. But this technique will not work for the language L, since for any sufficiently large n there must be a repetition among the states $i, ia, ia^2, \ldots, ia^n$. (And of course it was precisely because of this that we were able to prove in Theorem 2.2.6 that L is not in fact recognizable by an FSA.)

If a language such as L is to be recognizable, it must be by some kind of improved automaton with a memory facility capable of counting how many a's have been scanned. To get an idea of what this kind of improved automaton might be, consider a 'machine' with just three states i, q and t, and suppose that we also have a 'memory stack' which can contain words in an alphabet $M = \{\zeta, \alpha\}$. Instead of an action of A on Q (i.e., a partial map $\varphi : Q \times A \rightarrow Q$) we suppose that we have a partial map

$$\rho : Q \times M \times (A \cup \{1\}) \rightarrow Q \times M^*.$$

To be quite specific, suppose that

$$\rho(i, \zeta, a) = (i, \alpha\zeta) \quad \rho(i, \alpha, a) = (i, \alpha^2)$$
$$\rho(i, \alpha, b) = (q, 1) \quad \rho(q, \alpha, b) = (q, 1)$$
$$\rho(q, \zeta, 1) = (t, \zeta).$$

We may think of these as 'instructions' to our machine. The first two give the instruction:

   read a; add α to the memory stack.
The third and fourth instructions say:

read b; if the leftmost symbol in the memory stack is α remove it and go into, or remain in, state q.

The final instruction says:

read 1; if the leftmost symbol in the memory stack is ζ leave ζ alone and go into the terminal state t.

The symbol ζ is the *initial* memory stack symbol. A typical successful computation is

$$(i, \zeta, a^n b^n) \to (i, \alpha\zeta, a^{n-1} b^n) \to (i, \alpha^2\zeta, a^{n-2} b^n)$$
$$\to \cdots \to (i, \alpha^n \zeta, b^n) \to (q, \alpha^{n-1}\zeta, b^{n-1})$$
$$\to \cdots \to (q, \zeta, 1) \to (t, \zeta, 1).$$

By contrast, the computations

$$(i, \zeta, ab^2) \to (i, \alpha\zeta, b^2) \to (q, \zeta, b)$$

and

$$(i, \zeta, aba) \to (i, \alpha\zeta, ba) \to (q, \zeta, a)$$

both terminate without reaching the terminal state t and so are unsuccessful. It is not hard to be convinced that a computation beginning (i, ζ, w) can be successful if and only if $w = a^n b^n$ for some $n \geq 1$. Notice how the element α of the memory stack records how many initial a's have been scanned. After this the computation fails unless the remainder of the word consists of b's only and the number of b's is precisely the same as the number of initial a's.

5.1 Pushdown automata

The 'improved' machine constructed in the Introduction is too special in one important respect, in that a computation

$$(i, \zeta, w) \to \cdots \to (t, \zeta, 1)$$

recognizing a word w of length m must itself be of length precisely m. We shall want to allow for the possibility that our machine, as it scans from left to right, sometimes stops and adjusts its memory stack before continuing its rightwards scan. We shall also want at the outset to build indeterminacy into our new machines, and we shall discover that this is no longer the 'optional extra' that it was for FSAs, for the deterministic versions of our machines will recognize a smaller class of languages than the non-deterministic ones.

To make any further progress we now urgently require a formal definition.

Definition 5.1.1. A *pushdown automaton* (PDA) $\mathcal{P}$ is an octuple

$$\mathcal{P} = (Q, A, M, \mu, \rho, i, \zeta, T),$$

where
   Q is a finite set, the set of *states* of $\mathcal{P}$;
   $A \, (\neq \emptyset)$ is a finite set, the *alphabet* of $\mathcal{P}$;
   M is a finite set, the *memory stack alphabet* of $\mathcal{P}$;
   μ is a partial map from $Q \times M$ into the set $\mathcal{F}(Q \times M^*)$ of all finite subsets of $Q \times M^*$;
   ρ is a partial map from $Q \times M \times (A \cup \{1\})$ into $\mathcal{F}(Q \times M^*)$;
   $i \, (\in Q)$ is the *initial state*;
   $\zeta \, (\in M)$ is the *initial stack symbol*;
   $T \, (\subseteq Q)$ is the (non-empty) set of *terminal states*.

An *instantaneous description* (ID) of $\mathcal{P}$ is a triple (q, τ, w) in $Q \times M^* \times A^*$. If $\tau = \beta\tau'$ (with $\beta \in M$) and if the subset $\mu(q, \beta)$ contains the element (q', ξ) of $Q \times M^*$, then

$$(q, \tau, w) \rightarrow (q', \xi\tau', w) \tag{5.1.2}$$

is called a *transition* in $\mathcal{P}$. We can think of μ as giving the instruction:

   change q to q'; change β to ξ; leave w alone.

If $\tau = \beta\tau'$, $w = aw'$ (with $\beta \in M$, $a \in A$) and if the subset $\rho(q, \beta, a)$ contains the element (q', ξ) of $Q \times M^*$, then

$$(q, \tau, w) \rightarrow (q', \xi\tau', w') \tag{5.1.3}$$

is again called a *transition* in $\mathcal{P}$. Here we think of ρ as giving the instruction:

   change q to q'; change β to ξ; delete a.

The remaining type of transition arises only if the 'tape' containing the scanned word is blank: if $\rho(q, \beta, 1)$ contains the element (q', ξ) then

$$(q, \beta\tau', 1) \rightarrow (q', \xi\tau', 1) \tag{5.1.4}$$

is a *transition* in $\mathcal{P}$.

   If two IDs (q_1, τ_1, w_1) and (q_2, τ_2, w_2) are equal, or are linked by a finite sequence of transitions of these three kinds, we write

$$(q_1, \tau_1, w_1) \overset{*}{\rightarrow} (q_2, \tau_2, w_2).$$

Such a sequence of transitions is called a *computation* in $\mathcal{P}$, and we shall often want to refer to the individual transitions as *steps* in the computation. Steps of type (5.1.2) are called *stationary* — notice that the word w does not change during a stationary step — while a step of type (5.1.3), resulting as it does in the deletion of a letter of w, is called *progressive*.

A computation beginning with (i, ζ, w) and ending with a triple belonging to $T \times M^* \times \{1\}$ will be called *successful*, and a word w in A^* is said to be *recognized* by $\mathcal{P}$ if there is a successful computation beginning with (i, ζ, w). More formally, *the language $L(\mathcal{P})$ recognized by the PDA $\mathcal{P}$* is defined by

$$L(\mathcal{P}) = \left\{ w \in A^* : (\exists t \in T)(\exists \tau \in M^*)(i, \zeta, w) \overset{*}{\to} (t, \tau, 1) \right\}. \tag{5.1.5}$$

Remark 5.1.6. The name 'pushdown automaton' arises from a common (and quite helpful) visualization of the memory stack as being like a stack of plates in a cafeteria, in which only the top plate is accessible. In an ID (q, τ, w) the word τ is, as it were, a stack of plates. If $\tau = \beta \tau'$ then β is the top plate in the stack, and whether the transition is of type (5.1.2) or (5.1.3) or (5.1.4) only β, the top plate in the stack, has any influence on the immediate outcome.

Remark 5.1.7. In a computation

$$(i, \zeta, w) \overset{*}{\to} (t, \tau, 1)$$

we visualize the word w as being on a tape which is traversed irreversibly, but possibly with stops, from left to right. If the step is a stationary one then the scanner 'marks time' while the state and memory stack are adjusted; if it is a progressive one then again the state and the memory stack are adjusted, but this time the scanned letter is deleted and the scanner moves one space right.

Remark 5.1.8. If at any point in a computation the memory stack becomes empty, then the computation ceases, and is successful if and only if the outcome is $(t, 1, 1)$, where $t \in T$.

5.2 Examples

A series of simple examples should help to clarify the ideas introduced in the last section.

Example 5.2.1. Let $Q = \{i, q, t\}$, $A = \{a, b, c\}$, $M = \{\zeta, \alpha, \beta\}$, and define the partial map $\rho : Q \times M \times (A \cup \{1\}) \to \mathcal{F}(Q \times M^*)$ by

$$
\begin{aligned}
\rho(i, \zeta, a) &= \{(i, \alpha\zeta)\}, & \rho(i, \alpha, a) &= \{(i, \alpha^2)\}, \\
\rho(i, \beta, a) &= \{(i, \alpha\beta)\}, & \rho(i, \zeta, b) &= \{(i, \beta\zeta)\}, \\
\rho(i, \alpha, b) &= \{(i, \beta\alpha)\}, & \rho(i, \beta, b) &= \{(i, \beta^2)\}, \\
\rho(i, \zeta, c) &= \{(t, \zeta)\}, & \rho(i, \alpha, c) &= \{(q, \alpha)\}, \\
\rho(i, \beta, c) &= \{(q, \beta)\}, & \rho(q, \alpha, a) &= \{(q, 1)\}, \\
\rho(q, \beta, b) &= \{(q, 1)\}, & \rho(q, \zeta, 1) &= \{(t, \zeta)\}.
\end{aligned}
$$

Let μ be the empty map, and let $\mathcal{P}$ be the pushdown automaton

$$
(Q, A, M, \rho, \mu, i, \zeta, \{t\}).
$$

Then as an example of a successful computation we have

$$
(i, \zeta, a^2 bacaba^2) \overset{*}{\to} (i, \alpha\beta\alpha^2\zeta, caba^2) \to (q, \alpha\beta\alpha^2\zeta, aba^2)
$$

$$
\to (q, \beta\alpha^2\zeta, ba^2) \to (q, \alpha^2\zeta, a^2) \to (q, \alpha\zeta, a)
$$

$$
\to (q, \zeta, 1) \to (t, \zeta, 1).
$$

Here it is not hard to see that a successful computation beginning (t, ζ, w) is possible if and only if w is of the form zcz^R, where $z \in \{a, b\}^*$ and z^R is the reversed word of z. If $w = zcz'$, where $z \in \{a, b\}^+$, $z' \in \{a, b, c\}^*$, then the computation beginning (i, ζ, w) reaches $(i, z_M^R \zeta, cz')$ after $|z|$ steps, where for each u in $\{a, b\}^+$ we denote by u_M the word obtained from u by substituting α for a and β for b. The next step in the computation yields the ID $(q, z_M^R \zeta, z')$, and since both $\rho(q, \alpha, b)$ and $\rho(q, \beta, a)$ are undefined the computation fails unless z' contains no c's and is precisely z^R. Thus

$$
L(\mathcal{P}) = \{zcz^R : z \in \{a, b\}^*\}, \tag{5.2.2}
$$

the set of palindromic words in a, b and c having exactly one occurrence of c at the centre.

Before giving the next example we pause to record one important fact about PDAs, which is that in effect they do actually include FSAs. We have

Theorem 5.2.3. *Every rational language can be recognized by a pushdown automaton.*

Proof. Suppose that $L = L(\mathcal{A})$, where

$$\mathcal{A} = (Q, A, \varphi, i, T)$$

is an FSA. We define a PDA

$$\mathcal{P} = (Q, A, \{\varsigma\}, \mu, \rho, i, \varsigma, T),$$

where

$$\rho(q, \varsigma, a) = \{(\varphi(q, a), \varsigma)\} \quad (q \in Q, \ a \in A)$$

and μ is the empty map. Then an arbitrary successful path

$$i \to ia_1 \to \cdots \to ia_1 a_2 \ldots a_n \ (\in T)$$

(with $q_r = ia_1 a_2 \ldots a_r$) in $\mathcal{A}$ can be imitated by a successful computation

$$(i, \varsigma, a_1 a_2 \ldots a_n) \to (q_1, \varsigma, a_2 \ldots a_n)$$

$$\to (q_2, \varsigma, a_3 \ldots a_n) \to \cdots \to (q_n, \varsigma, 1)$$

in $\mathcal{P}$. Moreover, the only successful computations in $\mathcal{P}$ are precisely of this kind, and so $L(\mathcal{A}) = L(\mathcal{P})$. ∎

On the other hand we have already seen two examples, namely $\{a^n b^n : n \geq 1\}$ and $\{wcw^R : w \in \{a, b\}^*\}$, of languages that are not rational but are recognized by PDAs. (See Theorem 2.2.6 and Exercise 2.11.) Both these languages are context-free — see Example 4.3.1 and Exercise 4.5 — and we shall see that this is not an accident.

Both our examples of PDAs have been deterministic and μ has been the empty map. The next example is non-deterministic.

Example 5.2.4. Let $Q = \{i, q, t\}$, $A = \{a, b\}$, $M = \{\varsigma, \alpha, \beta\}$, and let

$$\rho : Q \times M \times (A \cup \{1\}) \to \mathcal{F}(Q \times M^*)$$

be given by

$$\rho(i, \varsigma, a) = \{(i, \alpha\varsigma)\}, \quad \rho(i, \beta, a) = \{(i, \alpha\beta)\},$$
$$\rho(i, \varsigma, b) = \{(i, \beta\varsigma)\}, \quad \rho(i, \alpha, b) = \{(i, \beta\alpha)\},$$
$$\rho(q, \alpha, a) = \{(q, 1)\}, \quad \rho(q, \beta, b) = \{(q, 1)\},$$

$$\rho(i, \alpha, a) = \{(i, \alpha^2), (q, 1)\},$$
$$\rho(i, \beta, b) = \{(i, \beta^2), (q, 1)\},$$
$$\rho(q, \varsigma, 1) = \{(t, \varsigma)\}.$$

Let μ be the empty map and let

$$\mathcal{P} = \big(Q, A, M, \rho, \mu, i, \zeta, \{t\}\big).$$

Here a typical successful computation is

$$\begin{aligned}(i,\zeta,a^2b^2a^2) &\to (i,\alpha\zeta,ab^2a^2) \to^{(1)} (i,\alpha^2\zeta,b^2a^2)\\ &\to (i,\beta\alpha^2\zeta,ba^2) \to^{(2)} (q,\alpha^2\zeta,a^2) \to (q,\alpha\zeta,a)\\ &\to (q,\zeta,1) \to (t,\zeta,1).\end{aligned}$$

At the stages marked (1) and (2) the indeterminacy comes into play: at (1) we take the first choice for $\rho(i,\alpha,a)$, while at (2) we take the second choice for $\rho(i,\beta,b)$.

Of course there is no sense in which the PDA 'knows' that these are the right choices. We are saying that $w \in L(\mathcal{P})$ if and only if one of the (possibly many) computations beginning with (i,ζ,w) is successful. We can think of the indeterminacy as causing a 'fanning out' of the computation. If we allow for all possibilities in computations beginning $(i,\zeta,a^2b^2a^2)$ we get a picture like Figure 5.2.5, in which the unique successful computation appears in the left hand column.

$$\begin{array}{ccccc}
(i,\zeta,a^2b^2a^2) & & & & \\
\downarrow & & & & \\
(i,\alpha\zeta,ab^2a^2) & \to & (q,\zeta,b^2a^2)\ \|\text{(fails)} & & \\
\downarrow & & & & \\
(i,\alpha^2\zeta,b^2a^2) & & & & \\
\downarrow & & & & \\
(i,\beta\alpha^2\zeta,ba^2) & \to & (i,\beta^2\alpha^2\zeta,a^2) & & \\
\downarrow & & \downarrow & & \\
(q,\alpha^2\zeta,a^2) & & (i,\alpha\beta\alpha^2\zeta,a) & \to & (q,\beta\alpha^2\zeta,1)\ \|\text{(fails)} \\
\downarrow & & \downarrow & & \\
(q,\alpha\zeta,a) & & (i,\alpha^2\beta\alpha^2\zeta,1)\ \|\text{(fails)} & & \\
\downarrow & & & & \\
(q,\zeta,1) & & & & \\
\downarrow & & & & \\
(t,\zeta,1) & & & &
\end{array}$$

Figure 5.2.5

We now show that

$$L(\mathcal{P}) = \{ zz^R : z \in \{a, b\}^+ \}, \tag{5.2.6}$$

the set of palindromes of even length in $\{a, b\}^+$.

First, if $w = zz^R$ for some non-empty word in the alphabet $\{a, b\}$ then we have a computation

$$(i, \zeta, zz^R) \overset{*}{\to} (i, z_M^R \zeta, z^R),$$

where as before z_M^R is the word obtained from z^R by substituting α for a and β for b. If we suppose without loss of generality that $z = ua$, with $u \in \{a, b\}^*$ (i.e., if we suppose that z ends in a) then $z^R = au^R$, $z_M^R = \alpha u_M^R$ and the next step in the computation is

$$(i, z_M^R \zeta, z^R) \to (q, u_M^R \zeta, u^R).$$

From this point on the computation is determinate, and at each stage the leftmost symbol in the scanned word, whether a or b, matches the leftmost symbol, whether α or β, in the memory stack. So we obtain

$$(q, u_M^R \zeta, u^R) \overset{*}{\to} (q, \zeta, 1) \to (t, \zeta, 1),$$

and the conclusion is that $w \in L(\mathcal{P})$.

Conversely, suppose that $w \in L(\mathcal{P})$. The computation

$$(i, \zeta, w) \overset{*}{\to} (t, \tau, 1)$$

is bound to have two main phases, one in which the state is i and the other in which the state is q. There is no way of changing i to t directly, and there is no way back from q to i or from t to q. It is possible to switch from phase 1 to phase 2 at any point where w has a repeated letter. So suppose (without loss of generality) that $w = ua^2v$ (with $u, v \in \{a, b\}^*$) and that the computation begins

$$(i, \zeta, w) \overset{*}{\to} (i, \alpha u_M^R \zeta, av) \to (q, u_M^R \zeta, v).$$

Thereafter the computation has no choice. If at any stage we arrive at an ID (q, τ, s) where τ begins with α and s with b, or where τ begins with β and s with a, then the computation fails, because $\rho(q, \alpha, b)$ and $\rho(q, \beta, a)$ are undefined. The middle coordinate can find its way back to ζ (which is something it has to do before t can put in an appearance) only if v has u^R as a left factor: $v = u^R v'$. But then we reach (q, ζ, v'), and the computation fails at this stage unless $v' = 1$. So $w = ua^2u^R = zz^R$ (with $z = ua$). ∎

The indeterminacy certainly seems to be an essential feature of this example. The automaton cannot 'know in advance' when it has reached the centre of the word, nor which (if any) of the several branches illustrated in Figure 5.2.5 will succeed. The point is that exactly one branch will succeed when (and only when) the word is a palindrome of even length.

We shall see eventually in Exercise 5.10 that the set of palindromes of even length cannot be recognized by a deterministic PDA.

We end this section with another example.

Example 5.2.7. Let $Q = \{i, q, t\}$, $A = \{a, b\}$, $M = \{\zeta, \alpha, \beta\}$, and let

$$\rho : Q \times M \times (A \cup \{1\}) \rightarrow \mathcal{F}(Q \times M^*)$$

be given by

$$\rho(i, \lambda, a) = \{(i, \alpha\lambda)\} \quad (\lambda = \alpha, \beta, \zeta),$$
$$\rho(i, \lambda, b) = \{(i, \beta\lambda)\} \quad (\lambda = \alpha, \beta, \zeta),$$

$$\rho(q, \alpha, a) = \rho(q, \beta, b) = \{(q, 1)\},$$

$$\rho(q, \zeta, 1) = (t, \zeta).$$

Let $\mu : Q \times M \rightarrow \mathcal{F}(Q \times M^*)$ be given by

$$\mu(i, \alpha) = \mu(i, \beta) = \{(q, 1)\}.$$

Let

$$\mathcal{P} = (Q, A, M, \rho, \mu, i, \zeta, \{t\}).$$

Then a typical successful computation is

$$(i, \zeta, a^3 b^2 ab^2 a^3) \xrightarrow{*} (i, \alpha\beta^2 \alpha^3 \zeta, b^2 a^3)$$
$$\rightarrow (q, \beta^2 \alpha^3 \zeta, b^2 a^3) \xrightarrow{*} (q, \zeta, 1) \rightarrow (t, \zeta, 1).$$

The indeterminacy of this PDA arises because of the choice between ρ and μ. The 'fanning out' we observed in Example 5.2.4 is more luxuriant here and for a word as long as $a^3 b^2 ab^2 a^3$ will not conveniently fit on a page. For the shorter word $ab^3 a$ the situation is as illustrated in Figure 5.2.8, where again the unique successful branch appears in the left hand column.

$$(i, \zeta, ab^3 a)$$

$$\downarrow$$

$$(i, \alpha\zeta, b^3 a) \quad \rightarrow \quad (q, \zeta, b^3 a)\| \text{ (fails)}$$

$$\downarrow$$

$$(i, \beta\alpha\zeta, b^2 a) \quad \rightarrow \quad (q, \alpha\zeta, b^2 a)\| \text{ (fails)}$$

$$\downarrow$$

$$(i, \beta^2\alpha\zeta, ba) \quad \rightarrow \quad (i, \beta^3\alpha\zeta, a) \quad \rightarrow \quad (i, \alpha\beta^3\alpha\zeta, 1)\| \text{ (fails)}$$

$$\downarrow \qquad\qquad\qquad\qquad \downarrow$$

$$(q, \beta\alpha\zeta, ba) \qquad (q, \beta^2\alpha\zeta, 1)\| \text{ (fails)}$$

$$\downarrow$$

$$(q, \alpha\zeta, a)$$

$$\downarrow$$

$$(q, \zeta, 1)$$

$$\downarrow$$

$$(t, \zeta, 1)$$

Figure 5.2.8

It is left as an exercise to the reader to show that

$$L(\mathcal{P}) = \{w \in \{a, b\}^+ : |w| \text{ is odd and } w^R = w\}, \qquad (5.2.9)$$

the set of palindromes of odd length in $\{a, b\}$.

5.3 Recognition by empty memory stack

As with finite state automata, there is of course a degree of arbitrariness about the definition of a pushdown automaton. Other definitions are possible, differing in this or that detail from the one we have adopted. The important issue is that the class of recognizable languages should not vary. One possible alternative approach is to have a different criterion for acceptance: instead of accepting a word w if there is a computation

$$(i, \zeta, w) \overset{*}{\rightarrow} (t, \zeta, 1)$$

in which t is a terminal state, we can drop all reference to terminal states and can choose to accept w if there is a computation

$$(i, \zeta, w) \overset{*}{\to} (q, 1, 1)$$

ending with an ID for which the memory stack is empty.

To exemplify this, let us make a small modification to Example 5.2.4.

Example 5.3.1. Let $Q = \{i, q\}$, $A = \{a, b\}$, $M = \{\zeta, \alpha, \beta\}$, and let

$$\rho : Q \times M \times (A \cup \{1\}) \to \mathcal{F}(Q \times M^*)$$

be given, almost as before, by

$$\rho(i, \zeta, a) = \{(i, \alpha\zeta)\}, \quad \rho(i, \beta, a) = \{(i, \alpha\beta)\},$$
$$\rho(i, \zeta, b) = \{(i, \beta\zeta)\}, \quad \rho(i, \alpha, b) = \{(i, \beta\alpha)\},$$
$$\rho(q, \alpha, a) = \{(q, 1)\}, \quad \rho(q, \beta, b) = \{(q, 1)\},$$

$$\rho(i, \alpha, a) = \{(i, \alpha^2), (q, 1)\},$$
$$\rho(i, \beta, b) = \{(i, \beta^2), (q, 1)\},$$
$$\rho(q, \zeta, 1) = \{(q, 1)\}.$$

(Only the last line is different.) Again let μ be the empty map. Write

$$\mathcal{P}' = (Q, A, M, \rho, \mu, i, \zeta),$$

(a septuple this time rather than an octuple, since no reference to terminal states is necessary). It is now fairly easy to see that a computation

$$(i, \zeta, 1) \overset{*}{\to} (q, 1, 1)$$

can take place if and only if $w = zz^R$ for some z in $\{a, b\}^+$.

This example suggests that a small modification will convert the one kind of PDA into the other, without disturbing the set of recognizable words. To state this properly we need a little notation. First, let us refer to our original type of PDA as a TSPDA (where TS stands for 'terminal state') and to the new kind of PDA as an ESPDA (where ES stands for 'empty stack'). If we have a TSPDA $\mathcal{P} = (Q, A, M, \rho, \mu, i, \zeta, T)$, let us temporarily write $T(\mathcal{P})$ for the set of words in A^* recognized by $\mathcal{P}$, i.e., the set of words w for which there is a successful computation

$$(i, \zeta, w) \overset{*}{\to} (t, \eta, 1)$$

for some t in T and some η in M^*. If we have an ESPDA $\mathcal{P} = (Q, A, M, \rho, \mu, i, \zeta)$ without terminal states, then, again temporarily, let $E(\mathcal{P})$ be the set of words w for which there is a successful computation

$$(i, \zeta, w) \overset{*}{\to} (q, 1, 1)$$

for some q in Q. We can now express the equivalence of the two kinds of PDA by means of the following theorem:

Theorem 5.3.2. *Let A be a finite alphabet and let L be a subset of A^*.*

(i) *If $L = T(\mathcal{P}_1)$ for some TSPDA $\mathcal{P}_1$, then there exists an ESPDA $\mathcal{P}_2$ such that $L = E(\mathcal{P}_2)$.*

(ii) *If $L = E(\mathcal{P}_2)$ for some ESPDA $\mathcal{P}_2$, then there exists a TSPDA $\mathcal{P}_1$ such that $L = T(\mathcal{P}_1)$.*

Proof. Suppose first that $L = T(\mathcal{P}_1)$ for some TSPDA $\mathcal{P}_1$, where

$$\mathcal{P}_1 = (Q_1, A, M_1, \rho_1, \mu_1, i_1, \zeta_1, T).$$

Let

$$\mathcal{P}_2 = (Q_2, A, M_2, \rho_2, \mu_2, i_2, \zeta_2),$$

where

$$Q_2 = Q_1 \cup \{i_2, q_2\}, \text{ with } i_2, q_2 \notin Q_1,$$
$$M_2 = M_1 \cup \{\zeta_2\}, \text{ with } \zeta_2 \notin M_1.$$

The map $\rho_2 : Q_2 \times M_2 \times (A \cup \{1\}) \to \mathcal{F}(Q_2 \times M_2^*)$ is given by

$$\rho_2(t, \alpha, 1) = \{(q_2, 1)\} \tag{5.3.3}$$

for all $t \in T$, $\alpha \in M_2$,

$$\rho_2(q_2, \alpha, 1) = \{(q_2, 1)\} \tag{5.3.4}$$

for all $\alpha \in M_2$, *and is otherwise the same as* ρ_1. The map $\mu_2 : Q_2 \times M_2 \to \mathcal{F}(Q_2 \times M_2^*)$ is likewise an extension of μ_1, the extra specification being

$$\mu(i_2, \zeta_2) = \{(i_1, \zeta_1\zeta_2)\}. \tag{5.3.5}$$

The idea is that whenever a computation of $\mathcal{P}_1$ reaches a successful conclusion the rules (5.3.3) and (5.3.4) come into play and empty the stack. The rule (5.3.5) places a new symbol at the bottom of the memory stack so that if the stack 'accidentally' empties in an (unsuccessful) computation in $\mathcal{P}_1$, arriving (say) at an ID $(q, 1, 1)$ with $q \notin T$, then $\mathcal{P}_2$ is not misled into acceptance.

To be more precise, suppose that $w \in T(\mathcal{P}_1)$, so that in $\mathcal{P}_1$ there is a computation

$$(i_1, \zeta_2, w) \xrightarrow{*} (t, \eta, 1) \tag{5.3.6}$$

for some $t \in T$, $\eta \in M^*$. Then there is a computation

$$(i_2, \zeta_2, w) \to (i_1, \zeta_1\zeta_2, w) \xrightarrow{*} (t, \eta\zeta_2, 1) \tag{5.3.7}$$

in $\mathcal{P}_2$. After the first step this computation is exactly like (5.3.6) except for the permanent presence of ζ_2 at the bottom of the memory stack. The new computation (5.3.7) then continues

$$(t, \eta\zeta_2, 1) \xrightarrow{*} (q_2, 1, 1)$$

using (5.3.3) and (5.3.4), and so $w \in E(\mathcal{P}_2)$.

We have shown that $T(\mathcal{P}_1) \subseteq E(\mathcal{P}_2)$. To show the reverse conclusion, suppose that $w \in E(\mathcal{P}_2)$, so that there is a computation

$$(i_2, \zeta_2, w) \xrightarrow{*} (q, 1, 1) \tag{5.3.8}$$

in $\mathcal{P}_2$ for some q in Q_2. This must begin with a step $(i_2, \zeta_2, w) \to (i_1, \zeta_1 \zeta_2, w)$ (using (5.3.5)). Because of the relentless presence of ζ_2 at the bottom of the memory stack in the computation that follows this initial step, the only way in which the stack can empty is if at some stage we arrive at an ID $(t, \eta \zeta_2, 1)$ in which t is a terminal state of $\mathcal{P}_1$. In such a case we then have that

$$(t, \eta \zeta_2, 1) \xrightarrow{*} (q_2, 1, 1)$$

in $\mathcal{P}_2$. During the phase

$$(i_1, \zeta_1 \zeta_2, w) \to (t, \zeta_2, 1)$$

there can be no recourse to the 'extra' specifications (5.3.3), (5.3.4) and (5.3.5) for ρ_2 and μ_2, and so there is a computation

$$(i_1, \zeta_1, w) \xrightarrow{*} (t, \eta, 1)$$

in $\mathcal{P}_1$. Thus $w \in T(\mathcal{P}_1)$.

To prove part (ii) of the theorem, let

$$\mathcal{P}_2 = (Q_2, A, M_2, \rho_2, \mu_2, i_2, \zeta_2)$$

be an ESPDA. Then let

$$\mathcal{P}_1 = (Q_1, A, M_1, \rho_1, \mu_1, \zeta_1, T),$$

where

$$Q_1 = Q_2 \cup \{i, t\}, \text{ with } i_1, t \notin Q_2,$$
$$M_1 = M_2 \cup \{\zeta_1\}, \text{ with } \zeta_1 \notin M_2,$$
$$T = \{t\}.$$

The map $\mu_1 : Q_1 \times M_1 \to \mathcal{F}(Q_1 \times M^*)$ is given by

$$\mu_1(i_1, \zeta_1) = \{(i_2, \zeta_2 \zeta_1)\} \tag{5.3.9}$$

and is otherwise the same as μ_2. The map ρ_1 is likewise an extension of ρ_2, the extra specifications being

$$\rho_1(q, \zeta_1, 1) = \{(t, \zeta_1)\} \tag{5.3.10}$$

for each $q \in Q_1$.

Suppose now that $w \in E(\mathcal{P}_2)$, so that there is a computation

$$(i_2, \zeta_2, w) \xrightarrow{*} (q, 1, 1)$$

in $\mathcal{P}_2$ for some q in Q. Then certainly there is a computation

$$(i_2, \zeta_2 \zeta_1, w) \xrightarrow{*} (q, \zeta_1, 1)$$

in $\mathcal{P}_1$, and so in $\mathcal{P}_1$ we have

$$(i_1, \zeta_1, w) \to (i_2, \zeta_2 \zeta_1, w) \xrightarrow{*} (q, \zeta_1, 1) \to (t, \zeta_1, 1).$$

Thus $w \in T(\mathcal{P}_1)$.

We have shown that $E(\mathcal{P}_2) \subseteq T(\mathcal{P}_1)$. To show the reverse inclusion, suppose that $w \in T(\mathcal{P}_1)$, so that there is a computation

$$(i_1, \zeta_1, w) \xrightarrow{*} (t, \tau, 1)$$

in $\mathcal{P}_1$. This must begin with

$$(i_1, \zeta_1, w) \to (i_2, \zeta_2 \zeta_1, w).$$

Moreover, the TSPDA cannot enter the state t except by means of (5.3.10), and this can come into play if and only if for some q in Q there is a computation

$$(i_2, \zeta_2 \zeta_1, w) \xrightarrow{*} (q, \zeta_1, 1)$$

in $\mathcal{P}_1$. Since no use can be made of (5.3.9) or (5.3.10) during this computation it is in effect a computation in $\mathcal{P}_2$. Precisely, we have

$$(i_2, \zeta_2, w) \to (q, 1, 1)$$

in $\mathcal{P}_2$. Thus $w \in E(\mathcal{P}_2)$ as required. ■

We distinguish between the two approaches by using the phrases 'recognition (or acceptance) by terminal state' and 'recognition (or acceptance) by empty memory stack'.

5.4   Pushdown automata and context-free languages

We have now encountered several languages (namely $\{a^n b^n : n \geq 1\}$, $\{ww^R : w \in \{a, b\}^+\}$, $\{w \in \{a, b\}^+ : |w|$ is odd and $w^R = w\}$) which are recognized by PDAs, are not regular, but are context-free. The main aim of this section will be to show that context-free languages bear the same relation to PDAs as regular languages do to FSAs. (See Kleene's Theorem (2.2.6).) To be precise, we have a theorem due to Schutzenberger and Chomsky:

Theorem 5.4.1. *Let A be a finite alphabet and let L be a non-empty subset of A^*. Then L is context-free if and only if $L = L(\mathcal{P})$ for some pushdown automaton $\mathcal{P}$.*

Proof. Suppose first that L is context-free, and in the first instance let us suppose also that $1 \notin L$. By Theorem 4.4.18 we may suppose that it is generated by a context-free grammar $\Gamma = (V, A, \pi, \sigma)$ in Chomsky Normal Form. We now construct a PDA

$$\mathcal{P} = \big(\{i, q, t\}, A, (V \backslash A) \cup \{\varsigma\}, \mu, \rho, i, \varsigma, \{t\}\big). \qquad (5.4.2)$$

First recall that the productions of Γ are either of type $\lambda \to \alpha\beta$ (with $\lambda, \alpha, \beta \in V \backslash A$) or of type $\lambda \to a$ (with $\lambda \in V \backslash A$, $a \in A$). Let

$$\mu(i, \varsigma) = \{(q, \sigma\varsigma)\} \qquad (5.4.3)$$

and for each production in Γ of type $\lambda \to \alpha\beta$ let

$$\mu(q, \lambda) \text{ contain } (q, \alpha\beta). \qquad (5.4.4)$$

Let

$$\rho(q, \varsigma, 1) = \{(t, \varsigma)\} \qquad (5.4.5)$$

and for each production in Γ of type $\lambda \to a$ let

$$\rho(q, \lambda, a) \text{ contain } (q, 1). \qquad (5.4.6)$$

Let us first establish

Lemma 5.4.7. *With the above definitions, $1 \notin L(\mathcal{P})$.*

Proof. Suppose, by way of contradiction, that there is a computation

$$(i, \varsigma, 1) \overset{*}{\to} (t, \tau, 1) \qquad (5.4.8)$$

in $\mathcal{P}$, where $\tau \in \big((V \backslash A) \cup \{\varsigma\}\big)^*$. This must begin with

$$(i, \varsigma, 1) \to (q, \sigma\varsigma, 1).$$

Thereafter, since the third coordinate of the ID must remain fixed at 1, there can be no steps based on (5.4.6). Steps based on (5.4.4) produce IDs of type $(q, \omega\varsigma, 1)$, where ω is a non-empty word in the alphabet $V \backslash A$, *not* involving ς. The instruction (5.4.5) can never come into play at all, since we never again reach an ID whose middle coordinate begins with ς. So $\mathcal{P}$ can never enter its terminal state t and a computation of type (5.4.8) is thus not possible. ∎

Before we prove anything further about the PDA $\mathcal{P}$, it may be helpful to mention that the basic strategy of our approach is that computations in $\mathcal{P}$ should simulate leftmost derivations in Γ. In the hope that it will not distract too much from the main business in hand, let us take time out to look at an example.

In Section 4.4 (see in particular Example 4.4.22) we established that the grammar

$$\Gamma = \left(\{\sigma, \lambda, \gamma, \delta, \beta_a, \beta_b, a, b\}, \{a, b\}, \pi, \sigma\right)$$

with productions

$$\sigma \to \beta_a\gamma, \qquad \gamma \to \lambda\beta_b, \qquad \sigma \to \beta_a\beta_b, \qquad \lambda \to \beta_a\delta,$$

$$\delta \to \lambda\beta_b, \qquad \lambda \to \beta_a\beta_b, \qquad \beta_a \to a, \qquad \beta_b \to b,$$

(in Chomsky Normal Form) generates the language $\{a^n b^n : n \geq 1\}$. A leftmost derivation of $a^3 b^3$ is

$$\sigma \Rightarrow \beta_a\gamma \Rightarrow a\gamma \Rightarrow a\lambda\beta_b \Rightarrow a\beta_a\delta\beta_b$$
$$\Rightarrow a^2\delta\beta_b \Rightarrow a^2\lambda\beta_b^2 \Rightarrow a^2\beta_a\beta_b^3 \Rightarrow a^3\beta_b^3 \qquad (5.4.9)$$
$$\Rightarrow a^3 b\beta_b^2 \Rightarrow a^3 b^2\beta_b \Rightarrow a^3 b^3.$$

In accordance with the instructions in the last paragraph we set up a PDA
$$\mathcal{P} = \left(\{i, q, t\}, \{a, b\}, \{\sigma, \lambda, \gamma, \delta, \beta_a, \beta_b, \zeta\}, \mu, \rho, i, \zeta, \{t\}\right)$$

where

$$\mu(i, \zeta) = \{(q, \sigma\zeta)\},$$
$$\mu(q, \sigma) = \{(q, \beta_a\gamma), (q, \beta_a\beta_b)\},$$
$$\mu(q, \lambda) = \{(q, \beta_a\delta), (q, \beta_a\beta_b)\},$$
$$\mu(q, \gamma) = \mu(q, \delta) = \{(q, \lambda\beta_b)\},$$

and

$$\rho(q, \beta_a, a) = \rho(q, \beta_b, b) = \{(q, 1)\},$$
$$\rho(q, \zeta, 1) = \{(t, \zeta)\}.$$

The derivation (5.4.9) in Γ is then simulated by a successful computation in $\mathcal{P}$ in the following way:

$$(i, \zeta, a^3 b^3) \to (q, \sigma\zeta, a^3 b^3) \to (q, \beta_a\gamma\zeta, a^3 b^3)$$
$$\to (q, \gamma\zeta, a^2 b^3) \to (q, \lambda\beta_b\zeta, a^2 b^3) \to (q, \beta_a\delta\beta_b\zeta, a^2 b^3)$$
$$\to (q, \delta\beta_b\zeta, ab^3) \to (q, \lambda\beta_b^2\zeta, ab^3) \to (q, \beta_a\beta_b^3\zeta, ab^3)$$
$$\to (q, \beta_b^3\zeta, b^3) \to (q, \beta_b^2\zeta, b^2) \to (q, \beta_b\zeta, b)$$
$$\to (q, \zeta, 1) \to (t, \zeta, 1).$$

Taking this example as encouragement that we are on the right track, let us now return to the task of showing that the language recognized by $\mathcal{P}$, as defined in (5.4.2) to (5.4.6), is precisely the language generated by the grammar $\Gamma = (V, A, \pi, \sigma)$. Suppose first that $w = b_1 b_2 \ldots b_m \in L \, (= L(\Gamma))$. For the moment we are assuming that $1 \notin L$, and so we have $m \geq 1$.

First, if the derivation of w in the grammar Γ is of length 1 then $w = b_1$ and we have

$$\sigma \Rightarrow b_1.$$

In this case we have a computation

$$(i, \zeta, b_1) \rightarrow (q, \sigma\zeta, b_1) \rightarrow (q, \zeta, 1) \rightarrow (t, \zeta, 1)$$

and so $b_1 \in L(\mathcal{P})$.

Otherwise the derivation begins

$$\sigma \Rightarrow \alpha\beta \quad (\alpha, \beta \in V \backslash A)$$

and we have a computation in $\mathcal{P}$ beginning

$$(i, \zeta, w) \rightarrow (q, \sigma\zeta, w) \rightarrow (q, \alpha\beta\zeta, w).$$

At a typical stage in a leftmost derivation of w in Γ we will have reached a word

$$b_1 b_2 \ldots b_k \delta_1 \delta_2 \ldots \delta_l,$$

where $b_1, b_2, \ldots, b_n \in A$, $\delta_1, \delta_2, \ldots, \delta_l \in V \backslash A$. Suppose inductively that there is a computation

$$(i, \zeta, b_1 b_2 \ldots b_m) \xrightarrow{*} (q, \delta_1 \delta_2 \ldots \delta_l \zeta, b_{k+1} \ldots b_m) \tag{5.4.10}$$

in $\mathcal{P}$. The next step in the leftmost derivation will involve either changing δ_1 to $\delta_1' \delta_1''$ or changing δ_1 to b_{k+1}. In the former case the computation (5.4.10) continues

$$(q, \delta_1 \delta_2 \ldots \delta_l \zeta, b_{k+1} \ldots b_m) \rightarrow (q, \delta_1' \delta_1'' \delta_2 \ldots \delta_l \zeta, b_{k+1} \ldots b_m),$$

while in the latter case it continues

$$(q, \delta_1 \delta_2 \ldots \delta_l, b_{k+1} \ldots b_m) \rightarrow (q, \delta_2 \ldots \delta_l, b_{k+2} \ldots b_m).$$

It is now clear that for all w in A^* the existence of a leftmost derivation $\sigma \xrightarrow{*} w$ in Γ implies the existence of a successful computation

$$(i, \zeta, w) \xrightarrow{*} (t, \zeta, 1)$$

in $\mathcal{P}$. We thus deduce that $L(\Gamma) \subseteq L(\mathcal{P})$.

Now let us suppose conversely that $w = b_1 b_2 \ldots b_m \in L(\mathcal{P})$, so that we have a computation

$$(i, \zeta, b_1 b_2 \ldots b_m) \overset{*}{\to} (t, \tau, 1) \tag{5.4.11}$$

for some τ in $((V \backslash A) \cup \{\zeta\})^*$. We may suppose that $m \geq 1$, since by Lemma 5.4.7 we know that $1 \notin L(\mathcal{P})$. The computation (5.4.11) must begin

$$(i, \zeta, b_1 b_2 \ldots b_m) \to (q, \sigma \zeta, b_1 b_2 \ldots b_m)$$

and the next step must be either

$$(q, \sigma \zeta, b_1 b_2 \ldots b_m) \to (q, \zeta, b_2 \ldots b_m)$$

(if $\sigma \to b_1$ is a production in Γ) or

$$(q, \sigma \zeta, b_1 b_2 \ldots b_m) \to (q, \alpha \beta \zeta, b_1 b_2 \ldots b_m)$$

(if $\sigma \to \alpha \beta$ is a production in Γ). In the former case the computation fails unless $w = b_1$; and in that case we have a derivation $\sigma \Rightarrow b_1$ in Γ. In the latter case we have a derivation $\sigma \Rightarrow \alpha \beta$.

Now suppose inductively that when after q (≥ 2) steps the computation (5.4.11) reaches the ID

$$(q, \delta_1 \delta_2 \ldots \delta_l \zeta, b_{k+1} \ldots b_m) \quad (l \geq 0)$$

then there is a derivation

$$\sigma \overset{*}{\Rightarrow} b_1 b_2 \ldots b_k \delta_1 \delta_2 \ldots \delta_l \tag{5.4.12}$$

in Γ. In effect, we anchored this induction in the previous paragraph. The next step in the computation will be based either on the fact that

$$(q, \delta_1' \delta_1'') \in \mu(q, \delta_1)$$

or on the fact that

$$(q, 1) \in \rho(q, \delta_1, b_{k+1}),$$

giving us (in the computation in $\mathcal{P}$) either

$$(q, \delta_1' \delta_1'' \delta_2 \ldots \delta_l, b_{k+1} \ldots b_m) \tag{5.4.13}$$

or

$$(q, \delta_2 \ldots \delta_l, b_{k+2} \ldots b_m). \tag{5.4.14}$$

In the former case $\delta \rightarrow \delta_1' \delta_2''$ is a production in Γ and the derivation (5.4.12) extends by one step to give

$$\sigma \overset{*}{\Rightarrow} b_1 b_2 \dots b_k \delta_1' \delta_1'' \delta_2 \dots \delta_l. \tag{5.4.15}$$

In the latter case $\delta_1 \rightarrow b_{k+1}$ is a production in Γ and the derivation (5.4.12) extends to give

$$\sigma \overset{*}{\Rightarrow} b_1 b_2 \dots b_{k+1} \delta_2 \dots \delta_l. \tag{5.4.16}$$

Now comparing (5.4.13) with (5.4.15) and (5.4.14) with (5.4.16) shows us that the inductive step is complete.

The computation (5.4.11) can enter the terminal state t only by invoking the instruction (5.4.5), and this instruction can be used only once. So the penultimate ID in (5.4.11) must be $(q, \zeta, 1)$. Applying the conclusion (5.4.12) to the case where $k = m$, we conclude that there is a derivation

$$\sigma \overset{*}{\Rightarrow} b_1 b_2 \dots b_m$$

in Γ. Thus $w = b_1 b_2 \dots b_m \in L(\Gamma)$.

We have now shown that every context-free language L not containing 1 is recognizable by a PDA. Before tackling the converse half of Theorem 5.4.1 we pause to consider the case where $1 \in L$. In this case our routine for putting the grammar into Chomsky Normal Form produces a grammar recognizing $L \backslash \{1\}$. By the argument just completed we produce a PDA $\mathcal{P}$ such that $L(\mathcal{P}) = L \backslash \{1\}$. Then if we simply add the instruction

$$\rho(i, \zeta, 1) = \{(t, \zeta)\}$$

to the instructions (5.4.2) to (5.4.6) for $\mathcal{P}$ we obtain a PDA $\mathcal{P}'$ with the property that $L(\mathcal{P}') = L$.

Suppose now that $L \ (\subseteq A^*)$ is recognized by a PDA. By Theorem 5.3.2 we may assume that the recognition is by empty memory stack. Thus we have a PDA

$$\mathcal{P} = (Q, A, M, \mu, \rho, i, \zeta) \tag{5.4.17}$$

(without terminal states) with the property that $w \in L$ if and only if there is a computation

$$(i, \zeta, w) \overset{*}{\rightarrow} (q, 1, 1)$$

for some q in Q. We must now define a context-free grammar $\Gamma = (V, A, \pi, \sigma)$ with the property that $L = L(\Gamma)$.

In fact we have to do something awkwardly abstract and define the set $V \backslash A$ to contain the Cartesian product $Q \times M \times Q$. To avoid confusion

with instantaneous descriptions (which are elements of $Q \times M^* \times A^*$)
we shall encase the triples that make up $Q \times M \times Q$ in square brackets:
$[p, \alpha, q]$. The only other element we require in $V \backslash A$ is a sentence symbol
σ. Thus, to summarize, we define

$$V \backslash A = (Q \times M \times Q) \cup \{\sigma\}. \tag{5.4.18}$$

The set π of productions of Γ is specified as follows. First, let π
contain a production

$$\sigma \rightarrow [i, \zeta, q] \tag{5.4.19}$$

for each q in Q. Secondly, whenever $\mu(q, \alpha)$ contains $(q', \beta_1 \beta_2 \ldots \beta_k)$, let
π contain a production

$$[q, \alpha, q_k] \rightarrow [q', \beta_1, q_1][q_1, \beta_2, q_2] \ldots [q_{k-1}, \beta_k, q_k] \tag{5.4.20}$$

for each choice of $q_1, q_2, \ldots, q_k$ in Q, provided $k \geq 1$. If $k = 0$, i.e., if
$\beta_1 \beta_2 \ldots \beta_k = 1$, let π contain the production

$$[q, \alpha, q'] \rightarrow 1. \tag{5.4.21}$$

Thirdly, whenever $\rho(q, \alpha, a)$ contains $(q', \beta_1 \beta_2 \ldots \beta_k)$, let π contain a
production

$$[q, \alpha, q_k] \rightarrow a[q', \beta_1, q_1][q_1, \beta_2, q_2] \ldots [q_{k-1}, \beta_k, q_k] \tag{5.4.22}$$

for each choice of $q_1, q_2, \ldots, q_k$ in Q, if $k \neq 0$; if $k = 0$, then let π contain

$$[q, \alpha, q'] \rightarrow a. \tag{5.4.23}$$

(We note in passing that the grammar we have defined is not in
Chomsky Normal Form, but it is context-free, and this is all that
matters.)

We want to show that (for all w in A^*) $w \in E(\mathcal{P})$ if and only if
$w \in L(\Gamma)$. (Here $E(\mathcal{P})$ is defined as in the paragraph preceding Theorem
5.3.2.) That is, we want to show that the existence of a computation
$(i, \zeta, w) \overset{*}{\rightarrow} (p, 1, 1)$ in $\mathcal{P}$ is equivalent to the existence of a derivation $\sigma \overset{*}{\Rightarrow} w$
in Γ. We shall use an inductive proof, and to make the induction work
it is convenient to consider a slightly more general situation. In fact we
prove:

Lemma 5.4.24. *Let $p, q \in Q$, $\alpha \in M$, $w \in A^*$. The following statements
are equivalent:*
   (i) *there is a computation*

$$(q, \alpha, w) \overset{*}{\rightarrow} (p, 1, 1)$$

in $\mathcal{P}$;

   (ii) *there is a derivation*

$$[q, \alpha, p] \overset{*}{\Rightarrow} w$$

in Γ.

Before proving this, we notice that it does give us the result we require. If there is a computation $(i, \zeta, w) \overset{*}{\rightarrow} (p, 1, 1)$ in $\mathcal{P}$ then there is a derivation

$$\sigma \Rightarrow [i, \zeta, p] \overset{*}{\Rightarrow} w$$

in Γ. Conversely, if there is a derivation $\sigma \overset{*}{\Rightarrow} w$ this must begin with a step $\sigma \Rightarrow [i, \zeta, p]$ for some p in Q and be followed by a derivation $[i, \zeta, p] \overset{*}{\Rightarrow} w$. By the lemma there must then exist a computation

$$(i, \zeta, w) \overset{*}{\rightarrow} (p, 1, 1)$$

in $\mathcal{P}$.

Proof of Lemma 5.4.24. We show first that (i) $\Rightarrow$ (ii). So suppose that there is a computation

$$(q, \alpha, w) \overset{*}{\rightarrow} (p, 1, 1) \tag{5.4.25}$$

in $\mathcal{P}$. If this computation involves just one step then either $w = 1$ and $\mu(q, \alpha)$ contains $(p, 1)$ or $w = b \in A$ and $\rho(q, \alpha, b)$ contains $(p, 1)$. In the former case there is a production $[q, \alpha, p] \rightarrow 1$ in Γ and in the latter case there is a production $[q, \alpha, p] \rightarrow b$. So certainly we have the required derivation $[q, \alpha, p] \overset{*}{\Rightarrow} w$.

Suppose now that the computation (5.4.25) involves n steps, where $n \geq 2$. The first step may be stationary (based on μ) or progressive (based on ρ); we cover both possibilities if we write it as

$$(q, \alpha, w) \rightarrow (q_1, \beta_1\beta_2 \ldots \beta_m, z),$$

where $w = az$ and $a \in A \cup \{1\}$. We may suppose that $m \geq 1$, for $m = 0$ would give either a successful computation with one step (if $z = 1$) or a failed computation (if $z \neq 1$). We now divide the remaining computation

$$(q_1, \beta_1\beta_2 \ldots \beta_m, z) \overset{*}{\rightarrow} (p, 1, 1)$$

into phases, according to the number of symbols in the memory stack. The first phase ends *when for the first time the memory stack contains fewer than m symbols*. During this phase the number of symbols in the memory stack may well have increased, but $\beta_2, \ldots \beta_m$ have had no influence on the computation. We arrive at an ID $(q_2, \beta_2 \ldots \beta_m, z_1)$, where $z = y_1 z_1$, with $y_1, z_1 \in A^*$. Since $\beta_2, \ldots \beta_m$ have had no influence on the computation there is in fact a computation

$$(q_1, \beta_1, y_1) \overset{*}{\to} (q_2, 1, 1)$$

in $\mathcal{P}$.

If $z_1 = 1$ and $m = 1$ then the successful computation

$$(q, \alpha, w) \to (q_1, \beta_1, z) \overset{*}{\to} (q_2, 1, 1)$$

is complete. If one of these happens without the other then the computation has failed, contrary to our assumption. So suppose that $z_1 \neq 1$ and $m \geq 2$. Then the next phase of the computation ends when for the first time the memory stack has fewer than $m - 1$ synbols. (Once again the memory stack may well grow during this phase.) We arrive at an ID

$$(q_3, \beta_3 \ldots \beta_m, z_2),$$

where $z_1 = y_2 z_2$ (with $y_2, z_2 \in A^*$) and since once again $\beta_3, \ldots \beta_m$ have had no influence at all on the computation, there exists a computation

$$(q_2, \beta_2, y_2) \overset{*}{\to} (q_3, 1, 1).$$

Continuing in this way, we obtain a phased computation

$$(q_1, \beta_1 \beta_2 \ldots \beta_m, z) \overset{*}{\to} (q_2, \beta_2 \ldots \beta_m, z_1)$$
$$\overset{*}{\to} (q_3, \beta_3 \ldots \beta_m, z_2) \overset{*}{\to} \cdots \overset{*}{\to} (q_m, \beta_m, z_m)$$
$$\overset{*}{\to} (q_{m+1}, 1, 1),$$

where

$$q_{m+1} = p, \ z = y_1 y_2 \ldots y_m, \ z_k = y_{k+1} \ldots y_m,$$

and where, for $i = 1, 2, \ldots, m$, there is a computation

$$(q_i, \beta_i, y_i) \overset{*}{\to} (q_{i+1}, 1, 1) \tag{5.4.26}$$

in $\mathcal{P}$.

Now each of the computations (5.4.26) involves fewer than n steps and so we may assume inductively that for $i = 1, 2, \ldots, m$ there is a derivation

$$[q_i, \beta_i, q_{i+1}] \overset{*}{\Rightarrow} y_i$$

in Γ. From (5.4.20) or (5.4.22) we also have a production

$$[q, \alpha, p] \rightarrow a[q_1, \beta_1, q_2][q_2, \beta_2, q_3] \ldots [q_m, \beta_m, p]$$

(with $a \in A \cup \{1\}$) corresponding to the first step of the computation (5.4.25). Hence there is a derivation

$$[q, \alpha, p] \overset{*}{\Rightarrow} a y_1 y_2 \ldots y_m \ (= w)$$

in Γ.

To prove (ii)$\Rightarrow$ (i) suppose that there is a derivation

$$[q, \alpha, p] \overset{*}{\Rightarrow} w \tag{5.4.27}$$

in Γ. We require to show that there is a computation

$$(q, \alpha, w) \overset{*}{\rightarrow} (p, 1, 1)$$

in $\mathcal{P}$. We shall use induction again, this time on the number of steps in the derivation (5.4.27). If there is just one step then we must have $w = a \in A$ or $w = 1$ and either $\rho(q, \alpha, a)$ contains $(p, 1)$ (in the former case) or $\mu(q, \alpha)$ contains $(p, 1)$ (in the latter case). In either case there is a single-step computation

$$(q, \alpha, w) \rightarrow (p, 1, 1)$$

in $\mathcal{P}$.

Suppose now that the derivation (5.4.27) has n steps, where $n \geq 2$. The first step must be

$$[q, \alpha, p] \Rightarrow a[q_1, \beta_1, q_2][q_2, \beta_2, q_3] \ldots [q_m, \beta_m, q_{m+1}],$$

where $q_{m+1} = p$ and $a \in A \cup \{1\}$. If $a \in A$ then $\rho(q, \alpha, a)$ contains $(q_1, \beta_1 \beta_2 \ldots \beta_m)$, while if $a = 1$ then $\mu(q, \alpha)$ contains $(q_1, \beta_1 \beta_2 \ldots \beta_m)$. The nature of the productions of Γ (see (5.4.19) – (5.4.23)) is such that the letter a, once printed at the left of the word, can never then be removed. So $w = az$ for some z in A^*. We may assume that $m \geq 1$, since $m = 0$ gives $[q, \alpha, p] \rightarrow a$, and this is just the case $n = 1$ already considered. We have a derivation

$$[q_1, \beta_1, q_2][q_2, \beta_2, q_3] \ldots [q_m, \beta_m, q_{m+1}] \overset{*}{\Rightarrow} z,$$

and since the grammar Γ is context-free it follows from Lemma 4.3.4 that z can be factorized as $y_1 y_2 \ldots y_m$ in such a way that there is a derivation

$$[q_i, \beta_i, q_{i+1}] \overset{*}{\Rightarrow} y_i$$

for $i = 1, 2, \ldots, m$. Since each of these derivations involves fewer than n steps we may assume inductively that for each i there is a computation

$$(q_i, \beta_i, y_i) \overset{*}{\to} (q_{i+1}, 1, 1)$$

in $\mathcal{P}$. Hence for each i there is a computation

$$(q_i, \beta_i \beta_{i+1} \ldots \beta_m, y_i y_{i+1} \ldots y_m) \overset{*}{\to} (q_{i+1}, \beta_{i+1} \ldots \beta_m, y_{i+1} \ldots y_m)$$

and so finally there is a computation

$$(q, \alpha, w) \to (q_1, \beta_1 \beta_2 \ldots \beta_m, z) \overset{*}{\to} (q_2, \beta_2 \ldots \beta_m, y_2 \ldots y_m)$$
$$\overset{*}{\to} (q_3, \beta_3 \ldots \beta_m, y_3 \ldots y_m) \overset{*}{\to} \cdots \overset{*}{\to} (p, 1, 1)$$

in $\mathcal{P}$. This completes the proof of Lemma 5.4.24, and so at last the connection between context-free languages and pushdown automata announced in Theorem 5.4.1 is established. ∎

5.5 The Pumping Lemma; non-context-free languages

The so-called Pumping Lemma (Theorem 2.2.7) of Chapter 2 was very useful in establishing the non-regularity of certain languages. The analogous lemma for context-free languages, while inevitably more complicated, can be used in a similar way.

Theorem 5.5.1 (The Pumping Lemma). *Let L be a context-free language. Then there is an integer N, depending only on L, with the property that if $z \in L$ and $|z| > N$ then z can be factorized as $z = uvwxy$, with $|vx| \geq 1$ and $|vwx| \leq N$, in such a way that L contains $uv^m xy^m z$ for all $m \geq 0$.*

Proof. By virtue of Theorem 4.4.18, we may assume that $L \backslash \{1\}$ is generated by a context-free grammar $\Gamma = (V, A, \pi, \sigma)$ in Chomsky Normal Form. Thus the productions in π are either of type

$$\lambda \to \alpha\beta \quad (\lambda, \alpha, \beta \in V \backslash A) \tag{5.5.2}$$

or of type

$$\lambda \to a \quad (\lambda \in V \backslash A, a \in A). \tag{5.5.3}$$

Let $N = 2^{p+1}$, where p is the number of productions of the former type.
In a typical derivation

$$\sigma \overset{*}{\Rightarrow} a_1 a_2 \ldots a_k \tag{5.5.4}$$

(where $a_1, a_2, \ldots, a_k \in A$) each a_i traces its ancestry back to σ along a unique 'ancestor path'. For example, if the grammar is that described in Example 4.4.22 then the derivation of $a^3 b^3$ described by (4.4.23) can be pictured as in Figure 5.5.5.

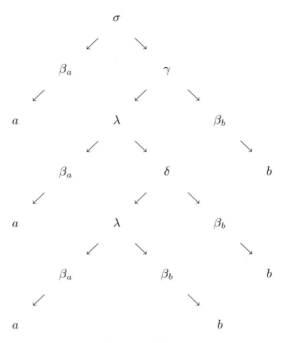

Figure 5.5.5

The ancestor path of the rightmost a is

$$\sigma \to \gamma \to \lambda \to \delta \to \lambda \to \beta_a \to a, \qquad (5.5.6)$$

of length 6, while the ancestor path of the leftmost a is

$$\sigma \to \beta_a \to a,$$

of length 2.

In general, the *depth* of a derivation (5.5.4) is the length of the longest path encountered. So, for example, the derivation (4.4.23) of $a^3 b^3$ has depth 6.

Changing our point of view, we can regard each a_i in (5.5.4) as a *descendant* of σ. The nature of the productions of Γ as given in (5.5.2) and (5.5.3) ensures that in a derivation of depth k the number of descendants of σ cannot exceed 2^k. So suppose now that in (5.5.4) the length k

of the word we obtain exceeds 2^p. Then at least one a_i has an ancestor path of length greater than p. Inevitably then , since p is defined as the number of productions of type (5.5.2) in Γ, some non-terminal symbol λ occurs more than once in the ancestor path of a_i. (The path (5.5.6), in which λ appears twice, provides an example of this phenomenon.)

Working back from a_i towards σ, we may in fact encounter λ several times, but we shall be concerned only with the first and the second encounters:

$$\sigma \to \cdots \to \lambda \to \cdots \to \lambda \to \cdots \to a_i.$$

Thus, while there may be several occurrences of λ in the phase $\sigma \to \cdots \to \lambda$, there are no 'unexpected' occurrences of λ in the phase

$$\lambda \to \cdots \to \lambda \to \cdots \to a_i, \tag{5.5.7}$$

and we may also assume that no other non-terminal symbol repeats during this phase. The length of the path (5.5.7) cannot therefore exceed $p+1$.

The derivation (5.5.4) may be represented as

$$\sigma \overset{*}{\Rightarrow} \zeta_1 \lambda \tau_1 \overset{*}{\Rightarrow} \zeta_2 \xi \lambda \eta \tau_2 \overset{*}{\Rightarrow} uvwxy,$$

where $\zeta_1 \overset{*}{\Rightarrow} \zeta_2$, $\tau_1 \overset{*}{\Rightarrow} \tau_2$, $\lambda \overset{*}{\Rightarrow} \xi \lambda \eta$, $\zeta_2 \overset{*}{\Rightarrow} u$, $\xi \overset{*}{\Rightarrow} v$, $\lambda \overset{*}{\Rightarrow} w$, $\eta \overset{*}{\Rightarrow} x$, $\tau_2 \overset{*}{\Rightarrow} y$ are all derivations in Γ. (See Lemma 4.3.4.) Here ζ_1 and τ_1 may be 1, and either ξ or η (but not both) may be 1. So we have $u, y \in A^*$, $|vx| \geq 1$. From the last paragraph we are moreover able to assume that the derivation

$$\lambda \overset{*}{\Rightarrow} \xi \lambda \eta \overset{*}{\Rightarrow} vwx$$

is of length not more than $p+1$. Hence $|vwx| \leq 2^{p+1} = N$.

Because of the 'self-embedding' property of λ we can introduce a loop into the derivation (5.5.4):

$$\sigma \overset{*}{\Rightarrow} \zeta_2 \xi \lambda \eta \tau_2 \overset{*}{\Rightarrow} \zeta_2 \xi^2 \lambda \eta^2 \tau_2 \overset{*}{\Rightarrow} uv^2 wx^2 y.$$

Indeed we now see that $uv^m wx^m y \in L(\Gamma)$ for all $m \geq 0$. ∎

As with the Pumping Lemma for FSAs, we can use this new lemma to demonstrate the existence of languages that are not context-free.

Example 5.5.8. Let

$$L = \{a^n b^n a^n : n \geq 1\}.$$

Suppose, by way of contradiction, that this language is context-free. Then for each sufficiently large n we can find u, v, w, x, y in $\{a, b\}^*$ such that $a^n b^n a^n = uvwxy$ and L contains $uv^m wx^m y$ for all $m \geq 0$.

Now the words in L all have the property that they have precisely two *boundaries*

$$aa \ldots a | bb \ldots b | aa \ldots a$$

marking the change from a to b and the change back from b to a. In general, if $z \in \{a, b\}^+$ has proper left factors $z_1, z_2, \ldots$, with $|z_i| = i$, and if $z = z_i t_i$ for $i = 1, 2, \ldots, |z| - 1$, then $B(z)$, the number of boundaries in z, is defined as the number of times when the last letter in z_i differs from the first letter in t_i. In the partitioning of $a^n b^n a^n$ into $uvwxy$ it is easy to see that if v or x contains a boundary then $B(uv^2wx^2y) > 2$ and so $uv^2wx^2y \notin L$. Hence $B(v) = B(x) = 0$, which is to say that each of v, x is a power of a single letter from $\{a, b\}$.

We can think of a word from L as being in three sections of equal length, Section 1 being the initial block of a's, Section 2 being the central block of b's and Section 3 being the final block of a's. Each of v, x lies in a single section (and they may conceivably be in the same section). If v and x lie in different sections then as m increases in the word $uv^m wx^m y$ the size of at least one of these two sections increases while the third remains unchanged, and so the word no longer belongs to w. If v and x lie in the same section then the situation is even worse, for the size of that section increases with m while the other two sections remain unchanged. There is no way in which all three sections can be made to grow equally. Accordingly

$$L = \{a^n b^n a^n : n \geq 1\} \text{ is not context-free.} \tag{5.5.9}$$

Inevitably, the new Pumping Lemma is harder to apply than the simpler lemma for regular languages. Another example may be helpful.

Example 5.5.10. Let $L = \{a^m b^n a^m b^n : m, n \geq 1\}$. Suppose by way of contradiction that this language is context-free. Then each sufficiently long word $a^m b^n a^m b^n$ in L can be partitioned into $uvwxy$ in such a way that $uv^k wx^k y \in L$ for all $k \geq 0$. The 'boundary' argument used in the last example applies here, and we may conclude that each of v and x is a power of a single letter. If (say) v is a positive power of a and x is a power of b then v must be a segment of one or other of the two blocks of a's in $a^m b^n a^m b^n$, and so of the two blocks of a's in $uv^2 wx^2 y$ one is longer than the other. Thus $uv^2 wx^2 y \notin L$. We conclude that we must either have $v = a^p$ and $x = a^q$, or $v = b^p$ and $x = b^q$; it will be sufficient to consider the former case. It is not hard to see by the argument we have used before that v must be extracted from the first block of a's and x from the second, and moreover we must have $p = q$, since otherwise the two blocks of a's increase at different rates as we increase k in $uv^k wx^k y$. So we have

$$a^m b^n a^m b^n = a^r.a^p.a^s b^n a^t.a^p.a^l b^n; \tag{5.5.11}$$

that is,

$$u = a^r, \ v = a^p, \ w = a^s b^n a^t, \ x = a^p, \ y = a^l b^n,$$

with

$$r, s, t, l \geq 0, \ p > 0, \ r + p + s = t + p + l = m.$$

Then

$$uv^k wx^k y = a^{m+(k-1)p} b^n a^{m+(k-1)p} b^n \in L$$

for all $k \geq 0$.

It would seem on the face of it that we have failed to prove that L is not context-free. But at this point the subtlest of the aspects of the Pumping Lemma (Theorem 5.5.1) comes into play, namely the stipulation that $|vwx| \leq N$. We have convinced ourselves that factorizations of the type (5.5.11) (along with the corresponding factorizations involving b's) are the only ones that ensure the inclusion of every $uv^k wx^k y$ in L. The number N referred to in the statement of the Pumping Lemma depends only on L, and we began our analysis of this example by looking at words $a^m b^n a^m b^n$ in which $2m + 2n > N$. If we make the stronger assumption that $n > N$ then

$$|vwx| = 2p + s + t + n > N,$$

and the factorization fails to satisfy the conditions of the Pumping Lemma.

Since the Pumping Lemma requires that a 'successful' factorization be possible for *all* sufficiently long words, we conclude that

$$L = \{a^m b^n a^m b^n : m, n \geq 1\} \text{ is not context-free.} \tag{5.5.12}$$

5.6 Closure properties of context-free languages

In Section 2.5 we saw that the class of regular languages is closed under the operations $\cup$, . and *. The same closure properties hold for context-free languages.

Theorem 5.6.1. *Let A be a finite alphabet.*

(i) *If L_1, L_2 are context-free languages in A^* then so are $L_1 \cup L_2$ and $L_1.L_2$.*

(ii) *If L is a context-free language in A^* then so is L^*.*

Proof. (i) For $i = 1, 2$ let $L_i = L(\Gamma_i)$, where

$$\Gamma_i = (V_i, A, \pi_i, \sigma_i)$$

is a context-free grammar. We may assume that $V_1 \backslash A$ and $V_2 \backslash A$ are disjoint. First, consider the grammar

$$\Gamma = (V_1 \cup V_2 \cup \{\sigma\}, A, \pi, \sigma),$$

where σ is a new symbol not in $V_1 \cup V_2$ and where π consists of the productions in $\pi_1 \cup \pi_2$ together with the extra productions

$$\sigma \to \sigma_1, \ \sigma \to \sigma_2.$$

Then $L(\Gamma) = L_1 \cup L_2$. To see this, note first that if $w \in L_1$ then there is a derivation $\sigma_1 \overset{*}{\Rightarrow} w$ in Γ_1 and hence a derivation

$$\sigma \Rightarrow \sigma_1 \overset{*}{\Rightarrow} w$$

in Γ. Thus $L_1 \subseteq L(\Gamma)$, and equally $L_2 \subseteq L(\Gamma)$.

Conversely, suppose that $w \in L(\Gamma)$, so that there exists a derivation $\sigma \overset{*}{\Rightarrow} w$. This must begin either with $\sigma \Rightarrow \sigma_1$ or with $\sigma \Rightarrow \sigma_2$. Let us assume, without loss of generality, that it begins with $\sigma \Rightarrow \sigma_1$. Then the fact that $V_1 \backslash A$ and $V_2 \backslash A$ are disjoint means that only productions from π_1 can from that point on be involved. Thus $\sigma_1 \overset{*}{\Rightarrow} w$ in Γ_1 and so $w \in L_1$. We have now shown that $L(\Gamma) = L_1 \cup L_2$, and it is clear that the grammar Γ is context-free.

To deal with $L_1.L_2$, let

$$\Gamma' = (V_1 \cup V_2 \cup \{\sigma\}, A, \pi', \sigma'),$$

where π' consists of the productions in $\pi_1 \cup \pi_2$ together with the extra production

$$\sigma' \to \sigma_1 \sigma_2.$$

If $w = w_1 w_2 \in L_1 L_2$, with $w_1 \in L_1$, $w_2 \in L_2$, then there are derivations $\sigma_1 \overset{*}{\Rightarrow} w_1$ and $\sigma_2 \overset{*}{\Rightarrow} w_2$ in Γ_1 and Γ_2 respectively, and so there is a derivation

$$\sigma' \Rightarrow \sigma_1 \sigma_2 \overset{*}{\Rightarrow} w_1 w_2$$

in Γ'. Thus $L_1 L_2 \subseteq L(\Gamma')$.

Conversely, suppose that $w \in L(\Gamma')$. A derivation $\sigma \overset{*}{\Rightarrow} w$ of w must begin with $\sigma \Rightarrow \sigma_1 \sigma_2$. By Lemma 4.3.4 we must have $w = w_1 w_2$, where $\sigma_1 \overset{*}{\Rightarrow} w_1$, $\sigma_2 \overset{*}{\Rightarrow} w_2$ are derivations in Γ'. But the disjointness of $V_1 \backslash A$ and $V_2 \backslash A$ ensures that the former derivation is in fact in Γ_1 and the latter in

Γ_2. Thus $w_1 \in L_1$, $w_2 \in L_2$ and so $w \in L_1 L_2$ as required. We have shown that $L_1 L_2 = L(\Gamma')$, and it is clear that Γ' is a context-free grammar.

(ii) Suppose that $L = L(\Gamma)$, where $\Gamma = (V, A, \pi, \sigma)$ is a context-free grammar. Let

$$\Gamma^* = \big(V \cup \{\tau\}, A, \pi^*, \tau\big),$$

where π^* consists of the productions in π together with the extra productions

$$\tau \to \sigma\tau, \ \tau \to 1.$$

Suppose now that $w = w_1 w_2 \ldots w_n \in L^+$, with $w_1, w_2, \ldots, w_n \in L$. Then there is a derivation $\sigma \overset{*}{\Rightarrow} w_i$ in Γ for each i and so there is a derivation

$$\tau \Rightarrow \sigma\tau \overset{*}{\Rightarrow} w_1 \tau \Rightarrow w_1 \sigma\tau \overset{*}{\Rightarrow} w_1 w_2 \tau$$
$$\overset{*}{\Rightarrow} \cdots \overset{*}{\Rightarrow} w_1 w_2 \ldots w_n \tau \Rightarrow w_1 w_2 \ldots w_n$$

in Γ^*. It follows that $L^+ \subseteq L(\Gamma^*)$, and the trivial observation that $\tau \Rightarrow 1$ is a derivation in Γ^* then gives the result that $L^* \subseteq L(\Gamma^*)$.

Conversely, suppose that $w \in L(\Gamma^*)$. The derivation $\tau \overset{*}{\Rightarrow} w$ must begin $\tau \Rightarrow \sigma\tau$, and the production $\tau \to \sigma\tau$ may be used several times before τ finally disappears. We shall establish the following lemma

Lemma 5.6.2. *Let $\tau \overset{*}{\Rightarrow} \zeta$ be a derivation in Γ^* in which the production $\tau \to \sigma\tau$ is used k times and the production $\tau \to 1$ is not used. Then $\zeta = \zeta_1 \zeta_2 \ldots \zeta_k \tau$, and there is a derivation $\sigma \overset{*}{\Rightarrow} \zeta_i$ in Γ for $i = 1, 2, \ldots, k$.*

Proof. If the production $\tau \to \sigma\tau$ is used once only then the derivation is

$$\tau \Rightarrow \sigma\tau \overset{*}{\Rightarrow} \zeta_1 \tau$$

and there is a derivation $\sigma \overset{*}{\Rightarrow} \zeta_1$ in Γ.

Suppose now that the production $\tau \to \sigma\tau$ is used k times, and consider the last use. We may suppose that the derivation takes the form

$$\sigma \overset{*}{\Rightarrow} \xi\tau \Rightarrow^{(1)} \xi\sigma\tau \overset{*}{\Rightarrow} \eta\zeta_k \tau = \zeta,$$

where the last use of $\tau \to \sigma\tau$ is at (1) and where $\xi \overset{*}{\Rightarrow} \eta$, $\sigma \overset{*}{\Rightarrow} \zeta_k$ are derivations in Γ. We may suppose inductively that $\xi = \zeta_1' \zeta_2' \ldots \zeta_{k-1}'$ and that there is a derivation $\sigma \overset{*}{\Rightarrow} \zeta_i'$ in Γ for $i = 1, 2, \ldots k-1$. By Lemma 4.3.4 the existence of a derivation $\xi = \zeta_1' \zeta_2' \ldots \zeta_{k-1}' \overset{*}{\Rightarrow} \eta$ implies that η can be expressed as $\zeta_1 \zeta_2 \ldots \zeta_{k-1}$ in such a way that there is a derivation $\zeta_i' \overset{*}{\Rightarrow} \zeta_i$ in Γ for $i = 1, 2, \ldots, k-1$. Thus $\zeta = \zeta_1 \zeta_2 \ldots \zeta_k \tau$, and there is a derivation $\sigma \overset{*}{\Rightarrow} \zeta_i$ in Γ for $i = 1, 2, \ldots, k$.

This completes the proof of the lemma. Returning now to the proof of the theorem, we may now suppose that the derivation $\tau \overset{*}{\Rightarrow} w$ in Γ^*, involving k uses of $\tau \to \sigma\tau$, takes the form

$$\tau \overset{*}{\Rightarrow} \zeta_1 \zeta_2 \ldots \zeta_k \tau \Rightarrow \zeta_1 \zeta_2 \ldots \zeta_k \overset{*}{\Rightarrow} z_1 z_2 \ldots z_k,$$

where, for each i,

$$\sigma \overset{*}{\Rightarrow} \zeta_i \overset{*}{\Rightarrow} z_i$$

is a derivation in Γ. Thus each z_i lies in L and so $w \in L^*$ as required. ∎

As we saw in Exercise 2.7, the class of regular languages is closed also under intersection and complementation. To see that the class of context-free languages is *not* closed under intersection, consider the following example. The language $L = \{a^m b^m : m \geq 1\}$ is known to be context-free. (See Example 4.3.1.) The language a^+, being regular, is certainly context-free and so by Theorem 5.6.1 the language

$$L_1 = L.a^+ = \{a^m b^m a^n : m, n \geq 1\}$$

is context-free. Similarly, since $L' = \{b^n a^n : n \geq 1\}$ is context-free, so is

$$L_2 = a^+.L' = \{a^m b^n a^n : m, n \geq 1\}.$$

But

$$L_1 \cap L_2 = \{a^n b^n a^n : n \geq 1\},$$

and we saw in (5.5.7) that this is not a context-free language. So we have

Theorem 5.6.3. *The class of context-free languages is NOT closed under intersection.* ∎

As a simple logical consequence we have another 'unfriendly' property of context-free languages:

Theorem 5.6.4. *The class of context-free languages is NOT closed under complementation.*

Proof. Suppose, by way of contradiction, that for every context-free language L in A^* the complement $A^* \backslash L$ is context-free. Let L_1, L_2 be arbitrary context-free languages. Then $A^* \backslash L_1$ and $A^* \backslash L_2$ are context-free and hence by Theorem 5.6.1 so is $(A^* \backslash L_1) \cup (A^* \backslash L_2)$. That is, by the De Morgan Law (1.1.1) $A^* \backslash (L_1 \cap L_2)$ is context-free. Hence, by the supposed property of closure under complementation, $L_1 \cap L_2$ is context-free. We have in fact shown that

$$L_1, L_2 \text{ context-free} \Rightarrow L_1 \cap L_2 \text{ context-free};$$

but we already know from Theorem 5.6.3 that this implication is false. Hence the class of context-free languages cannot after all be closed under complementation. ∎

There is nothing at all wrong with this indirect, 'logical' approach, but doubtless we shall all feel happier if we can find an actual example of a context-free language whose complement is not context-free. To this end we look yet again at $L = \{a^n b^n a^n : n \geq 1\}$, which by (5.5.7) we know to be not context-free, and we show that $\{a, b\}^* \backslash L$ is context-free. To see that this is so, first consider the context-free language

$$L_1 = \{a^q b^q : q \geq 1\}.$$

Since a^+ is regular and hence context-free, it follows by Theorem 5.6.1 that

$$L_2 = a^+ . L_1 = \{a^p b^q : p > q \geq 1\}$$

is context-free. Equally,

$$L_3 = L_1 . b^+ = \{a^p b^q : 1 \leq p < q\}$$

is context-free. Hence, by Theorem 5.6.1,

$$L_4 = L_2 \cup L_3 = \{a^p b^q : p, q \geq 1, \ p \neq q\}$$

is context-free. By the same token, so is

$$L_5 = \{b^q a^r : q, r \geq 1, \ q \neq r\}.$$

Hence

$$L_6 = L_4 . a^+ = \{a^p b^q a^r : p, q, r \geq 1, \ p \neq q\}$$

and

$$L_7 = a^+ . L_5 = \{a^p b^q a^r : p, q, r \geq 1, \ q \neq r\}$$

are both context-free.

The language $a^+ b^+ a^+$ is regular, and hence its complement

$$L_8 = \{a, b\}^* \backslash a^+ b^+ a^+,$$

being regular (by Exercise 2.7), is context-free. Now any word in $\{a, b\}^*$ that does *not* belong to $L \ (= \{a^n b^n a^n : n \geq 1\})$ is either outside $a^+ b^+ a^+$ or is of the form $a^p b^q a^r$ where either $p \neq q$ or $q \neq r$ (or both). That is,

$$\{a, b\}^* \backslash L = L_6 \cup L_7 \cup L_8;$$

hence, again by Theorem 5.6.1, $\{a, b\}^* \backslash L$ is context-free.

In the example we gave to establish Theorem 5.6.3 it was no accident that both L_1 and L_2 were not regular languages. Indeed we have

Theorem 5.6.5. *Let A be a finite alphabet and let L_1, L_2 be languages in A^* such that L_1 is regular and L_2 is context-free. Then $L_1 \cap L_2$ is context-free.*

Proof. Suppose that L_1 is recognized by an FSA

$$\mathcal{A}_1 = (Q_1, A, \varphi_1, i_1, T_1)$$

and that L_2 is recognized by a PDA

$$\mathcal{P}_2 = (Q_2, A, M_2, \mu_2, \rho_2, i_2, \zeta_2, T_2).$$

We construct a PDA to recognize $L_1 \cap L_2$ by coupling $\mathcal{A}_1$ and $\mathcal{P}_2$ 'in parallel'. Formally, let

$$\mathcal{P} = (Q_1 \times Q_2, A, M_2, \mu, \rho, \zeta, T_1 \times T_2),$$

where

$$\big((q_1, q_2'), \gamma\big) \in \mu\big((q_1, q_2), \lambda\big)$$

whenever $(q_2', \gamma) \in \mu(q_2, \lambda)$, and where

$$\big((q_1', q_2'), \gamma\big) \in \rho\big((q_1, q_2), \lambda, a\big)$$

whenever $\varphi(q_1, a) = q_1'$ and $(q_2', \gamma) \in \rho(q_2, \lambda, a)$.

We show that $L_1 \cap L_2 = L(\mathcal{P})$. The proof is rather awkward notationally, but is conceptually quite simple. We may visualize the 'coupled up' automaton $\mathcal{P}$ as computing simultaneously with $\mathcal{A}_1$ and $\mathcal{P}_2$. The FSA $\mathcal{A}_1$ simply 'marks time' during stationary, memory stack adjusting manœuvres in $\mathcal{P}_2$.

Suppose first that $w = a_1 a_2 \ldots a_m \in L_1 \cap L_2$. Then there is a computation

$$i_1 = q_{11} \to q_{12} \to \cdots \to q_{1,m+1} = t_1 \tag{5.6.6}$$

in $\mathcal{A}_1$, where $\varphi(q_{1i}, a_i) = q_{1,i+1}$ for $i = 1, 2, \ldots, m$, and there is a computation

$$(q_{21}, \zeta_2, w) \to (q_{22}, \tau_2, w_2) \to \cdots \to (q_{2n}, \tau_n, 1) \tag{5.6.7}$$

in $\mathcal{P}_2$, in which $q_{21} = i_2$ and $q_{2n} = t_2 \in T_2$. Notice that $n \geq m+1$, for the latter computation in general will involve some stationary steps as well as progressive steps. Suppose in fact that

$$w = w_2 = \cdots = w_{j_1},$$
$$w_{j_1+1} = \cdots = w_{j_2} = a_2 \ldots a_m,$$
$$\cdots \tag{5.6.8}$$
$$w_{j_{m-1}+1} = \cdots = w_{j_m} = a_m,$$
$$w_{j_m+1} = \cdots = w_n = 1.$$

We construct a computation in $\mathcal{P}$ beginning with $((i_1, i_2), \zeta_2, w)$ and ending with $((t_1, t_2), \tau_n, 1)$. It takes place in $m + 1$ phases. During the first phase both w and the state $q_{11} (= i_1)$ remain constant:

$$((i_1, i_2), \zeta_2, w) \xrightarrow{*} ((i_1, q_{2j_1}), \tau_{j_1}, w).$$

The next step is

$$((i_1, q_{2j_1}), \tau_{j_1}, w) \rightarrow ((q_{12}, q_{2, j_1+1}), \tau_{j_1+1}, a_2 \ldots a_m)$$

and is followed by a phase in which q_{12} and $a_2 \ldots a_m$ remain constant:

$$((q_{12}, q_{2, j_1+1}), \tau_{j_1+1}, a_2 \ldots a_m) \xrightarrow{*} ((q_{12}, q_{2j_2}), \tau_{j_2}, a_2 \ldots a_m).$$

In general we have phases

$$((q_{1l}, q_{2, j_{l-1}+1}), \tau_{j_{l-1}+1}, a_l \ldots a_m) \xrightarrow{*} ((q_{1l}, q_{2j_l}), \tau_{j_l}, a_l \ldots a_m)$$

followed by steps

$$((q_{1l}, q_{2j_l}), \tau_{j_l}, a_l \ldots a_m) \rightarrow ((q_{1, l+1}, q_{2, j_l+1}), \tau_{j_l+1}, a_{l+1} \ldots a_m)$$

in which a letter of w disappears. Recalling that $q_{1,m+1} = t_1$, we see that the final phase is

$$((t_1, q_{2, j_m+1}), \tau_{j_m+1}, 1) \xrightarrow{*} ((t_1, t_2), \tau_n, 1)$$

and so we have a successful computation

$$((i_1, i_2), \zeta, w) \xrightarrow{*} ((t_1, t_2), \tau_n, 1)$$

in $\mathcal{P}$. Thus $L_1 \cap L_2 \subseteq L(\mathcal{P})$.

Conversely, suppose that $w = a_1 a_2 \ldots a_m$ and that there is a successful computation

$$((q_{11}, q_{21}), \tau_1, w) \xrightarrow{*} ((q_{1j}, q_{2j}), \tau_j, w_j) \xrightarrow{*} ((q_{1n}, q_{2n}), \tau_n, 1) \qquad (5.6.9)$$

in $\mathcal{P}$, where

$$(q_{11}, q_{21}) = (i_1, i_2), \quad (q_{1n}, q_{2n}) = (t_1, t_2) \in T_1 \times T_2, \text{ and } \tau_1 = \zeta_2.$$

This computation divides naturally into $m + 1$ phases, divided by the steps in which a letter of w disappears. Suppose that the phases are as implied by the equations (5.6.8). Then, since changes in q_{1j} occur precisely when a letter of w disappears we must also have

$$q_{11} = \cdots = q_{1j_1} \ (= i_1)$$
$$q_{1,j_1+1} = \cdots = q_{1j_2} = \varphi(q_{11}, a_1)$$
$$\cdots$$
$$q_{1,j_k+1} = \cdots = q_{1,j_{k+1}} = \varphi(q_{1j_k}, a_k)$$
$$\cdots$$
$$q_{1,j_m+1} = \cdots = q_{1n} = t_1 = \varphi(q_{1j_m}, a_m)$$

and there is a computation

$$i_1 \to q_{1j_2} \to \cdots \to q_{1j_m} \to t_1$$

in $\mathcal{A}_1$. Hence $w \in L_1$.

Equally, for the computation (5.6.9) to take place, we require that

$$(q_{2,j+1}, \tau_{j+1}) \in \mu(q_{2j}, \tau_j, w_j)$$

when $j \notin \{j_1, j_2, \ldots j_m\}$, and

$$(q_{2,j_k+1}, \tau_{j_k+1}) \in \rho(q_{2j_k}, \tau_{j_k}, a_k)$$

for $k = 1, 2, \ldots, m$. But this implies that there is a computation

$$(i_2, \zeta_2, w) \overset{*}{\to} (t_2, \tau_n, 1)$$

in the PDA $\mathcal{P}_2$ and so $w \in L_2$. ∎

As an illustration of the use of this theorem, consider

Example 5.6.10. The language $L = \{w^2 : w \in \{a, b\}^+\}$ is not context-free. For suppose by way of contradiction that L is context-free. The language $L' = a^+b^+a^+b^+$ is regular and so by Theorem 5.6.5 the language

$$L \cap L' = \{a^m b^n a^m b^n : m, n \geq 1\}$$

is context-free. But by (5.5.8) this is not the case. Hence L is not context-free.

Exercises 5

5.1. Find a PDA recognizing each of the following context-free languages:

    (i) $\{a^n b^{2n} : n \geq 1\}$;

    (ii) $\{a^p b^{p+q} a^q : p, q \geq 1\}$;

    (iii) $\{a^m b^n : m, n \geq 1, m > n\}$;

    (iv) $\{w \in \{a, b\}^+ : |w|_a = |w|_b + 1\}$;

    (v) $\{wa^{|w|} : w \in \{a, b\}^+\}$.

5.2. Describe the language recognized by the PDA

$$\mathcal{P} = (Q, A, M, \mu, \rho, i, \zeta, T),$$

where $Q = \{i, q, q', t\}$, $A = \{a, b\}$, $M = \{\zeta, \alpha\}$, $T = \{t\}$, $\mu = \emptyset$, and

$$\begin{aligned}
\rho(i, \zeta, a) &= \{(i, \alpha^2 \zeta)\}, & \rho(i, \alpha, a) &= \{(i, \alpha^3)\}, \\
\rho(i, \alpha, b) &= \{(q, \alpha^2)\}, & \rho(q, \alpha, b) &= \{(q, \alpha^2)\}, \\
\rho(q, \alpha, a) &= \{(q', 1)\}, & \rho(q', \alpha, a) &= \{(q', 1)\}, \\
\rho(q', \zeta, 1) &= \{(t, \zeta)\}.
\end{aligned}$$

5.3. Describe the language recognized by the PDA

$$\mathcal{P} = (Q, A, M, \mu, \rho, i, \zeta, T),$$

where $Q = \{i, q, t\}$, $A = \{a, b\}$, $M = \{\zeta, \alpha, \beta\}$, $T = \{t\}$,

$$\begin{aligned}
\mu(q, \alpha) &= \{(i, 1)\}, & \mu(q, \zeta) &= \{(i, \beta \zeta)\}, \\
\rho(i, \zeta, a) &= \{(i, \alpha \zeta)\}, & \rho(i, \alpha, a) &= \{(i, \alpha^2)\}, \\
\rho(i, \beta, a) &= \{(i, 1)\}, & \rho(i, \zeta, b) &= \{(i, \beta^2 \zeta)\}, \\
\rho(i, \beta, b) &= \{(i, \beta^3)\}, & \rho(i, \alpha, b) &= \{(q, 1)\}, \\
\rho(i, \zeta, 1) &= \{(t, \zeta)\}.
\end{aligned}$$

5.4. (i) Find a PDA recognizing $L = \{w \in \{a, b\}^+ : |w|_a > |w|_b\}$.

(ii) Find a context-free grammar recognizing L.

5.5. Use the Pumping Lemma to show that the following subsets of a^* are not context-free: (i) $\{a^{n^2} : n \geq 1\}$; (ii) $\{a^p : p \text{ is prime}\}$.

5.6. Show that the following languages are not context-free:

    (i) $\{a^n b^n c^n : n \geq 1\}$;

    (ii) $\{a^n b^n c^n d^n : n \geq 1\}$;

    (iii) $\{a^n b^{2n} a^n : n \geq 1\}$.

5.7. Let $L = \{a^m b^n : m, n \geq 1, m \neq n\}$. Show that both L and $\{a, b\}^* \backslash L$ are context-free, but that neither is regular.

5.8. Show that the following languages are not context-free:

    (i) $\{w \in \{a, b, c\}^+ : |w|_a = |w|_b = |w|_c\}$;

    (ii) $\{ww^R w : w \in \{a, b\}^*\}$.

5.9. In Section 5.6 it was shown that L, the complement in $\{a, b\}^*$ of the language $\{a^n b^n a^n : n \geq 1\}$, is context-free. Find a context-free grammar generating L.

5.10. One possible approach to the definition of a DPDA (deterministic pushdown automaton) is to take it as a septuple

$$\mathcal{D} = (Q, A, M, \mu, \rho, i, \zeta),$$

where Q, A, M, i and ζ are as in Definition 5.1.1, except that μ is now a partial map from $Q \times M$ into $Q \times M^*$ and ρ is a partial map from $Q \times M \times (A \cup \{1\})$ into $Q \times M^*$, with the property that whenever $\mu(q, t)$ is defined then $\rho(q, \tau, a)$ is undefined for every a in $A \cup \{1\}$. Suppose that the DPDA recognizes by empty memory stack.

A *prefix code* in A^* is a non-empty subset of A^* with the property that $C \cap CA^+ = \emptyset$. Show that if L ($\subseteq A^*$) is recognized by a DPDA then L is a prefix code.

Deduce that $\{ww^R : w \in \{a, b\}^+\}$ and $\{a^m b^n : m \geq n \geq 1\}$, though context-free, are not recognizable by DPDAs.

5.11. Show that $L_1 = \{a^n b^n : n \geq 1\}$ and $L_2 = \{a^n b^{2n} : n \geq 1\}$ are recognizable by DPDAs, but that $L_1 \cup L_2$ is not.

5.12. Show that not every prefix code is context-free.

6 Turing machines

Introduction

In introducing the idea of a pushdown automaton in Chapter 5 we found it helpful to think of a word $w = a_1 a_2 \ldots a_n$ in A^* as printed on a tape

$$i$$

$$\downarrow$$

| a_1 | a_2 | $\ldots$ | a_n |

with the initial state i scanning the leftmost square. An FSA can be seen as acting on w by constantly moving right, deleting a symbol at each stage. A PDA similarly can never move left, but it can stand still temporarily in order to make adjustments to its memory stack. It can print symbols in its memory stack, but can never add any new symbols to the tape.

Part of the object of exploring a hierarchy of types of languages and machines is to examine the limits of computability. The FSA and the PDA can be seen as computing devices, and in the last few chapters we have explored their strengths and weaknesses. The next kind of machine to consider is the one devised by Alan Turing (1912–1954) and it is no accident that Turing was deeply involved in the development during the Second World War of the first computer, for with the *Turing machine* we reach the 'Rolls Royce' among the machines described in this book and acquire the kind of power that corresponds to that of a real computer.

In Chapter 2 we defined finite state automata (FSAs) and described the (rational or regular) languages recognized by such machines. The FSAs were originally taken as deterministic, but we saw that extending the definition to allow for non-determinism led to no extension of the class of recognizable languages.

157

In Chapters 4 and 5 the order of events was different. We defined context-free languages as being generated by a particular sort of grammar, and only then produced the notion of a pushdown automaton (PDA), giving us a class of machines recognizing precisely those languages. Our PDAs were intrinsically non-deterministic, and this was a necessary feature of their operation.

In this chapter we return to the earlier approach, defining the machine first and then considering the kind of grammar that will generate our new class of recognizable languages. Although our Turing machines will in the first instance be deterministic, we shall see in due course that the introduction of non-determinism (as in the case of FSAs) confers no advantage.

6.1 Definitions and examples

It will be much more helpful to defer further informal chat about Turing machines until after we have recorded a formal definition.

Definition 6.1.1. Let A be a finite alphabet. A *Turing machine* (TM) $\mathcal{T}$ is a septuple

$$\mathcal{T} = (Q, A, M, \Delta, \theta, i, T),$$

where

Q is a finite set, the set of *states* of $\mathcal{T}$;

$M \supseteq A$ is a finite set, the set of *tape symbols* of $\mathcal{T}$;

$\Delta \ (\in M \backslash A)$ is the *blank* symbol;

$i \ (\in Q)$ is the *initial* state;

T, a non-empty subset of Q, is the set of *terminal* states;

θ is a partial map from $Q \times M$ into $Q \times M \times \{L, R\}$.

(Here L, R are symbols standing for *left* and *right* respectively.)

An *instantaneous description* (ID) of $\mathcal{T}$ is a word $w_1 q w_2$, where $w_1 \in M^*$, $w_2 \in M^+$ and $q \in Q$.

We now define a *computation* in $\mathcal{T}$. We think of the state symbol q in the ID $w_1 q w_2$ as scanning the leftmost letter of w_2. Thus if $w_2 = \alpha w_2'$ (with $\alpha \in M$, $w_2' \in M^*$) then the behaviour of the machine depends on the value of $\theta(q, \alpha)$. To be precise, consider first an ID

$$z_1 \beta q \alpha \gamma z_2,$$

with $\alpha, \beta, \gamma \in M$, $z_1, z_2 \in M^*$. Then:

(i) if $\theta(q, \alpha) = (q', \alpha', R)$ we write

$$z_1 \beta q \alpha \gamma z_2 \rightarrow z_1 \beta \alpha' q' \gamma z_2; \qquad (6.1.2)$$

that is, the scanned symbol α changes to α', the state q changes to q' and the scanner moves one space to the right;

(ii) if $\theta(q, \alpha) = (q', \alpha', L)$ we write

$$z_1 \beta q \alpha \gamma z_2 \rightarrow z_1 q' \beta \alpha' \gamma z_2; \qquad (6.1.3)$$

that is, the scanned symbol α changes to α', the state q changes to q' and the scanner moves one space to the left.

If q is near the edge of the ID there are some special provisions. Precisely, if we have an ID $z_1 q \alpha$ (with $z_1 \in M^*$, $\alpha \in M$) and if $\theta(q, \alpha) = (q', \alpha', R)$ then we write

$$z_1 q \alpha \rightarrow z_1 \alpha' q' \Delta; \qquad (6.1.4)$$

that is, the machine behaves as in (6.1.2) but prints another 'blank' symbol on the right. By contrast, if we have a similar situation on the left-hand side, that is, if we have an ID $q \alpha z_2$ (with $\alpha \in M$, $z_2 \in M^*$) and if $\theta(q, \alpha) = (q', \alpha', L)$ then the machine stops. It is as if $\theta(q, \alpha)$ were undefined. In effect we are regarding the tape as extendible on the right but not on the left.

If two IDs ω_1 and ω_2 are connected by a finite sequence of changes of types (6.1.2), (6.1.3) and (6.1.4) then we refer to the sequence as a *computation* in T (or as a T-*computation* and write $\omega_1 \overset{*}{\rightarrow} \omega_2$.

If $w \in A^*$ then we say that w is *recognized* by T if there is a computation

$$iw \overset{*}{\rightarrow} \zeta_1 t \zeta_2,$$

in which i is the initial state, t is a terminal state, and $\zeta_1 \in M^*$, $\zeta_2 \in M^+$. The *language $L(T)$ recognized by T* is defined as the set of all w in A^* recognized by T.

A TM, unlike a PDA, does not have a separate memory stack, but it has the power to scan both left and right and it can print symbols from $M \backslash A$ at any point on the tape. It will turn out that TMs recognize all context-free languages as well as some languages that are not context-free. However, it will be useful first to look at some examples.

First we show how to construct a TM to recognize the subset $L = \{a^n b^n : n \geq 1\}$ of $\{a, b\}^*$. This language is known to be context-free (Example 4.3.1) but not regular (Theorem 2.2.6).

Example 6.1.5. Let

$$Q = \{i, q_1, q_2, q_3, q_4, t\}, \ M = \{a, b, \Delta, \lambda, \mu\}, \ T = \{t\}.$$

Let

$$\theta(i, a) = (q_1, \lambda, R)$$
$$\theta(q_1, a) = (q_1, a, R)$$
$$\theta(q_1, \mu) = (q_2, \mu, R)$$
$$\theta(q_2, b) = (q_3, \mu, L)$$
$$\theta(q_2, \mu) = (q_3, \mu, L)$$
$$\theta(q_1, b) = (q_3, \mu, L)$$

(Change i to q_1 and the initial a to λ; move right until b is encountered; change b to μ and begin to move left.)

$$\theta(q_3, \mu) = (q_3, \mu, L)$$
$$\theta(q_3, a) = (q_3, a, L)$$
$$\theta(q_3, \lambda) = (i, \lambda, R)$$

(Move left until λ is encountered; return to state i and prepare to repeat.)

$$\theta(i, \mu) = (q_4, \mu, R)$$
$$\theta(q_4, \mu) = (q_4, \mu, R)$$
$$\theta(q_4, \Delta) = (t, \Delta, R)$$

(When all a's from the initial block are gone, scan right looking for a's and b's; if none found, enter terminal state.)

To understand the operation of this machine, let us look at the demonstration that $a^3 b^3 \in L(\mathcal{T})$. We have a computation

$$ia^3 b^3 \rightarrow \lambda q_1 a^2 b^3 \xrightarrow{*} \lambda a^2 q_1 b^3$$
$$\rightarrow \lambda a q_3 a \mu b^2 \xrightarrow{*} q_3 \lambda a^2 \mu b^2 \rightarrow \lambda i a^2 \mu b^2$$
(end of first loop)
$$\rightarrow \lambda^2 q_1 a \mu b^2 \xrightarrow{*} \lambda^2 a \mu q_2 b^2 \rightarrow \lambda^2 a q_3 \mu^2 b$$
$$\xrightarrow{*} \lambda q_3 \lambda a \mu^2 b \rightarrow \lambda^2 i a \mu^2 b$$
(end of second loop)
$$\rightarrow \lambda^3 q_1 \mu^2 b \xrightarrow{*} \lambda^3 \mu^2 q_2 b \rightarrow \lambda^3 \mu q_3 \mu^2$$
$$\xrightarrow{*} \lambda^2 q_3 \lambda \mu^3 \rightarrow \lambda^3 i \mu^3$$
(end of third loop)
$$\rightarrow \lambda^3 \mu q_4 \mu^2 \xrightarrow{*} \lambda^3 \mu^3 q_4 \Delta \rightarrow \lambda^3 \mu^3 \Delta t \Delta.$$

It is clear that this computation does not depend upon having precisely 3 a's and 3 b's. In fact for every $n \geq 1$ there is a computation

$$ia^n b^n \xrightarrow{*} \lambda^n \mu^n \Delta t \Delta.$$

Thus $L \subseteq L(\mathcal{T})$.

To show the reverse inclusion, note that the complement of L in A^* (where $A = \{a, b\}$) is the union of the sets

$$bA^*, \ a^*, \ \{a^m b^n z : z \notin bA^*, \ m > n\},$$

$$\{a^m b^n z : z \in A^*, \ m < n\}, \ \{a^n b^n z : z \in aA^*\}.$$

That is, if $w \notin L$ then it either: (1) begins with b; or (2) consists entirely of a's; or (3) begins with a block of a's followed by a shorter block of b's; or (4) begins with a block of a's followed by a longer block of b's; or (5) is of the form wz, where $w \in L$ and $z \in aA^*$.

In Case (1) the computation beginning with iw fails at the first step, since $\theta(i, b)$ is undefined.

In Case (2), with $w = a^k$ (say), the computation fails at the first step if $k = 0$. If $k \geq 1$ we have a computation

$$ia^k \rightarrow \lambda q_1 a^{k-1} \overset{*}{\rightarrow} \lambda a^{k-1} q_1 \Delta,$$

and the computation then fails.

In Case (3) the looping process takes place as we have observed, except that the supply of b's runs out before the supply of a's. We get

$$ia^m b^n z \overset{*}{\rightarrow} \lambda^n i a^{m-n} \mu^n z \rightarrow \lambda^{n+1} q_1 a^{m-n-1} \mu^n z \overset{*}{\rightarrow} \lambda^{n+1} a^{m-n-1} \mu^n q_2 z,$$

and whether $z = 1$ or $z \in aA^*$ the computation then fails.

In Case (4) the supply of a's runs out first. We get

$$ia^m b^n z \overset{*}{\rightarrow} \lambda^m i \mu^m b^{n-m} z \rightarrow \lambda^m \mu q_4 \mu^{m-1} b^{n-m} z \overset{*}{\rightarrow} \lambda^m \mu^m q_4 b^{n-m} z,$$

and the computation fails because $\theta(q_4, b)$ is undefined.

Finally, in Case (5) we get

$$ia^n b^n z \overset{*}{\rightarrow} \lambda^n \mu^n q_4 z,$$

and again the computation fails. We conclude that $L(\mathcal{T}) = L$.

We shall see eventually that *every* context-free language is recognizable by a TM, but this is by no means obvious. What we can say at this stage is that every *regular* (or *rational*) language is recognizable by a TM. We show this in effect by devising a TM to simulate an arbitrary FSA.

Theorem 6.1.6. *Every rational language is recognizable by a Turing machine.*

Proof. Suppose that L is a rational language. Then by Kleene's Theorem (2.5.2) we may suppose that L is recognized by an FSA

$$\mathcal{A} = (Q, A, \varphi, i, T).$$

Define a TM

$$\mathcal{T} = (Q \cup \{h\}, A, A \cup \{\Delta\}, \theta, i, \{h\}),$$

where
$$\theta(q, a) = (q', \Delta, R) \text{ whenever } \varphi(q, a) = q';$$
$$\theta(t, \Delta) = (h, \Delta, R) \text{ for every } t \text{ in } T.$$

Then every successful computation

$$i \xrightarrow{a_1} q_1 \xrightarrow{a_2} q_2 \xrightarrow{a_3} \cdots \xrightarrow{a_{n-1}} q_{n-1} \xrightarrow{a_n} t$$

in $\mathcal{A}$, where

$$\varphi(i, a_1) = q_1, \ \varphi(q_j, a_{j+1}) = q_{j+1} \ (j = 1, \ldots, n-2)$$

and $\varphi(q_{n-1}, a_n) = t$, corresponds to a successful computation

$$ia_1 a_2 \ldots a_n \to \Delta q_1 a_2 \ldots a_n \to \cdots$$
$$\cdots \to \Delta^{n-1} q_{n-1} a_n \to \Delta^n t \Delta \to \Delta^{n+1} h \Delta$$

in T, and so $L = L(\mathcal{A}) \subseteq L(T)$.

Conversely, if $w = a_1 a_2 \ldots a_n \in L(T)$ then there is a computation

$$ia_1 a_2 \ldots a_n \xrightarrow{*} z_1 h z_2 \tag{6.1.7}$$

in T. The state h cannot be reached unless a terminal state t of $\mathcal{A}$ scans a blank square; so in fact the computation (6.1.7) is of the form

$$ia_1 a_2 \ldots a_n \xrightarrow{*} z_1 t \Delta \to z_1 \Delta h \Delta.$$

The TM, moreover, can only move right during the computation, and so in fact the computation must be

$$ia_1 a_2 \ldots a_n \xrightarrow{*} \Delta q_1 a_2 \ldots a_n \xrightarrow{*} \Delta^n t \Delta \to \Delta^{n+1} h \Delta,$$

where

$$\varphi(i, a_1) = q_1, \ \varphi(q_j, a_{j+1}) = q_{j+1} \ (j = 1, \ldots, n-2)$$

and $\varphi(q_{n-1}, a_n) = t$. It follows that $w \in L(\mathcal{A})$. ∎

The simulation of an FSA by a TM is easy, as we have now seen, but the simulation of a PDA, because of its intrinsic indeterminacy, causes more difficulty, and we shall defer it until Section 5. For the moment let us look at another example, one which gives a hint of the extra power of Turing machines, since the language involved is known (see Example 5.5.8) not to be context-free.

Example 6.1.8. Let

$$L = \{a^n b^n a^n : n \geq 1\}.$$

Let

$$\mathcal{T} = (Q, \{a, b\}, M, \Delta, \theta, i, \{t\}),$$

where

$$Q = \{i, q_1, q_2, q_3, q_4, t\}, \quad M = \{a, b, \lambda, \mu, \nu, \Delta\},$$

and where

$\theta(i, a) = (q_1, \lambda, R)$	(Move right, changing first a
$\theta(q_1, a) = (q_1, a, R)$	to λ, first b to μ,
$\theta(q_1, \mu) = (q_1, \mu, R)$	first a of second block of a's
$\theta(q_1, b) = (q_2, \mu, R)$	to ν, and prepare to move left.)
$\theta(q_2, b) = (q_2, b, R)$	
$\theta(q_2, \nu) = (q_3, \nu, R)$	
$\theta(q_3, \nu) = (q_3, \nu, R)$	
$\theta(q_2, a) = (q_4, \nu, L)$	
$\theta(q_3, a) = (q_4, \nu, L)$	

$\theta(q_4, \nu) = (q_4, \nu, L)$	(Move left until λ
$\theta(q_4, b) = (q_4, b, L)$	is encountered; return to
$\theta(q_4, \mu) = (q_4, \mu, L)$	initial state and prepare to
$\theta(q_4, a) = (q_4, a, L)$	begin again.)
$\theta(q_4, \lambda) = (i, \lambda, R)$	

$\theta(i, \mu) = (q_5, \mu, R)$	(When first block of a's is exhausted,
$\theta(q_5, \mu) = (q_5, \mu, R)$	move right until Δ is encountered;
$\theta(q_5, \nu) = (q_5, \nu, R)$	then enter terminal state.)
$\theta(q_5, \Delta) = (t, \Delta, R)$	

A typical successful computation is

$$i a^n b^n a^n \rightarrow \lambda q_1 a^{n-1} b^n a^n \overset{*}{\rightarrow} \lambda a^{n-1} q_1 b^n a^n$$

$$\rightarrow \lambda a^{n-1} \mu q_2 b^{n-1} a^n \overset{*}{\rightarrow} \lambda a^{n-1} \mu b^{n-1} q_2 a^n$$

$$\rightarrow \lambda a^{n-1} \mu b^{n-2} q_4 b \nu a^{n-1} \overset{*}{\rightarrow} q_4 \lambda a^{n-1} \mu b^{n-1} \nu a^{n-1}$$

$$\rightarrow \lambda i a^{n-1} \mu b^{n-1} \nu a^{n-1}$$

(end of first loop)

$$\overset{*}{\rightarrow} \lambda^2 i a^{n-2} \mu^2 b^{n-2} \nu^2 a^{n-2} \overset{*}{\rightarrow} \cdots \overset{*}{\rightarrow} \lambda^n i \mu^n \nu^n$$

(repeating the process, replacing the a's and b's)

$$\rightarrow \lambda^n \mu q_5 \mu^{n-1} \nu^n \overset{*}{\rightarrow} \lambda^n \mu^n \nu^n q_5 \Delta \rightarrow \lambda^n \mu^n \nu^n \Delta t \Delta.$$

It is now easy to see that if a word w in $\{a, b\}^*$ is not of the form $a^n b^n a^n$ then the computation beginning with iw must fail. If $w \in bA^*$ then the computation cannot even begin, since $\theta(i, b)$ is undefined. Certainly for a computation to continue w must have a block of a's followed by a block of b's followed by another block of a's. If these blocks are of unequal length or if there are any letters following the second block of a's then the computation fails. For example,

$$ia^2ba \rightarrow \lambda q_1 aba \rightarrow \lambda a q_1 ba \rightarrow \lambda a \mu q_2 a$$
$$\rightarrow \lambda a q_4 \mu \nu \overset{*}{\rightarrow} q_4 \lambda a \mu \nu \rightarrow \lambda i a \mu \nu$$
$$\rightarrow \lambda^2 q_1 \mu \nu \rightarrow \lambda^2 \mu q_1 \nu \parallel (\text{fails});$$

$$iabab \rightarrow \lambda q_1 bab \rightarrow \lambda \mu q_2 ab \rightarrow \lambda q_4 \mu \nu b$$
$$\overset{*}{\rightarrow} q_4 \lambda \mu \nu b \rightarrow \lambda i \mu \nu b \rightarrow \lambda \mu q_5 \nu b$$
$$\rightarrow \lambda \mu \nu q_5 b \parallel (\text{fails}).$$

The examples we have considered in this section, while giving a hint of the power of Turing machines, also indicate how lengthy and tedious the process of writing the 'program' (for that is what the specification of θ is, in effect) can be. It is necessary to develop, as in real computer programming, some standard subroutines. This process will be begun in the next section.

6.2. Turing machines as text-processors

The word-processor, with its screen editing facility, is by now a familiar object. Some of the standard Turing machine subroutines we shall want to use can be presented as editing procedures. Certainly we can always think of a piece of text as a finite word $a_1 a_2 \ldots a_n$ (in an alphabet A which includes spaces, punctuation marks, etc., as well as letters of the ordinary kind). The first requirement in any editing system is to be able to

6.2.1. *Move the cursor.*

If the TM is in state q scanning the symbol a_i then we can instruct it to move r places to the right by means of the subroutine

$$\theta(q, a) = (q_1, a, R) \quad (a \in A),$$
$$\theta(q_j, a) = (q_{j+1}, a, R) \quad (a \in A, \ j = 1, 2, \ldots, r-1).$$

Clearly a move of r spaces to the left can be achieved just as easily.

Having arrived at the right place, we may want to

6.2.2. *Substitute one letter for another.*

Suppose that the TM is in state q, scanning a_i, and that we wish to change a_i to a_i'. The subroutine given by

$$\theta(q, a_i) = (q', a_i', R), \quad \theta(q', a_{i+1}) = (q, a_{i+1}, L)$$

(where $a_{i+1} = \Delta$ if $i = n$) then gives

$$a_1 \ldots a_{i-1} q a_i \ldots a_n \rightarrow a_i \ldots a_{i-1} a_i' q' a_{i+1} \ldots a_n \rightarrow a_i \ldots a_{i-1} q a_i' a_{i+1} \ldots a_n.$$

It is easy to generalize this so as to

6.2.3. *Substitute one piece of text for another of the same length.*
Suppose that we wish to change $a_1 a_2 \ldots a_n$ into

$$a_1 \ldots a_{i-1} b_i \ldots b_j a_{j+1} \ldots a_n$$

(where $1 \leq i < j \leq n$.) Then the subroutine

$$\theta(q, a_i) = (q_1, b_i, R)$$
$$\theta(q_k, a_{i+k}) = (q_{k+1}, b_{i+k}, R) \quad (k = 1, \ldots, j - i - 1)$$
$$\theta(q_{j-i}, a_j) = (q_{j-i-1}', b_j, L)$$
$$\theta(q_k', b_{i+k}) = (q_{k-1}', b_{i+k}, L) \quad (k = 2, \ldots, j - i - 1)$$
$$\theta(q_1', b_{i+1}) = (q, b_{i+1}, L)$$

gives a computation

$$a_1 \ldots a_{i-1} q a_i \ldots a_n \xrightarrow{*} a_1 \ldots a_{i-1} q b_i \ldots b_j a_{j+1} \ldots a_n.$$

A more ambitious piece of editing is to

6.2.4. *Move a piece of text one space to the right, inserting a given symbol*
Suppose that we want to change $a_1 a_2 \ldots a_n$ to $a_1 \ldots a_i c a_{i+1} \ldots a_n$ by moving $a_{i+1} \ldots a_n$ one space to the right and inserting c. Suppose that we have a state symbol q_a for each a in A and a tape symbol $\lambda \notin A$. Let

$$\theta(q, a) = (q_a, \lambda, R) \quad (a \in A)$$
$$\theta(q_a, b) = (q_b, a, R) \quad (a, b \in A)$$
$$\theta(q_a, \Delta) = (q', a, L) \quad (a \in A)$$
$$\theta(q', a) = (q', a, L) \quad (a \in A)$$
$$\theta(q', \lambda) = (q'', c, R).$$

Then, for example, if $A = \{a, b, c\}$,

$$acbqabac \rightarrow acb\lambda q_a bac \rightarrow acb\lambda aq_b ac$$
$$\rightarrow acb\lambda abq_a c \rightarrow acb\lambda abaq_c \Delta \rightarrow acb\lambda abq' ac$$
$$\overset{*}{\rightarrow} acbq'\lambda abac \rightarrow acbcq'' abac.$$

If the insertion is to be made at the extreme right of the word (something that scarcely counts as editing at all) then a simpler subroutine suffices. If, for example, we wish to change aba to $abac$ we let

$$\theta(q, \Delta) = (q', c, R)$$

and obtain

$$abaq\Delta \rightarrow abacq'\Delta.$$

The reverse process can also be useful:

6.2.5. *Delete a symbol and close up the space.*

The subroutine given by

$$\theta(q, a) = (q', \epsilon, R) \quad (a \in A)$$
$$\theta(q', a) = (q_a, a, L) \quad (a \in A)$$
$$\theta(q_a, \epsilon) = (q, a, R) \quad (a \in A)$$
$$\theta(q', \Delta) = (q'', \Delta, L)$$
$$\theta(q'', \epsilon) = (q'', \Delta, R)$$

then gives (for example, if $A = \{a, b, c\}$)

$$abcqbabc \rightarrow abc\epsilon q' abc \rightarrow abcq_a \epsilon abc$$
$$\rightarrow abcaqabc \rightarrow abca\epsilon q' bc \rightarrow abcaq_b \epsilon bc$$
$$\rightarrow abcabqbc \rightarrow abcabeq' c \rightarrow abcabq_c \epsilon c$$
$$\rightarrow abcabc\epsilon q' \Delta \rightarrow abcabcq'' \epsilon\Delta \rightarrow abcabc\Delta q'' \Delta.$$

That is, b (the scanned symbol) has been deleted, and the space is closed up. There is of course no difficulty about moving the cursor back if necessary. It is indeed sometimes convenient to begin by using (6.2.4) to insert a 'marker' symbol μ to the left of q, obtaining

$$abc\mu qbabc.$$

Then we can extend the instructions for the deletion routine by specifying

$$\theta(q'', a) = (q'', a, L) \quad (a \in A \cup \{\Delta\})$$
$$\theta(q'', \mu) = (\bar{q}, \mu, R)$$

and thus obtain the word $abc\mu\bar{q}abc$.

The next editing procedure to be considered is

6.2.6. *Duplicate a piece of text.*

Here we need states q_a $(a \in A)$ and two disjoint sets λ_a $(a \in A)$ and μ_a $(a \in A)$ of tape symbols. Let

$$\theta(q, a) = (q_a, \lambda_a, R) \quad (a \in A)$$
$$\theta(q_a, b) = (q_a, b, R) \quad (a, b \in A)$$
$$\theta(q_a, \mu_b) = (q_a, \mu_b, R) \quad (a, b \in A)$$
$$\theta(q_a, \Delta) = (q', \mu_a, L) \quad (a \in A)$$
$$\theta(q', x) = (q', x, L) \quad (x \in A \cup \{\mu_a : a \in A\})$$
$$\theta(q', \lambda_a) = (q, a, R) \quad (a \in A)$$
$$\theta(q, \mu_a) = (q'', a, R) \quad (a \in A)$$
$$\theta(q'', \mu_a) = (q'', a, R) \quad (a \in A).$$

Then, for example,

$$qaba \to \lambda_a q_a ba \xrightarrow{*} \lambda_a baq_a \Delta \to \lambda_a bq' a\mu_a$$
$$\xrightarrow{*} q' \lambda_a ba\mu_a \to aqba\mu_a \to a\lambda_b q_b a\mu_a$$
$$\xrightarrow{*} a\lambda_b a\mu_a q_b \Delta \to a\lambda_b aq' \mu_a \mu_b \xrightarrow{*} aq' \lambda_b a\mu_a \mu_b$$
$$\to abqa\mu_a \mu_b \to ab\lambda_a q_a \mu_a \mu_b \xrightarrow{*} ab\lambda_a \mu_a \mu_b q_a \Delta$$
$$\to ab\lambda_a \mu_a q' \mu_b \mu_a \xrightarrow{*} abq' \lambda_a \mu_a \mu_b \mu_a \to abaq\mu_a \mu_b \mu_a$$
$$\to abaaq'' \mu_b \mu_a \xrightarrow{*} abaabaq'' \Delta.$$

Given a word $a_1 a_2 \ldots a_n$ we can certainly describe a Turing machine that will interchange a_i and a_j. We simply use (6.2.5) twice to remove a_i and a_j and then use (6.2.4) to insert them in their new places. And once we accept that all interchanges of this kind (what we call *transpositions*) are possible, we deduce that any rearrangement at all of the letters $a_1, a_2, \ldots a_n$ can be achieved by means of a Turing machine, for any rearrangement can be carried out by a succession of transpositions.

The final piece of editing we shall examine is more akin to the spelling checks available in most word-processing systems.

6.2.7. *Check whether two words are identical.*

Here we envisage a word $w_1 \# w_2$ made up of two words w_1 and w_2 separated by a marker symbol #. We make use of state symbols q_a, q'_a $(a \in A), q, q', q''$ and t (terminal) and of tape symbols λ_a, μ_a $(a \in A)$. For all a, b in A and all x in

$$A \cup \{\mu_a : a \in A\} \cup \{\#\},$$

let

$$\theta(q, a) = (q_a, \lambda_a, R)$$ (Mark the first letter
$$\theta(q_a, b) = (q_a, b, R)$$ and move right
$$\theta(q_a, \#) = (q'_a, \#, R)$$ to see whether the first
$$\theta(q'_a, \mu_b) = (q'_a, \mu_b, R)$$ letter after # is
$$\theta(q'_a, a) = (q', \mu_a, L)$$ is the same.)

$$\theta(q', x) = (q', x, L)$$ (Move left to the marked letter and
$$\theta(q', \lambda_a) = (q, a, R)$$ prepare to start again.)

$$\theta(q, \#) = (q'', \#, R)$$ (When all letters to the left of # are
$$\theta(q'', \mu_a) = (q'', a, R)$$ scanned, move right to check that no
$$\theta(q'', \Delta) = (t, \Delta, R).$$ letters remain on the right of #.)

Then, for example, we have a successful computation

$$qab\#ab \to \lambda_a q_a b\#ab \overset{*}{\to} \lambda_a b\#q'_a ab$$
$$\to \lambda_a b q' \#\mu_a b \overset{*}{\to} q' \lambda_a b\#\mu_a b \to aqb\#\mu_a b$$

(end of first loop)

$$\to a\lambda_b q_b \#\mu_a b \overset{*}{\to} a\lambda_b \#\mu_a q'_b b \to a\lambda_b \#q' \mu_a \mu_b$$
$$\overset{*}{\to} aq' \lambda_b \#\mu_a \mu_b \to abq\#\mu_a \mu_b$$

(end of second loop; entering final phase)

$$\to ab\#q'' \mu_a \mu_b \overset{*}{\to} ab\#abq'' \Delta \to ab\#ab\Delta t\Delta.$$

By contrast, the computation

$$qab\#ba \overset{*}{\to} \lambda_a b\#q'_a ba$$

fails, since $\theta(q'_a, b)$ is undefined. Also, the computation

$$qab\#aba \overset{*}{\to} abq\#\mu_a \mu_b a \to ab\#q'' \mu_a \mu_b a \overset{*}{\to} ab\#abq'' a$$

again fails, since $\theta(q'', a)$ is undefined. Only when w_1, w_2 are identical does a computation beginning with $qw_1\#w_2$ succeed in reaching the terminal state t.

6.3. Alternative Turing machines

There is a fair degree of arbitrariness about the definition of a Turing machine. Indeed the definition we have adopted is not quite Turing's definition. We have envisaged a 'semi-infinite' tape with a definite left-hand end, whereas Turing — and several later authors — allowed a tape infinite in both directions. On the face of it, a 'left infinite Turing machine' (LITM) might give rise to a larger class of recognizable words than a TM, but in fact this is not the case. (See Exercise 6.5.) We shall indeed see that some quite ambitious generalizations of a TM lead to no extension at all of the class of recognizable words, which suggests that this class of words has an importance well beyond its association with a particular kind of machine.

Many 'improved' Turing machines have been proposed and we shall not attempt an exhaustive survey. But as an example let us look at what we might call a *multi-tape* TM, where we envisage finitely many tapes laid parallel to each other and where a whole column of squares is scanned at once.

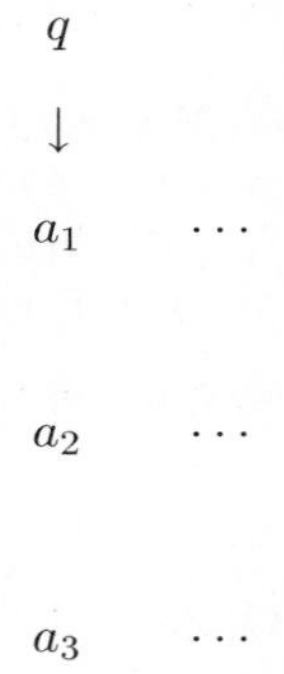

$$q$$

$$\downarrow$$

$$a_1 \quad \cdots$$

$$a_2 \quad \cdots$$

$$a_3 \quad \cdots$$

Figure 6.3.1. A 3-TM

Formally, an *n-Turing machine* (briefly, an *n*-TM) is defined in exactly the same way as a TM except that θ is now a partial map from $Q \times M^n$ into $Q \times M^n \times \{L, R\}$. We can think of the state q as scanning a column vector

$$\mathbf{a} = \begin{pmatrix} a_1 \\ a_2 \\ \vdots \\ a_n \end{pmatrix}.$$

If $\theta(q, \mathbf{a}) = (q', \mathbf{a}', R)$ (for example) then a changes to $\mathbf{a}'$, q changes to q' and the scanner moves one space to the right.

An ID is now a string

$$\mathbf{a}^{(1)} \ldots \mathbf{a}^{(j-1)} q \mathbf{a}^{(j)} \ldots \mathbf{a}^{(n)} \tag{6.3.2}$$

of column vectors, with a state symbol q scanning the column vector $\mathbf{a}^{(j)}$. We envisage a computation starting with a word $w = b_1 b_2 \ldots b_k$

from A^* printed on the first tape, with the other tapes being blank; thus the initial ID is

$$i\hat{\mathbf{b}}_1\hat{\mathbf{b}}_2\ldots\hat{\mathbf{b}}_k,$$

where, for $j = 1, 2, \ldots, k$,

$$\hat{\mathbf{b}}_j = \begin{pmatrix} b_j \\ \Delta \\ \vdots \\ \Delta \end{pmatrix}. \tag{6.3.3}$$

If a computation beginning this way reaches an ID (6.3.2) in which q is a terminal state, we say that the computation is successful and that the n-TM *recognizes* w.

One might expect that an n-TM would have some advantage over an ordinary TM, in that one or more of the extra tapes could be used as an additional memory or control facility. We shall in fact see that the extra tapes confer no advantage at all. First, and unsurprisingly, we have

Theorem 6.3.4. *Every language recognized by a* TM *is recognized by an* n-TM.

Proof. Let $L = L(T)$, where

$$T = (Q, A, M, \Delta, \theta, i, T)$$

is a TM. Define an n-TM

$$T' = (n, Q, A, M, \Delta, \theta', i, T).$$

Using the notation $\hat{u}$ as in (6.3.3) we define

$$\theta'(q, \hat{\mathbf{u}}) = (q', \hat{\mathbf{v}}, R)$$

whenever $\theta(q, u) = (q, v, R)$ and

$$\theta'(q, \hat{\mathbf{u}}) = (q', \hat{\mathbf{v}}, L)$$

whenever $\theta(q, u) = (q, v, L)$. In effect we are simply filling the extra tapes with unchanging Δs. It is then clear that a successful computation

$$ia_1a_2\ldots a_m \overset{*}{\rightarrow} u_1\ldots u_{i-1}tu_i\ldots u_k$$

in T can be simulated by a successful computation

$$i\hat{\mathbf{a}}_1\hat{\mathbf{a}}_2\ldots\hat{\mathbf{a}}_m \overset{*}{\rightarrow} \hat{\mathbf{u}}_1\ldots\hat{\mathbf{u}}_{i-1}t\hat{\mathbf{u}}_i\ldots\hat{\mathbf{u}}_k$$

in T'. Furthermore, since there is no mechanism in T' for introducing symbols other than Δ to the tapes numbered from 2 to n, there can be no successful computation in T' that is not derived in the obvious way from a successful computation in T. Thus $L(T') = L(T)$. ∎

More surprisingly, we also have

Theorem 6.3.5. *Every language recocognized by an n-TM can be recognized by a TM.*

Proof. Consider an n-TM

$$T = (n, Q, A, M, \Delta, \theta, i, T).$$

We shall construct a TM

$$T'' = (Q', A, M', \Delta, \theta', i', T)$$

which simulates its action. Notice that T' has the same set of terminal symbols as T and we suppose that $Q' \supseteq Q$, $M' \supseteq M$. A typical initial ID of T is (in the notation of (6.3.3)) $i\hat{a}_1\hat{a}_2 \ldots \hat{a}_m$, where $a_1 a_2 \ldots a_m \in A^*$. We shall want to begin the corresponding computation in T' with $i'a_1 a_2 \ldots a_m$, and the first stage will be to change this into an ID in which the column vectors $\hat{a}_1, \hat{a}_2, \ldots, \hat{a}_m$ are laid on their sides, giving

$$i'a_1\Delta \ldots \Delta a_2\Delta \ldots \Delta \ldots a_m\Delta \ldots \Delta.$$

This is done by using the insertion subroutine (6.2.4) $(m-1)(n-1)$ times to insert $n-1$ Δs following each a_j. For convenience we shall also insert a marker symbol $\#$ before each a_j and at the end, arriving at

$$i\#a_1\Delta^{n-1}\#a_2\Delta^{n-1}\# \ldots a_m\Delta^{n-1}\#.$$

Of course the initial state of T' is i', but at this stage we may suppose that we have arranged to enter the state i so that the simulation of T can begin.

Consider now a typical instruction of T, of the form

$$\theta(q, \mathbf{u}) = (q', \mathbf{u}', R),$$

where

$$\mathbf{u} = \begin{pmatrix} u_1 \\ u_2 \\ \vdots \\ u_n \end{pmatrix}, \quad \mathbf{u}' = \begin{pmatrix} u'_1 \\ u'_2 \\ \vdots \\ u'_n \end{pmatrix} \tag{6.3.6}$$

For each such instruction we introduce new states $p_1, \ldots, p_{n+1}, p'$ to T' and specify

$$\begin{aligned}
\theta'(q, \#) &= (p_1, \#, R) \\
\theta'(p_j, u_j) &= (p_{j+1}, u'_j, R) \quad (j = 1, 2, \ldots, n) \\
\theta'(p_{n+1}, \#) &= (p', \#, L) \\
\theta'(p_{n+1}, \Delta) &= (p', \#, L) \\
\theta'(p', u'_n) &= (q', u'_n, R).
\end{aligned} \tag{6.3.7}$$

Then corresponding to the change

$$\ldots q\mathbf{uv}\ldots \to \ldots \mathbf{u}'q'\mathbf{v}\ldots$$

in a T-computation we have a computation

$$\ldots q\#u_1\ldots u_n\#v_1\ldots v_n\#\ldots \to \ldots \#p_1u_1\ldots u_n\#v_1\ldots v_n\#\ldots$$
$$\overset{*}{\to}\ldots \#u_1'\ldots u_n'p_{n+1}\#v_1\ldots v_n\#\ldots \to \ldots \#u_1'\ldots u_{n-1}'p'u_n'\#v_1\ldots v_n\#\ldots$$
$$\to \ldots \#u_1'\ldots u_n'q'\#v_1\ldots v_n\#\ldots$$

in T'.

In the computation just described no use was made of the instruction

$$\theta'(p_{n+1}, \Delta) = (p', \#, L).$$

This comes into play only when in the T-computation the cursor reaches the right-hand edge and there is an instruction

$$\theta(q, \Delta) = (q', \mathbf{u}', R)$$

giving rise to a transition

$$\omega q \to \omega\mathbf{u}'q'$$

for some ω in M^*. We envisage T as printing Δ (the column vector consisting entirely of Δs) and then replacing it by $\mathbf{u}'$. The corresponding computation in T' is then based on (6.3.7), with $u_1 = u_2 = \cdots = u_n = \Delta$, and takes place as follows:

$$\ldots q\# \to \ldots \#p_1\Delta \overset{*}{\to} \ldots \#u_1'u_2'\ldots u_n'p_{n+1}\Delta$$
$$\to \ldots \#u_1'u_2'\ldots u_{n-1}'p'u_n'\# \to \ldots \#u_1'u_2'\ldots u_n'q'\#.$$

For an instruction

$$\theta(q, \mathbf{u}) = (q', \mathbf{u}', L)$$

of T the simulating subroutine in T' has to be a little more complicated. For each such instruction we introduce new state symbols $s_1, \ldots, s_{n+1}, s', s'', \bar{s}$. Then let

$$\theta'(q, \#) = (s_1, \#, R)$$
$$\theta'(s_j, u_j) = (s_{j+1}, u_j', R) \quad (j = 1, 2, \ldots, n)$$
$$\theta'(s_{n+1}, \#) = (s', \#, L)$$
$$\theta'(s_{n+1}, \Delta) = (s', \#, L)$$
$$\theta'(s', u_j') = (s', u_j', L) \quad (j = 1, 2, \ldots, n - 1)$$
$$\theta'(s', \#) = (s'', \#, L)$$
$$\theta'(s'', \zeta) = (s'', \zeta, L) \text{ for every tape symbol } \zeta \text{ except } \#$$
$$\theta'(s'', \#) = (\bar{s}, \#, R)$$
$$\theta'(\bar{s}, \zeta) = (q', \zeta, L) \text{ for every tape symbol } \zeta.$$

Then corresponding to the change

$$\ldots \mathbf{v}q\mathbf{u} \ldots \to \ldots q'\mathbf{v}\mathbf{u}' \ldots$$

in $\mathcal{T}$ we have a computation

$$\ldots \#v_1 \ldots v_n q \# u_1 \ldots u_n \# \ldots \to \#v_1 \ldots v_n \# s_1 u_1 \ldots u_n \# \ldots$$
$$\xrightarrow{*} \ldots \#v_1 \ldots v_n \# u_1' \ldots u_n' s_{n+1} \# \ldots \to \ldots \#v_1 \ldots v_n \# u_1' \ldots u_{n-1}' s' u_n' \# \ldots$$
$$\xrightarrow{*} \ldots \#v_1 \ldots v_n s' \# u_1' \ldots u_n' \# \ldots \to \ldots v_1 \ldots v_{n-1} s'' v_n \# u_1' \ldots u_n' \# \ldots$$
$$\xrightarrow{*} \ldots s'' \# v_1 \ldots v_n \# u_1' \ldots u_n' \# \ldots \to \ldots \# \bar{s} v_1 \ldots v_n \# u_1' \ldots u_n' \# \ldots$$
$$\to \ldots q' \# v_1 \ldots v_n \# u_1' \ldots u_n' \# \ldots .$$

Again the instruction $\theta'(s_{n+1}, \Delta) = (s', \#, L)$ comes into play when the cursor reaches the right-hand edge and we require to simulate a change

$$\ldots \mathbf{v}q\Delta \to \ldots q'\mathbf{v}\mathbf{u}'$$

by a computation

$$\ldots v_1 \ldots v_n q\Delta \xrightarrow{*} \ldots q' \# v_1 \ldots v_n \# u_1' \ldots u_n' \#.$$

From the way we have constructed $\mathcal{T}'$ it is clear that we can simulate every $\mathcal{T}$-computation by a $\mathcal{T}'$-computation. Thus $L(\mathcal{T}) \subseteq L(\mathcal{T}')$.

Conversely, suppose that

$$w = a_1 a_2 \ldots a_m \in L(\mathcal{T}'),$$

so that there is a computation beginning with $i'w$ and ending in a terminal state. (Recall that the terminal states of $\mathcal{T}'$ are just those of $\mathcal{T}$.) The computation cannot enter any state at all of $\mathcal{T}$ without going through the initial phase

$$i'w \xrightarrow{*} i \# a_1 \Delta^{n-1} \# a_2 \Delta^{n-1} \# \ldots \# a_m \Delta^{n-1} \#.$$

The $\mathcal{T}'$-computation will then have a number of 'excursions' during which states, labelled p and s, from $Q' \backslash Q$ make an appearance. Since the computation is successful these must subsequently disappear, and whenever this happens the effect of the excursion has been to simulate a step in a $\mathcal{T}$-computation. There is no way in which the $\mathcal{T}'$-computation can change the feature that precisely n tape symbols lie between successive occurrences of $\#$, and it is also the case that state symbols from Q will always appear in a position scanning $\#$. So a typical ID when T' is in state $q\ (\in Q)$ is

$$\#u_{11}\ldots u_{n1}\#\ldots\#u_{1,j-1}\ldots u_{n,j-1}q\#u_{1j}\ldots u_{nj}\#\ldots\#u_{1k}\ldots u_{nk},$$

and this can be reached if and only if there is a T-computation

$$i\hat{\mathbf{a}}_1\hat{\mathbf{a}}_2\ldots\hat{\mathbf{a}}_m\overset{*}{\to}\mathbf{u}_1\ldots\mathbf{u}_{j-1}q\mathbf{u}_j\ldots\mathbf{u}_k,$$

where

$$\mathbf{u}_i = \begin{pmatrix} u_{1i} \\ u_{2i} \\ \vdots \\ u_{ni} \end{pmatrix} \qquad (i = 1, 2, \ldots, k).$$

In particular, if q is terminal, this is a successful T-computation, and so $a_1 a_2 \ldots a_m \in L(T)$. Thus $L(T') \subseteq L(T)$. ∎

6.4. Non-deterministic Turing machines

In the study of finite state automata we gained no advantage by allowing non-determinacy, but in pushdown automata, where we allowed non-determinacy from the outset, the assumption of determinacy proved to be a real restriction. (See Exercise 5.10.) In this section we shall see that non-determinacy brings no advantage to Turing machines, in the sense that every language recognized by a non-deterministic Turing machine (NDTM) is recognized also by a (deterministic) Turing machine (TM).

It is fairly clear what we must mean by an NDTM T. We simply modify Definition 6.1.1 so that θ becomes a partial map from $Q \times M$ into $\mathcal{P}(Q \times M \times \{L, R\})$, the set of all subsets of the finite set $Q \times M \times \{L, R\}$. The description of a computation in T then begins with phrases like 'if $\theta(q, \alpha)$ contains (q', α', R)', and we can imagine all possible computations as 'fanning out' into a tree with a branch point wherever the relevant instruction $\theta(q, \alpha)$ presents the machine with a choice. The pattern of choices can be represented in a tree diagram, as exemplified by Figure 6.4.1, and any particular path through the tree will be thought of as representing 'a computation' in T. If among the computations beginning with iw there is a computation $iw\overset{*}{\to}u_1 t u_2$ in which $t \in T$, then w is said to be *recognized by* T.

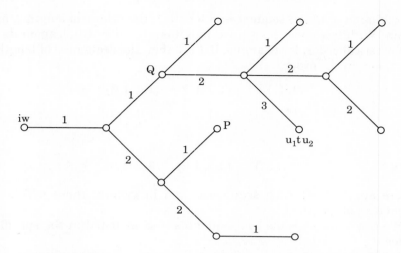

Figure 6.4.1. A schema of a possible set of computations of an NDTM

Our purpose is to prove

Theorem 6.4.2. *Let A be a finite alphabet and let $L \subseteq A^*$. If L is recognized by an NDTM T then there exists a TM T' recognizing L.*

Proof. Let

$$T = (Q, A, M, \Delta, \theta, i, T)$$

be an NDTM, with $L(T) = L$. The stategy of the proof is to construct a (deterministic) TM T' which for a given initial ID iw carries out in succession all possible computations of T beginning with iw. If there is a successful computation in T then one of the 'subcomputations' in T' will succeed. Otherwise none of the subcomputations of T' will succeed and the whole T'-computation fails.

First we need a way of describing the various possible computations in the NDTM T. For a given q in Q and α in M we can in an arbitrary way label the members of the set $\theta(q, \alpha)$:

$$\theta(q, \alpha) = \{\theta_1(q, \alpha), \theta_2(q, \alpha), \dots \theta_s(q, \alpha)\}. \tag{6.4.3}$$

The size of this set will normally vary with q and α, but we may suppose that there is a number d such that $s \leq d$ in all cases. In Figure 6.4.1 we have made the listing implied by (6.4.3) correspond to a labelling of the edges. Thus the route from iw to $u_1 t u_2$ can be labelled $(1, 1, 2, 3)$. In general, a particular computation can be labelled by a finite sequence of integers drawn from the set $\{1, 2, \dots, d\}$.

Accordingly, let us consider the set S of all sequences (of all possible lengths) of integers from $\{1, 2, \dots, d\}$. We can suppose that S is listed by taking first all the sequences of length 1, then those of length 2, and so on. Within a given length we may take them in what amounts to alphabetical order: for example, if $d = 3$ then the sequences of length 4 are taken in the order

$$(1, 1, 1, 1), \quad (1, 1, 1, 2), \quad (1, 1, 1, 3),$$

$$(1, 1, 2, 1), \quad (1, 1, 2, 2), \quad (1, 1, 2, 3),$$

$$\cdots$$

$$(3, 3, 3, 1), \quad (3, 3, 3, 2), \quad (3, 3, 3, 3).$$

There are $3^4 = 81$ such sequences, and in general there will be d^l sequences of length l.

We can thus envisage the sequences in S as listed in the specified order

$$S_1, \ S_2, \ S_3, \ldots$$

where

$$S_1 = (1), \ldots, S_d = (d), \ S_{d+1} = (1, 1) \text{ etc.},$$

and we can think of each of the members of S as labelling a particular sequence of choices in a computation of T.

Of course it will normally be the case that not every sequence corresponds to an actual computation. If we return once more to Figure 6.4.1 (where $d = 3$) we see that the sequence $(1, 2, 1, 2)$ does not correspond to any computation since there is no exit from the point labelled P, and the sequence $(1, 1, 3)$ likewise fails to correspond to any computation because there are only two choices at the point labelled Q. This will not turn out to be a serious difficulty.

The first task our TM T' must perform is to generate the members of S. We describe a subroutine into which T' will go at regular intervals.

6.4.4. *Subroutine to change S_j to S_{j+1}.*

Suppose that we begin with an ID $q \lambda S_j$, where we think of λ as a left marker symbol, and where in writing S_j we leave out all commas and brackets. Then let

$\theta(q, \lambda) = (q, \lambda, R)$	(Move right
$\theta(q, s) = (q, s, R) \ (s = 1, \ldots, d)$	until Δ is
$\theta(q, \Delta) = (q_1, \Delta, L)$	encountered.)
$\theta(q_1, d) = (q_1, d, L)$	(Reading from
$\theta(q_1, s) = (q_2, s + 1, L) \ (s = 1, \ldots, d - 1)$	the right, replace
$\theta(q_2, d) = (q_2, 1, R)$	the first $s \neq d$
$\theta(q_2, \Delta) = (q_3, \Delta, L)$	by $s + 1$; then move right
$\theta(q_3, s) = (q_3, s, L) \ (s = 1, \ldots, d)$	replacing each d

$\theta(q_3, \lambda) = (q_4, \lambda, R)$

$\theta(q_4, s) = (q', s, L)\ (s = 1, \ldots, d)$

(by 1; then return the cursor to the left.)

$\theta(q_1, \lambda) = (q_4, \lambda, R)$

$\theta(q_5, s) = (q_5, 1, R)$

$\theta(q_5, \Delta) = (q_6, 1, L)$

(If no $s \neq d$ is encountered, move right, printing 1s.)

$\theta(q_6, 1) = (q_6, 1, L)$

$\theta(q_6, \lambda) = (q_7, \lambda, R)$

$\theta(q_7, 1) = (q', 1, L)$

(Return the cursor to the left.)

To see the operation of this subroutine, consider for example $d = 3$ and $S_j = 123$. Then

$$q\lambda 123 \overset{*}{\to} \lambda 123 q\Delta \to \lambda 12 q_1 3 \to \lambda 1 q_1 23$$
$$\to \lambda 13 q_2 3 \to \lambda 131 q_2 \Delta \to \lambda 13 q_3 1 \overset{*}{\to} q_3 \lambda 131$$
$$\to \lambda q_4 131 \to q' \lambda 131.$$

The latter part of the subroutine comes into play if $S_j = dd \ldots d$. For example, with $d = 3$,

$$q\lambda 333 \overset{*}{\to} \lambda 333 q\Delta \to \lambda 33 q_1 3 \overset{*}{\to} q_1 \lambda 333$$
$$\to \lambda q_5 333 \overset{*}{\to} \lambda 111 q_5 \Delta \to \lambda 11 q_6 11$$
$$\to q_6 \lambda 1111 \to \lambda q_7 1111 \to q' \lambda 1111,$$

just as required.

We are envisaging that a given sequence in S will function as a 'control' sequence, in effect telling the TM T' which choice it should make at each stage among the possibly many choices offered by the NDTM T. To demonstrate how this is done, we consider an ID

$$ia_1 a_2 \ldots a_n \# S$$

where S is some sequence in S and $\#$ is a marker symbol. Introducing new state and tape symbols as necessary, we define

$\theta'(q, \alpha) = (p_1, \bar{\alpha}, R)$

$\theta'(p_1, \zeta) = (p_1, \zeta, R)$ for all $\zeta \neq \#$

$\theta'(p_1, \#) = (p_2, \#, R)$

$\theta'(p_2, \bar{k}) = (p_2, \bar{k}, R)\ (k = 1, \ldots, d)$

$\theta'(p_2, k) = (u_k, \bar{k}, L)\ (k = 1, \ldots, d)$

(Mark α and then move right for instructions from the control sequence, marking the instruction being used.)

$\theta'(u_k, \bar{l}) = (u_k, \bar{l}, L)\ (k, l \in \{1, \ldots, d\})$

$\theta'(u_k, \zeta) = (u_k, \zeta, L)$ for all $\zeta \in M \cup \{\#\}$

$\theta'(u_k, \bar{\alpha}) = \theta_k(q, \alpha)\ (k = 1, \ldots, s)$

(Move left and make the appropriate change.)

So if, for example, we take $S = 1223$, then after two stages in the T'-computation we might have reached an ID $\zeta_1 q \alpha \zeta_2 \# \bar{1} \bar{2} 23$, and the computation now proceeds

$$\zeta_1 q \alpha \zeta_2 \# \bar{1} \bar{2} 23 \overset{*}{\rightarrow} \zeta_1 \bar{\alpha} \zeta_2 \bar{1} \bar{2} p_2 23$$
$$\rightarrow \zeta_1 \bar{\alpha} \zeta_2 \# \bar{1} u_2 \bar{2} \bar{2} 3 \overset{*}{\rightarrow} \zeta_1 u_2 \bar{\alpha} \zeta_2 \bar{1} \bar{2} \bar{2} 3.$$

Then if $\theta_2(q, \alpha) = (q', \gamma, R)$, say, the next step in the T'-computation gives the ID

$$\zeta_1 \gamma q' \zeta_2 \bar{1} \bar{2} \bar{2} 3.$$

The computation ceases when either u_k scans an $\bar{\alpha}$ in a case where $k > s = |\theta(q, \alpha)|$ or when the sequence S is exhausted. (In the latter case p_2, moving right, encounters a blank square and receives no instruction.)

We are now ready to tack together the various bits and pieces we have been examining and to produce a proper (if somewhat informal) description of the TM T' we are seeking. Using the subroutines described in Section 6.2 we can certainly duplicate a given word $a_1 a_2 \ldots a_n$ in A^* and put a marker λ in between. So we may suppose that for every $a_1 a_2 \ldots a_n$ in A^* the deterministic T'-computation begins

$$i a_1 a_2 \ldots a_n \overset{*}{\rightarrow} q a_1 a_2 \ldots a_n \lambda a_1 a_2 \ldots a_n.$$

The next step is to print $\#1\rho$ at the end. The symbol ρ here is simply a right marker, so in effect we are printing $\#S_1$. (Recall that in the listing $S_1, S_2, \ldots$ of S the sequence S_1 was (1).) We then use 1 as a control to the right of λ and carry out the (deterministic) T-computation corresponding to the sequence S_1. We end with

$$a_1 a_2 \ldots a_n \lambda \zeta \# \bar{S}_1,$$

where ζ is an ID of T.

The next stage is to delete everything between λ and $\#$, change $\bar{S}_1$ to S_2, duplicate $a_1 a_2 \ldots a_n$ between λ and $\#$, and adjust the position and state of the cursor. We obtain

$$a_1 \ldots a_n \lambda i a_1 \ldots a_n \# S_2,$$

and are ready to begin again, this time using S_2 as the control sequence.

This process continues. Of course, as we have seen, it is perfectly possible that the phase of the T'-computation beginning at

$$a_1 \ldots a_n \lambda i a_1 \ldots a_n \# S_j$$

fails at some stage because the sequence S_j does not correspond to a legitimate sequence of choices within the T-computation. To cover this eventuality we simply insert a 'premature escape' facility into the control sequence subroutine. If $k > |\theta(q, \alpha)|$ then we can specify $\theta'(u_k, \bar{\alpha})$ in such a way that we move straight into the routine leading to

$$a_1 \ldots a_n \lambda i a_1 \ldots a_n \# S_{j+1}.$$

If $a_1 \ldots a_n \in L(T)$, so that there is a computation

$$i a_1 \ldots a_n \xrightarrow{*} \zeta_1 t \zeta_2$$

in T, with t a terminal state, then for some sequence S_m the computation in T' will reach

$$a_1 \ldots a_n \lambda \zeta_1 t \zeta_2 \# \bar{S}_m.$$

If we suppose that T' has the same terminal states as T we thus conclude that $a_1 \ldots a_n \in L(T')$. Thus $L(T) \subseteq L(T')$.

Conversely, if $a_1 \ldots a_n \notin L(T)$ then no choice of the sequence S_m will lead the T'-computation into a terminal state. So $a_1 \ldots a_n \notin L(T')$. We have now shown that $L(T) = L(T')$, as required. ∎

6.5. Turing machines and context-free languages

Once we had the text-processing routines in Section 6.2 there was no great difficulty in simulating the action of a *deterministic* PDA by means of a TM. The difficulty was in building in the essential non-determinacy of PDAs. However, in view of Theorem 6.4.2 it is now enough to simulate a PDA using an NDTM, and this is relatively easy. By Theorem 5.4.1 a language is recognized by a PDA if and only if it is context-free. So we can state the main result of this section as follows:

Theorem 6.5.1. *Every context-free language is recognized by a Turing machine.*

Proof. Let L be a context-free language. Then $L = L(\mathcal{P})$, where

$$\mathcal{P} = (Q, A, M, \mu, \rho, i, \zeta, T)$$

is a PDA, recognizing by terminal state. (See Sections 5.2, 5.3.) We create an NDTM

$$T = (Q', A, M', \Delta, \theta, i, T),$$

where $Q \subseteq Q'$, $A \cup M \subseteq M'$. The IDs of $\mathcal{P}$ are triples (q, τ, w), with $\tau \in M^+$, $w \in A^*$, and the initial ID, if we are testing a word w for membership of $L(\mathcal{P})$, is (i, ζ, w). If we are testing w for membership of $L(T)$ we want to begin with iw, and we suppose that T has an initial (deterministic) subroutine (see Section 6.2) which changes this to $iw\#\zeta$, where $\#$ as usual is a marker symbol not in M. In general we shall want to make the ID (q, τ, w) of $\mathcal{P}$ correspond to an ID $qw\#\tau$ of T.

We now indicate how T can be constructed so that every transition in $\mathcal{P}$ can be simulated by a computation in T.

There are three cases to consider, corresponding to the three types of transition described in (5.1.2), (5.1.3) and (5.1.4). First, suppose that $\tau = \beta\tau'$ ($\beta \in M$, $\tau' \in M^*$) and that $\mu(q, \beta)$ contains (q', ξ), where $\xi = \alpha_1 \ldots \alpha_m \in M^*$, so that there is a transition

$$(q, \tau, w) \rightarrow (q', \xi\tau', w)$$

in $\mathcal{P}$. The corresponding ID of T is $qw\#\tau$, and we suppose that

$$\theta(q, a) \text{ contains } (q_1, \bar{a}, R) \quad \text{(for all } a \in A)$$
$$\theta(q_1, b) \text{ contains } (q_1, b, R) \quad \text{(for all } b \in A)$$
$$\theta(q_1, \#) \text{ contains } (q_1, \#, R).$$

This marks the original place (by changing a to $\bar{a}$) and then moves right to meet β. At this point a standard TM subroutine deletes β and replaces it by $\xi = \alpha_1 \ldots \alpha_m$ before going into a state q_β (depending on q and β). This then moves left until it encounters the marked symbol $\bar{a}$. It then changes to state q' (in accordance with the specification of $\mu(q, \beta)$), while replacing $\bar{a}$ by a.

The second case is where $\tau = \beta\tau'$ (as before), $w = aw'$ ($a \in A$, $w \in A^*$) and $\rho(q, \beta, a)$ contains (q', ξ). Again we start with an ID $qw\#\tau$ and we suppose that

$$\theta(q, a) \text{ contains } (q_1, \bar{a}, R)$$
$$\theta(q_1, b) \text{ contains } (q_1, b, R) \quad \text{(for all } b \text{ in } A)$$
$$\theta(q_1, \#) \text{ contains } (q_1, \#, R).$$

This gives rise to the ID $\bar{a}w'\#q_1\beta\tau'$, and as in the previous case we can insert standard subroutines to change β into ξ, to go into a state q_β (depending on both q and β) and to move left until $\bar{a}$ is encountered. This gives rise to the ID

$$q_\beta\bar{a}w'\#\xi\tau',$$

from which point it is easy to change into the state q' (as determined by $\rho(q, \beta, a)$), delete $\bar{a}$ and move right. Precisely, the instruction

$$\theta(q_\beta, \bar{a}) = (q', \Delta, R)$$

gives the ID $\Delta q'w'\#\xi\tau'$. There is no harm in retaining the leftmost Δ, but if desired there is a standard routine to remove it and to shunt the whole ID one space left, obtaining $q'w'\#\xi\tau'$.

The final case, where $w = 1$ and where $\rho(q, \beta, 1)$ contains (q', ξ), is handled in essentially the same way , and the transition

$$(q, \beta\tau', 1) \to (q', \xi\tau', 1)$$

in $\mathcal{P}$ is simulated by a computation

$$q\#\beta\tau' \xrightarrow{*} q'\#\xi\tau'$$

in $\mathcal{T}$. The details are omitted.

From the construction of $\mathcal{T}$ it is evident that $L(\mathcal{P}) \subseteq L(\mathcal{T})$. The reverse inclusion is also true. We may certainly assume that the extra states involved in simulating each of the instructions of $\mathcal{P}$ are specific to that instruction and are outside Q. A computation in $\mathcal{T}$ begins in a state i belonging to Q and (if successful) ends in a terminal state t belonging to Q. Every 'excursion' outside Q must eventually return to Q, and when this happens the TM $\mathcal{T}$ has simulated a step in a $\mathcal{P}$-computation. The TM $\mathcal{T}$, though non-deterministic, is constructed so that the only successful computations are those that simulate successful $\mathcal{P}$-computations. Thus $L(\mathcal{T}) \subseteq L(\mathcal{P})$. ∎

6.6. Turing machines and phrase structure grammars

We have seen that the languages recognized by finite state automata are exactly those generated by regular grammars. Equally, the languages recognized by pushdown automata coincide with those generated by context-free grammars. When, in Chapter 4, we first encountered the idea of a grammar the most general notion of all was that of a phrase structure grammar. We can now show that the languages generated by such grammars are exactly those recognized by Turing machines.

However, before giving a formal statement of this result we pause to fill in a gap in our battery of text-editing subroutines.

6.6.1. *Find the first occurrence, reading from the left, of a given sub-word.*

In a way this subroutine properly belongs in Section 6.2, but we have deferred it until now because it is a little inconvenient to carry it out using a deterministic Turing machine. This is because of the possibility of 'false starts' when we are looking for a subword with repeated letters.

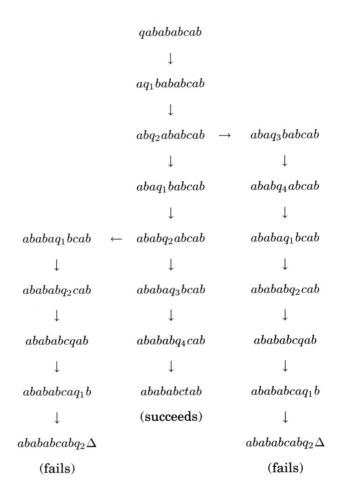

Figure 6.6.2.
Searching for $ababc$

For example, if we look for the first occurrence of $ababc$ in $abababcab$ then in scanning from the left we might appear to be quite close to success when we reach the second b, when in fact it is a second attempt, beginning at the second occurrence of a, that is destined to be successful.

Let us envisage a computation as starting from an ID qw, where w may or may not contain one or more occurrences of the word $\alpha_1\alpha_2\ldots\alpha_k$. Then let

$\theta(q, x) = (q, x, R)\ \big(x \in M\backslash\{\alpha_1\}\big)$ (Move right until
$\theta(q, \alpha_1) = (q_1, \alpha_1, R)$ α_1 is found.)

For $i = 1, 2, \ldots, k - 2$ let

$\theta(q_i, \alpha_{i+1}) = (q_{i+1}, \alpha_{i+1}, R)$ if $\alpha_{i+1} \neq \alpha_1$ (Test the letters following α_1.)

$\theta(q_i, \alpha_{i+1}) = \{(q_{i+1}, \alpha_{i+1}, R), (q_1, \alpha_1, R)\}$ if $\alpha_{i+1} = \alpha_1$
$\theta(q_i, \alpha_1) = (q_1, \alpha_1, R)$ if $\alpha_{i+1} \neq \alpha_1$

$\theta(q_i, x) = (q, x, R)$ if $x \neq \alpha_1, \alpha_{i+1}$ (Resume search.)
$\theta(q_{k-1}, \alpha_k) = (t, \alpha_k, R)$ (Enter terminal state having found $\alpha_1 \alpha_2 \ldots \alpha_k$.)

$\theta(q_{k-1}, \alpha_1) = (q_1, \alpha_1, R)$ if $\alpha_1 \neq \alpha_k$
$\theta(q_{k-1}, x) = (q, x, R)$ if $x \neq \alpha_1, \alpha_k$ (Resume search.)

For the example already mentioned, that of finding the leftmost occurrence of $ababc$ in $abababcab$, the computation proceeds as in Figure 6.6.2.

It is easy to modify the routine so as to continue searching for second and subsequent occurrences of $\alpha_1 \alpha_2 \ldots \alpha_k$. The non-determinacy of the machine, which enables several tests to take place simultaneously, is particularly useful if occurrences overlap, as for example if we look for occurrences of aba in $abababa$.

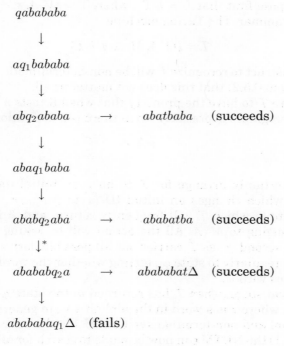

Figure 6.6.3.

To achieve the desired effect we simply modify the last three instructions of the subroutine to

$$\theta(q_{k-1}, \alpha_k) = \begin{cases} (t, \alpha_k, R) & \text{if } \alpha_k \neq \alpha_1, \\ \{(t, \alpha_k, R), (q_1, \alpha_1, R)\} & \text{if } \alpha_k = \alpha_1; \end{cases}$$

$$\theta(q_{k-1}, \alpha_1) = (q_1, \alpha_1, R) \text{ if } \alpha_k \neq \alpha_1,$$

$$\theta(q_{k-1}, x) = (q, x, R) \text{ if } x \neq \alpha_k, \alpha_1.$$

The operation of this subroutine in finding occurrences of aba in $abababa$ then proceeds as in Figure 6.6.3. There are three successful endpoints to the computation and in each case the occurrence of aba that has been found lies immediately to the left of t.

Once an occurrence of $\alpha_1 \alpha_2 \ldots \alpha_k$ has been found, we can delete it, or duplicate it, or replace it by another word, using routines already developed.

We are now ready to prove

Theorem 6.6.4. *Let A be a finite alphabet and let L be a subset of A^*. Then L is recognizable by a Turing machine if and only if it is generated by a phrase structure grammar.*

Proof. Suppose first that $L = L(\Gamma)$, where $\Gamma = (V, A, \pi, \sigma)$ is a phrase structure grammar. The Turing machine

$$\mathcal{T} = (Q, A, M, \Delta, \theta, i, T)$$

we shall construct to recognize L will be non-deterministic, but we have seen (Theorem 6.5.2) that this does not matter at all.

We require $\mathcal{T}$ to have the property that when it tests a typical word w in A^* it recognizes w precisely when there is a derivation

$$\sigma \overset{*}{\Rightarrow} w$$

of w in Γ.

We can certainly arrange for $\mathcal{T}$ to have an initial (deterministic) subroutine which changes an initial ID iw to $\#wq_0\#\sigma$. Here q_0 is a 'central' state to which $\mathcal{T}$ will frequently return, and $\#$ is a 'marker' symbol belonging to $M \backslash A$. All the action will be taking place to the right of the second $\#$, as $\mathcal{T}$ carries out all possible derivations from σ and returns regularly to state q_0, testing whether the word on the right of $\#$ coincides with w.

At a typical stage, when $\mathcal{T}$ has returned to the state q_0, the ID will be $\#wq_0\#\tau$, where τ is a word in the alphabet V, in general containing both terminal and non-terminal symbols. Using the subroutine developed in (6.6.1) the NDTM can now be made to search for all occurrences as subwords of τ of all words appearing on the left of arrows in the productions of Γ. Whenever such a subword u is found (say with $\tau = \zeta_1 u \zeta_2$, where $u \to v$ is a production of Γ) then $\mathcal{T}$ will change u to v, move the cursor back and enter a 'duplication check' state q_d, giving

$$\#w q_d \# \zeta_1 v \zeta_2.$$

At this point $\mathcal{T}$ will check whether $\zeta_1 v \zeta_2$ is identical to w, using the routine (6.2.7). If so then it will enter a terminal state and the word w is accepted. If not, then the machine returns to the ID

$$\#w q_0 \# \zeta_1 v \zeta_2$$

and the generation of words in $L(\Gamma)$ continues.

This is a fairly complex and highly non-deterministic process, and even a fairly simple example is hard to describe in detail. It may, however, be helpful to illustrate the process in the case of a grammar in which $A = \{a, b\}$ and with productions

$$\sigma \to 1, \ \sigma \to a \sigma b.$$

(Here $L = \{a^n b^n : n \geq 0\}$.) The demonstration that $a^2 b^2 \in L(\mathcal{T})$ proceeds as shown in Figure 6.6.5.

$$i a^2 b^2$$

$$\downarrow^*$$

$$\#a^2 b^2 q_0 \# \sigma \quad \overset{*}{\to} \quad \#a^2 b^2 q_d \# 1 \quad \overset{*}{\to} \quad \#a^2 b^2 q_0 \# 1$$

$$\downarrow^* \qquad\qquad\qquad\qquad\qquad\qquad \text{(dead end)}$$

$$\#a^2 b^2 q_d \# a \sigma b$$

$$\downarrow^*$$

$$\#a^2 b^2 q_0 \# a \sigma b \quad \overset{*}{\to} \quad \#a^2 b^2 q_d \# ab \quad \overset{*}{\to} \quad a^2 b^2 q_0 \# ab$$

$$\downarrow^* \qquad\qquad\qquad\qquad\qquad\qquad \text{(dead end)}$$

$$\#a^2 b^2 q_d \# a^2 \sigma b^2 \quad \overset{*}{\to} \quad \#a^2 b^2 q_d \# a^2 b^2 \quad \overset{*}{\to} \quad \#a^2 b^2 t a^2 b^2$$

$$\downarrow^* \qquad\qquad\qquad\qquad\qquad\qquad \text{(succeeds)}$$

$$\#a^2 b^2 q_d \# a^3 \sigma b^3$$

(continues)

Figure 6.6.5.

On the right of the second $\#$ the NDTM $\mathcal{T}$ is generating all possible words derived from σ in the grammar Γ and is checking each against the word w held in store between the $\#$'s. It is clear that $w \in L(\mathcal{T})$ if and only if $w \in L(\Gamma)$.

Suppose now that $L = L(\mathcal{T})$, where

$$\mathcal{T} = (Q, A, M, \Delta, \theta, i, T)$$

is a (deterministic) TM. We must find a phrase structure grammar Γ such that $L = L(\Gamma)$. Our stategy will be to arrange for Γ to generate each word in A^*, suitably 'polluted' for the moment with non-terminal symbols, to test the word by simulating the action of $\mathcal{T}$, and finally to purge it of non-terminal symbols if and only if it is recognized by $\mathcal{T}$. There is a difficulty in carrying out this strategy, in that the action of $\mathcal{T}$ on a word w in A^* will normally mutilate w beyond recognition. So it is necessary to hold unchanged a second copy of w while testing the first copy. If the test on the first copy is successful then the grammar deletes everything except the untouched second copy.

The most convenient way to produce the required second 'control' copy using a phrase structure grammar is to arrange for each $w = a_1 a_2 \ldots a_m$ to appear in a duplicated form as

$$[a_1 a_1][a_2 a_2] \ldots [a_m a_m],$$

where we may regard $[$ and $]$ as non-terminal symbols. More precisely, we let

$$\Gamma = (V, A, \pi, \sigma),$$

where

$$V = M \cup Q \cup \{[,], \epsilon, \sigma, \lambda, \mu\}.$$

The productions of Γ fall into three groups as follows.

(Γ1) $\sigma \to i\lambda$, $\lambda \to [aa]\lambda$ $(a \in A)$, $\lambda \to \mu$, $\mu \to [\Delta\epsilon]\mu$, $\mu \to 1$.

(Γ2) Whenever $\theta(q, \beta) = (q', \gamma, R)$ in $\mathcal{T}$, there is a production

$$q[\beta a] \to [\gamma a]q'$$

in Γ for every a in $A \cup \{\epsilon\}$. Whenever $\theta(q, \beta) = (q', \gamma, L)$ in $\mathcal{T}$, there is a production

$$[\tau b]q[\beta a] \to q'[\tau b][\gamma a]$$

for every a, b in $A \cup \{\epsilon\}$ and for every τ in M.

(Γ3) For each terminal state t of $\mathcal{T}$, there are productions

$$t[\alpha\beta] \to t[\alpha\beta]t, \quad [\alpha\beta]t \to t[\alpha\beta]t$$

for each α, β in M. Also, there is a production

$$t[\zeta a] \to a$$

for each ζ in M and each a in A and a production

$$t[\zeta \epsilon] \to 1$$

for each ζ in M.

These three groups correspond to the three phases we indicated when announcing the strategy. The productions in Group ($\Gamma 1$) set up the word, those in Group ($\Gamma 2$) simulate $\mathcal{T}$, while those in Group ($\Gamma 3$) carry out the clearing up operation.

Consider now a word $w = a_1 a_2 \ldots a_n$ in A^* recognized by $\mathcal{T}$, and suppose that in the successful $\mathcal{T}$-computation beginning with $i a_1 a_2 \ldots a_n$ the computation uses k (≥ 0) squares of the tape beyond the right-hand end of the word $a_1 a_2 \ldots a_n$. In the grammar Γ it is clear that the productions listed in ($\Gamma 1$) give

$$\sigma \overset{*}{\Rightarrow} i[a_1 a_1][a_2 a_2] \ldots [a_n a_n][\Delta \epsilon]^k.$$

From a word of this type there is no escape except by means of the productions ($\Gamma 2$).

We now show

Lemma 6.6.6. *If there is a $\mathcal{T}$- computation*

$$i a_1 a_2 \ldots a_n \overset{*}{\to} \zeta_1 \ldots \zeta_{r-1} q \zeta_r \ldots \zeta_m \tag{6.6.7}$$

with $\zeta_1, \ldots, \zeta_m \in M$, then there is a derivation

$$
\begin{aligned}
i[a_1 a_1][a_2 a_2] &\ldots [a_n a_n][\Delta \epsilon]^k \\
&\overset{*}{\Rightarrow} [\zeta_1 a_1] \ldots [\zeta_{r-1} a_{r-1}] q [\zeta_r a_r] \ldots [\zeta_{n+k} a_{n+k}]
\end{aligned}
\tag{6.6.8}
$$

in Γ, where

$$a_{n+1} = a_{n+2} = \cdots = a_{n+k} = \epsilon, \quad \zeta_{m+1} = \zeta_{m+2} = \cdots \zeta_{m+k} = \Delta.$$

Proof. The proof is by induction on the number of steps in the computation (6.6.7), the result being clear if there are no steps at all.

Suppose inductively that we have a $\mathcal{T}$-computation given by (6.6.7), and that there is a corresponding Γ-derivation given by (6.6.8). Consider the next step of the $\mathcal{T}$-computation. If $\theta(q, \zeta_r) = (q', \zeta_r', R)$ then

$$\zeta_1 \ldots \zeta_{r-1} q \zeta_r \ldots \zeta_m \to \zeta_1 \ldots \zeta_{r-1} \zeta_r' q' \zeta_{r+1} \ldots \zeta_m$$

if $r < m$, and

$$\zeta_1 \ldots \zeta_{r-1} q \zeta_r \ldots \zeta_m \rightarrow \zeta_1 \ldots \zeta_{m-1} \zeta'_m q' \Delta$$

if $r = m$. From the definition of the productions in (Γ2) we see that there is a corresponding step

$$[\zeta_1 a_1] \ldots [\zeta_{r-1} a_{r-1}] q [\zeta_r a_r] \ldots [\zeta_{n+k} a_{n+k}]$$
$$\Rightarrow [\zeta_1 a_1] \ldots [\zeta_{r-1} a_{r-1}] [\zeta'_r a_r] q' [\zeta_{r+1} a_{r+1}] \ldots [\zeta_{n+k} a_{n+k}]$$

(valid in both cases) in the Γ-derivation. Equally, if $\theta(q, \zeta_r) = (q', \zeta'_r, L)$ then the next step of the $\mathcal{T}$-computation gives the ID

$$\zeta_1 \ldots \zeta_{r-2} q' \zeta_{r-1} \zeta'_r \zeta_{r+1} \ldots \zeta_m$$

and there is a corresponding step in the Γ-derivation leading to

$$[\zeta_1 a_1] \ldots [\zeta_{r-2} a_{r-2}] q' [\zeta_{r-1} a_{r-1}] [\zeta'_r a_r] [\zeta_{r+1} a_{r+1}] \ldots [\zeta_{n+k} a_{n+k}].$$

Thus the Lemma is proved.

We are assuming that $w = a_1 a_2 \ldots a_n \in L(\mathcal{T})$, and so there is a computation

$$i a_1 a_2 \ldots a_n \overset{*}{\rightarrow} \gamma_1 \ldots \gamma_{r-1} t \gamma_r \ldots \gamma_m,$$

where t is a terminal state. By Lemma 6.6.6 there is a derivation

$$\sigma \overset{*}{\Rightarrow} i [a_1 a_1] [a_2 a_2] \ldots [a_n a_n] [\Delta \epsilon]^k$$
$$\overset{*}{\Rightarrow} [\gamma_1 a_1] \ldots [\gamma_{r-1} a_{r-1}] t [\gamma_r a_r] \ldots [\gamma_{n+k} a_{n+k}].$$

in Γ. At this point the productions (Γ3) come into play. The first stage is to spread t through the whole ID, giving

$$t [\gamma_1 a_1] t \ldots t [\gamma_{r-1} a_{r-1}] t [\gamma_r a_r] t \ldots t [\gamma_{n+k} a_{n+k}].$$

Then finally we make use of the productions $t[\zeta a] \rightarrow a$ and $t[\zeta \epsilon] \rightarrow 1$, obtaining the word

$$a_1 a_2 \ldots a_n.$$

We have now shown that $L(\mathcal{T}) \subseteq L(\Gamma)$. To show the reverse inclusion, suppose that $w \in L(\Gamma)$. The derivation of w can only begin with $\sigma \rightarrow i\lambda$, which introduces a symbol from Q. Now a symbol from Q can be removed only by means of the productions (Γ3), which apply only to terminal symbols, and so the middle phase of the derivation must in effect simulate a successful $\mathcal{T}$-computation. If at the beginning of this phase the ID is

$$i [a_1 a_1] [a_2 a_2] \ldots [a_n a_n] [\Delta \epsilon]^k,$$

then the only word in the terminal alphabet that can be reached by a Γ-derivation is $a_1 a_2 \ldots a_n$, for the second symbols inside the brackets [,] remain constant. So in fact the derivation must take the form

$$\sigma \overset{*}{\Rightarrow} i[a_1 a_1][a_2 a_2] \ldots [a_n a_n][\Delta \epsilon]^k$$
$$\overset{*}{\Rightarrow} [\delta_1 a_1] \ldots [\delta_{r-1} a_{r-1}] t[\delta_r a_r] \ldots [\delta_{n+k} a_{n+k}]$$
$$(\text{with } a_{n+1} = \cdots a_{n+k} = 1)$$
$$\overset{*}{\Rightarrow} a_1 a_2 \ldots a_n,$$

and the middle phase simulates a successful $\mathcal{T}$-computation

$$i a_1 a_2 \ldots a_n \overset{*}{\rightarrow} \delta_1 \ldots \delta_{r-1} t \delta_r \ldots \delta_{n+k},$$

in which $w = a_1 a_2 \ldots a_n$. Thus $w \in L(\mathcal{T})$.   ∎

6.7. The Chomsky hierarchy

We have examined two ways of specifying types of subsets of A^*. First, we can specify a type of subset in terms of its recognizability by this or that kind of machine. Secondly, we can specify it in terms of its generation by this or that kind of grammar. We have now established some correspondences between the two approaches:

Regular grammars   :   Finite state automata
Context-free grammars   :   Pushdown automata
Phrase structure grammars   :   Turing machines

The languages recognized by Turing machines are called *recursively enumerable*. If we denote the classes of rational languages (in A^*), context-free languages and recursively enumerable languages respectively by **Rat**, **CF** and **RE**, then we can summarize our present knowledge by the containments

$$\mathcal{F}(A^*) \subset \mathbf{Rat} \subset \mathbf{CF} \subset \mathbf{RE} \subseteq \mathcal{P}(A^*).$$

The hierarchy of language types represented by this formula is part of what is often called the *Chomsky hierarchy*. (Languages generated by so-called *context-sensitive* grammars also form part of this hierarchy, but these are less central to our main message and have not been discussed.)

We have seen by examples that the containments **Rat** $\subset$ **CF** and **CF** $\subset$ **RE** are proper, and it is of course pertinent to ask whether the containment **RE** $\subseteq \mathcal{P}(A^*)$ is also proper. To put the question another way, we ask whether there exist subsets of A^* which are *not* recognizable by a Turing machine. This question is answered in the next section.

Our treatment of Turing machines (it is only fair to mention) has skated over one of the most significant questions, the so-called Halting Problem. If we start a TM T on and ID iw (with $w \in A^*$) it may reach a terminal state (in which case $w \in L(T)$) or it may reach a dead end in some non-terminal state (in which case $w \notin L(T)$.) But the computation may alternatively go on and on, and in such a way that we will not know whether it will ever stop. In the introduction to Chapter 2 we promised that the treatment of Turing machines would be 'modest', and so it has turned out. So for questions concerning the Halting Problem and related ideas the reader is referred to a more specialized book, such as Davis (1958) or Hopcroft and Ullman (1979).

6.8 Sets that are not TM-recognizable

The Pumping Lemmas (Theorems 2.2.7 and 5.5.1) were efficacious in demonstrating the existence first of non-rational and then of non-context-free languages. While there is no corresponding result for Turing machines, it is possible to show that there are languages that cannot be recognized by a TM, that is to say, languages that are not recursively enumerable. This is an important result with many ramifications which we shall not be exploring: suffice it to say that the fact that there are questions that cannot be resolved by any realistic algorithmic technique has had a huge influence on the course and on the style of twentieth century mathematics. The argument is one of great beauty and ingenuity, though it can cause a certain discomfort at a first reading.

We begin by showing that the information contained in the definition of a TM

$$T = (Q, A, M, \Delta, \theta, i, T),$$

with $a, b \in A$, can be encoded using only the letters a and b. The precise way in which this encoding is carried out is arbitrary; the essence is that we should be able to decode the appropriate words in $\{a, b\}^*$ so as to recover the TM. Precisely what we mean by an encoding will emerge as we make our definitions.

It will be convenient from this point on to assume that we are dealing with a TM

$$T = (Q, A, M, \Delta, \theta, i, T)$$

in which $i \notin T$. This is no real restriction: given a TM T in which $i \in T$ we can always find a TM T' for which $i \notin T$ and such that $L(T') = L(T)$. (See Exercise 6.6.)

We can regard a typical TM instruction

$$\theta(q, \alpha) = (q', \alpha', R)$$

as an element $(q, \alpha, q', \alpha', R)$ of $Q \times M \times Q \times M \times \{L, R\}$. Suppose now that

$$Q = \{q_1, \ldots, q_k, t_1, \ldots, t_l\},$$

with $i = q_1$ and $T = \{t_1, \ldots, t_l\}$. Suppose also that $A = \{a_1, \ldots, a_m\}$, with $a_1 = a$, $a_2 = b$, and that $M \backslash A = \{\alpha_1, \ldots, \alpha_n\}$, with $\alpha_1 = \Delta$. Define $c_1 : Q \to \{a, b\}^*$ and $c_2 : M \to \{a, b\}^*$ by

$$c_1(q_i) = ba^{2i+1}b \ (i = 1, \ldots, k), \qquad c_1(t_j) = ba^{2j}b \ (j = 1, \ldots, l),$$
$$c_2(a_r) = ba^{2r+1}b \ (r = 1, \ldots, m), \qquad c_2(\alpha_s) = ba^{2s}b \ (s = 1, \ldots, n);$$

also, define $c_3 : \{L, R\} \to \{a, b\}^*$ by

$$c_3(L) = bab, \ c_3(R) = ba^2b$$

We then define a map

$$c : Q \times M \times Q \times M \times \{L, R\} \to \{a, b\}^*$$

by the rule that

$$c(q, \alpha, q', \alpha', X) = c_1(q)bc_2(\alpha)bc_1(q')bc_2(\alpha')bc_3(X).$$

(Here $q, q' \in Q$, $\alpha, \alpha' \in M$, and $X \in \{L, R\}$.) Thus, for example, the instruction

$$\theta(i, b) = (q_4, \Delta, R)$$

(in which q_4 is not terminal) is encoded as

$$ba^3b^3a^5b^3a^9b^3a^2b^3a^2b,$$

while the instruction

$$\theta(q_2, \Delta) = (t_2, \alpha_3, L)$$

(in which t_2 is terminal and $\alpha_3 \notin A$) is encoded as

$$ba^5b^3a^2b^3a^4b^3a^6b^3ab.$$

Now, a Turing machine is specified in essence by a sequence of (say) p instructions of the type

$$\theta(q, \alpha) = (q', \alpha', X).$$

We define a map

$$K : \left(Q \times M \times Q \times M \times \{L, R\}\right)^P \to \{a, b\}^*$$

by

$$K(z_1, z_2, \ldots, z_p) = c(z_1)b^2 c(z_2)b^2 \ldots b^2 c(z_p),$$

where $z_1, z_2, \ldots, z_p \in Q \times M \times Q \times M \times \{L, R\}^*$.

We are justified in calling K an *encoding,* for it is one-one and so a codeword in A^* corresponds to a unique TM. For example, if we take a codeword

$$ba^3 b^3 a^3 b^3 a^5 b^3 a^3 b^3 ab^4 a^5 b^3 a^2 b^3 a^2 b^3 a^5 b^3 a^2 b, \qquad (6.8.1)$$

then we may deduce that this encodes a TM

$$\left(\{i, q_2, t_1\},\ \{a, b\},\ \{a, b, \Delta\},\ \Delta,\ \theta,\ i,\ \{t_1\}\right),$$

where

$$\theta(i, a) = (q_2, a, L),\ \theta(q_2, \Delta) = (t_1, b, R).$$

(Notice that the presence of b^4 heralds the end of one instruction and the start of the next.)

In fact it is not quite accurate to describe K as a function from the set of Turing machines into $\{a, b\}^*$, since the order in which we specify the TM instructions is arbitrary. Thus, for example, the word

$$ba^5 b^3 a^2 b^3 a^2 b^3 a^5 b^3 a^2 b^4 a^3 b^3 a^3 b^3 a^5 b^3 a^3 b^3 ab$$

represents the same TM as the word (6.8.1). We can, however, talk of *the set of codewords* of a Turing machine $\mathcal{T}$, and the set W of elements of $\{a, b\}^*$ which are codewords of Turing machines is now a reasonable, if somewhat complicated, concept.

Each codeword w determines a unique Turing machine $\mathcal{T}(w)$, and it is then possible to feed this Turing machine with its own codeword. If $w \in L\big(\mathcal{T}(w)\big)$ then we shall call it a *good* codeword. Let G be the subset of W consisting of all the good codewords.

Theorem 6.8.2. *The set $D = \{a, b\}^* \backslash G$ is not recursively enumerable.*

Proof. Suppose by way of contradiction that D is recursively enumerable. Then $D = L(\mathcal{T})$ for some Turing machine $\mathcal{T}$. Let w be a codeword of $\mathcal{T}$, so that $\mathcal{T} = \mathcal{T}(w)$, and suppose first that w is a good codeword. Then on the one hand we have $w \in G$, while on the other hand the definition of 'good' gives us that $w \in L(\mathcal{T}(w)) = D$. This is a contradiction.

Suppose alternatively that w is not a good codeword. Then on the one hand we have $w \in D$, while on the other hand we have $w \notin L(\mathcal{T}(w)) = D$. Again we have a contradiction.

Since both the possible assumptions about w lead to contradictions we are force to conclude that there can be no Turing machine recognizing D. ∎

It is of course reasonable to ask whether a set such as D might be recognizable by a substantially improved type of machine. The trouble is that even if we could find such a machine then if it has any reasonable 'finitary' property we could almost certainly encode it as we have done with Turing machines, and a revised definition of 'good' would lead to exactly the same situation. We are forced to conclude that there are subsets of $\{a, b\}^*$ whose listing is beyond the capacity of any reasonable algorithmic process.

Exercises 6

6.1. Construct Turing machines to recognize
    (i) $\{a^n b^{2n} : n \geq 1\}$;
    (ii) $\{w \in A^* : w^R = w\}$, the set of palindromes, where A is an arbitrary finite alphabet.

6.2. Construct Turing machines to recognize the languages
    (i) $\{a^n b^n c^n : n \geq 1\}$;
    (ii) $\{a^n b^n c^n d^n : n \geq 1\}$;
    (iii) $\{a^n b^{2n} a^n : n \geq 1\}$. (None of these languages is context-free, by Exercise 5.6.)

6.3 Construct Turing machines to recognize the languages
    (i) $\{w \in \{a, b, c\}^+ : |w|_a = |w|_b = |w|_c\}$;
    (ii) $\{w w^R w : w \in \{a, b\}^*\}$. (Neither of these languages is context-free, by Exercise 5.8.)

6.4. Let $\mathcal{T}$ be the Turing machine $(Q, A, M, \Delta, \theta, i, \{t\})$, where

$$Q = \{i, i', i_2, q_1, q_2, q_3, q_4, q_1', q_2', t\} \quad A = \{a, b\}, \quad M = \{a, b, \bar{a}, \bar{b}, a^*\},$$

and let

$$\theta(i, a) = (q_1, a^*, R)$$
$$\theta(q_1, a) = (q_1, a, R)$$
$$\theta(q_1, \bar{b}) = (q_2, \bar{b}, R) \qquad\qquad \theta(i', a) = (q_1', \bar{a}, R)$$
$$\theta(q_1, b) = (q_3, \bar{b}, L) \qquad\qquad \theta(q_1', a) = (q_1', a, R)$$
$$\theta(q_2, \bar{b}) = (q_2, \bar{b}, R) \qquad\qquad \theta(q_1', \bar{b}) = (q_2', \bar{b}, R)$$
$$\theta(q_2, b) = (q_3, \bar{b}, L) \qquad\qquad \theta(q_2', b) = (q_3, \bar{b}, L)$$

$$\theta(q_1, \Delta) = (t, \Delta, R) \qquad\qquad \theta(q_2, \Delta) = (t, \Delta, R)$$

$$\theta(q_3, \bar{b}) = (q_3, \bar{b}, L) \qquad\qquad \theta(i', \bar{b}) = (q_4, \bar{b}, L)$$
$$\theta(q_3, a) = (q_3, a, L) \qquad\qquad \theta(q_4, a) = (q_4, a, L)$$
$$\theta(q_3, \bar{a}) = (i', a, R) \qquad\qquad \theta(q_4, a^*) = (q_1, a^*, R).$$
$$\theta(q_3, a^*) = (i', a^*, R)$$

(i) Show that

$$L(T) = \{a^m b^n : m \text{ divides } n\}.$$

(ii) Show how to modify T so that it recognizes

$$\{a^m b^n : m \text{ does not divide } n\}.$$

6.5. Define an LITM to be a Turing machine, but with a tape infinitely extendible to the left as well as to the right. Thus if we have an ID

$$q\alpha_1\alpha_2 \ldots \alpha_k$$

and the instruction is $\theta(q, \alpha_1) = (q', \beta, L)$ then the ID changes to $q'\Delta\beta\alpha_2 \ldots \alpha_k$. Show how an LITM can be simulated by a TM.

6.6. Show how a TM

$$T = (Q, A, M, \Delta, \theta, i, T)$$

in which $i \in T$ can be simulated by a TM

$$T' = \left(Q \cup \{i', q\}, A, M, \Delta, \theta', i', T\right)$$

in which the initial state is not terminal.

6.7. Devise a TM subroutine to reverse a piece of text.

6.8. Devise a TM subroutine to transform $b_1 b_2 \ldots b_n \# c_1 c_2 \ldots c_n$ into the interleaved word $b_1 c_1 b_2 c_2 \ldots b_n c_n$.

6.9. Devise a TM subroutine to convert $\beta^k \# a_1 a_2 \ldots a_n$ into

$$(a_1 a_2 \ldots a_n)^k.$$

(Regard β^k as a 'control' word indicating how often the word $a_1 a_2 \ldots a_n$ is to be duplicated.

6.10. Describe a TM that will recognize $\{a^{n^2} : n \geq 1\}$.

6.11. Describe a TM that will recognize $\{a^p : p \text{ is prime}\}$.

6.12. Show that if L_1 and L_2 are recursively enumerable subsets of A^* then so are $L_1 \cup L_2$ and $L_1.L_2$. (Hint: construct phrase structure grammars generating $L_1 \cup L_2$ and $L_1.L_2$.)

6.13. Show that if L is a recursively enumerable subset of A^* then so is L^*.

7 Varieties

Introduction

In Chapter 3 we explored some important links between rational languages and abstract algebra, the key notion being that of the *syntactic monoid* of a language. In this chapter we explore these links further. The central result is the Variety Theorem (7.3.11) due to Eilenberg (1975). The deepest result (Theorem 7.4.9), due to Schutzenberger (1965) establishes a connection between aperiodic monoids and so-called 'star-free' languages.

It is necessary to begin the chapter with some further algebraic ideas.

7.1 Some more algebra

The notion of a *congruence* on a monoid was introduced in Section 1.3. If S is a monoid and ρ_i $(i \in I)$ are congruences on S then the relation

$$\rho = \bigcap_{i \in I} \rho_i$$

is again a congruence on S. The verification of this is routine: for example,

$$(a, a'), (b, b') \in \rho$$
$$\Rightarrow (\forall i \in I) \, (a, a'), (b, b') \in \rho_i$$
$$\Rightarrow (\forall i \in I) \, (ab, a'b') \in \rho_i$$
$$\Rightarrow (ab, a'b') \in \rho.$$

If $\mathbf{R}$ is a relation on S (that is, any subset of $S \times S$), consider the set of all congruences on S containing $\mathbf{R}$. (There is necessarily at least one such congruence, namely the *universal* congruence $S \times S$.) The intersection of all these congruences is again a congruence containing **R**, and is in fact the unique smallest congruence containing **R**. We call it the congruence on S *generated by* **R** and write it as $\mathbf{R}^\#$. The key property of $\mathbf{R}^\#$ is as follows:

$$\text{If } \rho \text{ is a congruence and } \mathbf{R} \subseteq \rho, \text{ then } \mathbf{R}^\# \subseteq \rho. \qquad (7.1.1)$$

The following theorem will prove useful when we come to prove a crucial result on varieties in Section 7.2.

Theorem 7.1.2. *Let A be a finite alphabet and let ρ be a congruence on A^* such that A^*/ρ is finite. Then there exists a finite subset $\mathbf{T}$ of $A^* \times A^*$ such that $\mathbf{T}^\# = \rho$.*

Proof. For each ρ-class $w\rho$ we define

$$l(w\rho) = \min\{|z| : z \in w\rho\}.$$

Thus, if

$$k = 1 + \max\{l(w\rho) : w \in A^*\},$$

we can be sure that every ρ-class contains a word of length strictly less than k. Let

$$\mathbf{T} = \{(u, v) \in A^* \times A^* : u \, \rho \, v, \ |u| \le k, \ |v| \le k\},$$

and let us denote by τ the congruence $\mathbf{T}^\#$ on A^* generated by $\mathbf{T}$. Certainly $\mathbf{T}$ is finite, since the total number of words in A^* of length not exceeding k is

$$1 + |A| + \cdots + |A|^k.$$

It is also clear that $\mathbf{T} \subseteq \rho$ and so by (7.1.1) we have that $\tau \subseteq \rho$.

We aim to show that $\tau = \rho$. So let us suppose by way of contradiction that there exist u, v in A^* such that

$$(u, v) \in \rho, \quad (u, v) \notin \tau. \qquad (7.1.3)$$

We may in fact suppose that u and v have been chosen so that $|u| + |v|$ is as small as possible, and clearly there is no harm in assuming that $|u| \ge |v|$. If $|u| < k$ then by definition of $\mathbf{T}$ we have $(u, v) \in \mathbf{T} \subseteq \tau$, in contradiction to (7.1.3). So $|u| \ge k$ and we may write $u = zu'$, where $|u'| = k$. By the definition of k we can find v' in A^* such that $|v'| < k$ and $(u', v') \in \rho$. Thus

$$(u', v') \in \mathbf{T} \subseteq \tau \subseteq \rho.$$

Hence from

$$(v, u) \in \rho, \ u = zu' \text{ and } (zu', zv') \in \rho$$

we deduce that $(v, zv') \in \rho$. On the other hand, if we had $(v, zv') \in \tau$ then from

$$u = zu', \ (zu', zv') \in \tau \text{ and } (zv', v) \in \tau$$

we would deduce that $(u, v) \in \tau$, and we are assuming that this is not the case. So we have

$$(v, zv') \in \rho, \quad (v, zv') \notin \tau.$$

Since

$$|v| + |zv'| < |v| + |zu'| = |u| + |v|,$$

we now have a contradiction, for we had chosen $|u| + |v|$ to be as small as possible. We deduce that $\rho = \mathbf{T}^{\#}$. ∎

An *ideal* of a monoid S is a non-empty subset I of S with the property that

$$a \in I, \ s \in S \quad \Rightarrow \quad sa \in I, \ as \in I.$$

More compactly, I is an ideal if

$$SI \subseteq I, \quad IS \subseteq I. \tag{7.1.4}$$

Ideals are an important source of congruences. Given an ideal I of S, we define a relation ρ_I on S by the rule that $(a, b) \in \rho_I$ if and only if either $a = b$ or a and b both belong to I. It is routine to verify the congruence properties and to show that the ρ_I-classes are the singleton sets $\{a\}$, where $a \in S \backslash I$, together with the set I. The quotient monoid S/ρ_I is usually written simply as S/I. Its set of elements consists of I together with the singleton sets $\{a\}$ (with $a \in S \backslash I$), and multiplication is given by

$$\{a\}\{b\} = \begin{cases} \{ab\} & \text{if } ab \notin I \\ I & \text{if } ab \in I \end{cases} \tag{7.1.5}$$
$$\{a\}I = I\{a\} = II = I.$$

The element I acts as a zero of S/I, and it is frequently useful to think of S/I as consisting of the elements of $S \backslash I$ together with a zero, where the product of two non-zero elements a and b in $S \backslash I$ is simply their product in S if this lies outside I, and is zero otherwise.

This particular kind of congruence is called a *Rees* congruence (after the pioneering work of David Rees (1940)) and the quotient monoid S/I is called a *Rees* quotient.

We now give a very brief account of certain crucially important equivalence relations on a monoid. These relations, concerned with mutual divisibility, were introduced by J. A. Green (1951). Given a monoid M and elements a, b in M we write $a \mathcal{R} b$ if a and b are mutually right divisible, i.e., if there exist u, v in M such that

$$au = b, \quad bv = a. \tag{7.1.6}$$

Dually, we write $a \mathcal{L} b$ if there exist x, y in M such that

$$xa = b, \quad yb = a. \tag{7.1.7}$$

A two-sided analogue is provided by the relation $\mathcal{J}$, defined by the rule that $a \mathcal{J} b$ if there exist u, v, x, y in M such that

$$uav = b, \quad xby = a. \tag{7.1.8}$$

It is routine to show that $\mathcal{R}$, $\mathcal{L}$ and $\mathcal{J}$ are all equivalence relations. In general they are *not* congruence relations.

For a full account of these relations see Howie (1976), or Clifford and Preston (1961), or Lallement (1979). We confine ourselves here to one or two key properties that will be of use later in the chapter.

First, we have

Theorem 7.1.9. *Let a, b be elements of a finite monoid M and suppose that there exist u, v, x in M such that $a = ubv$, $b = ax$. Then $a \mathcal{R} b$.*

Proof. From the given relations we have $a = uaxv$. We can repeat this expansion process and obtain

$$a = u^n a(xv)^n \quad (n = 1, 2, \ldots).$$

Now by Theorem 1.4.8 we know that some power of xv is idempotent. Suppose therefore that we have chosen n so that $(xv)^n = (xv)^{2n}$. Then

$$a = u^n a(xv)^{2n} = \big(u^n a(xv)^n\big)x(vx)^{n-1}v$$
$$= ax(vx)^{n-1}v = b(vx)^{n-1}v.$$

This equality, together with the given equation $ax = b$, implies that $a \mathcal{R} b$.

We have a left-right dual result as follows:

Theorem 7.1.10. *Let a, b be elements of a finite monoid M and suppose that there exist u, v, x in M such that $a = ubv$, $b = xa$. Then $a \mathcal{L} b$.* ∎

Next, recall from Section 1.4 that for a given idempotent e in a semigroup or monoid S the group H_e is defined by

$$H_e = \{a \in S : ea = ae = a, \quad (\exists x \in S)\, ax = xa = e\}.$$

Theorem 7.1.11. *Let a be an element of a finite monoid M. If $a \mathcal{R} 1$ then $a \in H_1$.*

Proof. Since $a \, \mathcal{R} \, 1$ there exists u in M such that $au = 1$. Hence

$$a^2 u^2 = a(au)u = au = 1,$$

and more generally we have

$$a^n u^n = 1 \quad (n = 1, 2, \ldots).$$

By the finiteness of M we have integers $m \geq 0$, $r \geq 1$ such that $a^m = a^{m+r}$. (See equation (1.4.7).) Hence

$$a^r = a^r . a^m u^m = a^{m+r} u^m = a^m u^m = 1.$$

We now have
$$1a = a1 = a, \quad a.a^{r-1} = a^{r-1}.a = 1,$$

and so $a \in H_1$ as required.   ∎

Finally, and for use in Section 7.4, let us consider an element z of a finite monoid M and define

$$W(z) = \{y \in M : z \notin MyM\}. \tag{7.1.12}$$

It is of course possible that $W(z) = \emptyset$. For example, if M is a group then, for all y, z in M,

$$z = y^{-1}yz \in MyM;$$

thus $W(z) = \emptyset$ for all z in M. However, if M has a zero element 0 then for all $z \neq 0$ in M we have $z \notin M0M$ and so $0 \in W(z)$. It follows that $W(z) \neq \emptyset$.

We also have the following result:

Theorem 7.1.13. *With the above definitions:*
  (i) *if $W(z) \neq \emptyset$ then $W(z)$ is the largest ideal of M not containing z;*
  (ii) *if $|W(z)| = 1$ then M has a zero element 0 and $W(z) = \{0\}$;*
  (iii) *for all z, t in M, $W(z) = W(t)$ if and only if $z \, \mathcal{J} \, t$.*

Proof. (i) If $a \in W(z)$ and s is an arbitrary element of M then

$$MsaM \subseteq MaM, \quad MasM \subseteq MaM;$$

so from $z \notin MaM$ we immediately deduce that

$$z \notin MsaM, \quad z \notin MasM,$$

i.e., that $sa, as \in W(z)$. So $W(z)$ is an ideal. Clearly $z \in MzM$ and so $z \notin W(z)$.

To establish the maximality of $W(z)$, consider an arbitrary ideal I of M not containing z, and let $a \in I$. Then $MaM \subseteq MIM \subseteq I$ and so from $z \notin I$ we can deduce that $z \notin MaM$. Thus $a \in W(z)$. We have shown that $I \subseteq W(z)$, exactly as required.

(ii) Suppose that $W(z) = \{u\}$. Then $\{u\}$ is an ideal of M and so for all s in M we have

$$s\{u\} \subseteq \{u\}, \quad \{u\}s \subseteq \{u\}.$$

That is to say, $su = us = u$; thus u is the zero element of M.

(iii) Suppose first that $W(z) = W(t)$. Then $z \notin W(t)$ and $t \notin W(z)$, from which we deduce that

$$t \in MzM, \quad z \in MtM,$$

i.e., that $z \mathcal{J} t$. Conversely, suppose that $z \mathcal{J} t$, so that $z \in MtM$, $t \in MzM$. Consider now an element $u \notin W(z)$. Then $z \in MuM$ and so

$$t \in MzM \subseteq M(MuM)M \subseteq MuM.$$

Thus $u \notin W(t)$. We have shown that

$$u \notin W(z) \Rightarrow u \notin W(t).$$

The opposite implication follows in the same way and so $W(z) = W(t)$ as required. ∎

7.2 Varieties and F-varieties

The algebraic notion of a *variety* goes back to Birkhoff (1935) and is discussed in some detail in (Cohn 1965). We shall be concerned here only with monoids and semigroups; so to be definite and to avoid unnecessary abstraction let us consider Birkhoff's definition as it applies to the (reasonably representative) case of monoids. A class **V** of monoids is called a *variety* if it is closed under the taking of submonoids, quotient monoids and direct products. That is to say, **V** is a variety if:

(7.2.1) whenever $A \in \mathbf{V}$ and B is a submonoid of A then $B \in \mathbf{V}$;
(7.2.2) whenever $A \in \mathbf{V}$ and ρ is a congruence on A then $A/\rho \in \mathbf{V}$;
(7.2.3) whenever $A_i \in \mathbf{V}$ $(i \in I)$ then

$$\prod_{i \in I} A_i \in \mathbf{V}.$$

Examples of varieties abound. The class **Com** of *commutative* monoids forms a variety, and so does the class **BM** of *band monoids,* by which we mean the class of monoids in which every element is *idempotent.* By contrast, the class **Gp** of groups is *not* a variety of monoids, since (for example) the submonoid $\{1, a, a^2, \ldots\}$ of the infinite cyclic group $\langle a \rangle$ is not a group.

Finiteness is strongly built into the concept of an automaton and emerges in the finiteness of the syntactic monoid associated with a rational language. For this reason we find it useful to make a crucial change in the Birkhoff definition. It is the clause (7.2.3) in the definition that allows infinities to creep into the classical notion of a variety. So, while keeping clauses (7.2.1) and (7.2.2), we replace (7.2.3) by:

(7.2.4) whenever $A, B \in \mathbf{V}$ then $A \times B \in \mathbf{V}$,

and agree that a class $\mathbf{V}$ of *finite* monoids satisfying (7.2.1), (7.2.2) and (7.2.4) will be called an F-variety. (The term 'pseudo-variety' has been used by some authors, and several authors have preferred simply to use the term 'variety'.)

Remark 7.2.5. It is clear that the closure property given by (7.2.4) extends to arbitrary finite direct products: if $A_1, A_2, \ldots, A_n \in \mathbf{V}$ then $A_1 \times A_2 \times \cdots \times A_n \in \mathbf{V}$.

Remark 7.2.6. It is sometimes convenient to describe the closure properties (7.2.1) and (7.2.2) together:

(7.2.7) if $A \in \mathbf{V}$ and B divides A then $B \in \mathbf{V}$.

Clearly the class of all finite monoids within a variety $\mathbf{V}$ is an F-variety. However, this is by no means the end of the story, for the class **FGp** of *finite* groups is an F-variety of monoids, and we saw above that the class **Gp** of *all* groups is not a variety of monoids. (The F-variety property holds since every submonoid of a finite group is a subgroup – see Exercise 1.8.)

Suppose now that we have a collection of F-varieties $\mathbf{V}_i$ ($i \in I$), where I may be finite or infinite, but is certainly non-empty. Let

$$\mathbf{V} = \bigcap_{i \in I} \mathbf{V}_i;$$

then $\mathbf{V}$ is an F-variety. The verification is routine: for example,

$$
\begin{aligned}
A, B \in \mathbf{V} &\Rightarrow A, B \in \mathbf{V}_i \text{ for every } i \text{ in } I \\
&\Rightarrow A \times B \in \mathbf{V}_i \text{ for every } i \text{ in } I \\
&\Rightarrow A \times B \in \mathbf{V}.
\end{aligned}
$$

Now let us consider a collection of finite monoids M_j ($j \in J$), where again J may be finite or infinite, but is certainly non-empty. Let $\mathbf{V}$ be the intersection of all the F-varieties containing every M_j. (Notice that the collection of F-varieties containing every M_j is non-empty, since at the very least it contains the F-variety of *all* finite monoids.) From the result of the last paragraph we see that $\mathbf{V}$ is an F-variety containing every M_j, and is the *smallest* F-variety with this property, in the sense that every F-variety containing every M_j must contain $\mathbf{V}$. We refer to $\mathbf{V}$ as the F-variety *generated by* the monoids M_j, and write

$$\mathbf{V} = \mathbf{V}\langle M_j : j \in J \rangle.$$

We have:

Theorem 7.2.8. *With the above definitions, $A \in \mathbf{V}\langle M_j : j \in J \rangle$ if and only if there exist $j_1, j_2, \ldots, j_n$ in J such that A divides $M_{j_1} \times M_{j_2} \times \cdots \times M_{j_n}$.*

Proof. Suppose first that A divides $M_{j_1} \times M_{j_2} \times \cdots \times M_{j_n}$. Now $\mathbf{V}\langle M_j : j \in J \rangle$ is an F-variety containing $M_{j_1}, M_{j_2}, \ldots M_{j_n}$, and so by Remarks 7.2.5 and 7.2.6 contains every monoid dividing $M_{j_1} \times M_{j_2} \times \cdots \times M_{j_n}$. Thus $A \in \mathbf{V}\langle M_j : j \in J \rangle$.

Conversely, let $\mathbf{C}$ be the set of all monoids dividing $M_{j_1} \times M_{j_2} \times \cdots \times M_{j_n}$ for some n and for some $j_1, j_2, \ldots, j_n$ in J. Then $\mathbf{C}$ is an F-variety. The closure properties (7.2.1) and (7.2.2) are clear. To prove (7.2.4), suppose that $A, B \in \mathbf{C}$, so that

$$A \text{ divides } M_{j_1} \times \cdots \times M_{j_n}, \quad B \text{ divides } M_{k_1} \times \cdots \times M_{k_p}.$$

Then $A \times B$ divides $M_{j_1} \times \cdots \times M_{j_n} \times M_{k_1} \times \cdots \times M_{k_p}$ and so $A \times B \in \mathbf{C}$.

We have shown that $\mathbf{C}$ is an F-variety containing every M_j. Consequently we must have

$$\mathbf{C} \subseteq \mathbf{V}\langle M_j : j \in J \rangle.$$

The proof of Theorem 7.2.8 is now complete. ∎

Let $X = \{x_1, x_2, \ldots\}$ be an infinite alphabet and let u, v be distinct elements of X^*. We say that a monoid M *satisfies the equation* $u = v$ if $\varphi(u) = \varphi(v)$ for *every* morphism $\varphi : X^* \to M$. Intuitively we may think of u and v as words in the 'variables' $x_1, x_2, \ldots$. Any particular morphism φ in effect substitutes elements of M for the variables, and M satisfies the equation $u = v$ if and only if u and v give rise to the same element of M for every possible substitution. For example, a monoid M is commutative if and only if it satisfies the equation $x_1 x_2 = x_2 x_1$; it is a *band* (a semigroup consisting of idempotents) if and only if it satisfies the equation $x^2 = x$.

If (u_n, v_n) is a (finite or infinite) sequence of pairs of distinct elements in X^*, then we may consider the class **C** of all monoids satisfying the equations $u_n = v_n$ $(n = 1, 2, \ldots)$. We say that the class **C** is *equationally defined,* or more specifically that **C** is *defined by the equations* $u_n = v_n$ $(n = 1, 2, \ldots)$. It is not hard to see that **C** is a variety; we denote it by

$$\mathbf{V}[u_1 = v_1,\ u_2 = v_2, \ldots] \text{ or by } \mathbf{V}[u_n = v_n\ (n \geq 1)].$$

Birkhoff (1935) showed the more difficult converse result that every variety is equationally defined. Notice in particular that the class **Com** of commutative monoids and the class **B** of band monoids are given, respectively, by

$$\mathbf{Com} = \mathbf{V}[xy = yx], \quad \mathbf{B} = \mathbf{V}[x^2 = x].$$

For F-varieties we require a slightly more sophisticated notion. Certainly if the F-variety **V** consists of the finite members of a variety $\mathbf{W} = \mathbf{V}[u_1 = v_1,\ u_2 = v_2, \ldots]$ then it consists of all finite monoids satisfying the equations $u_1 = v_1, u_2 = v_2, \ldots$, and so may be said to be equationally defined; but we have already seen that not every F-variety arises in this way. However, it will be helpful, temporarily at least, to use the notation $\mathbf{V}_F[u_1 = v_1,\ u_2 = v_2, \ldots]$ for the F-variety consisting of the finite members of a variety $\mathbf{V}[u_1 = v_1,\ u_2 = v_2, \ldots]$.

If we have an infinite sequence (u_n, v_n) $(n = 1, 2, \ldots)$ of pairs of distinct elements in X^* then a class **C** of finite monoids is said to be *ultimately defined* by the equations $u_n = v_n$ $(n \geq 1)$ if it consists precisely of those finite monoids M which satisfy the equations $u_n = v_n$ *for all sufficiently large* n. To put the matter more precisely, suppose that, for $n = 1, 2, \ldots$,

$$\mathbf{V}_n = \mathbf{V}_F[u_n = v_n],$$

the F-variety defined by the single equation $u_n = v_n$, and let

$$\mathbf{W}_m = \bigcap_{n \geq m} \mathbf{V}_n \quad (m \geq 1).$$

Thus $\mathbf{W}_m = \mathbf{V}_F[u_m = v_m,\ u_{m+1} = v_{m+1}, \ldots]$, and evidently

$$\mathbf{W}_1 \subseteq \mathbf{W}_2 \subseteq \cdots .$$

The class $\mathbf{V}[[u_n = v_n\ (n \geq 1)]]$ *ultimately defined by the equations* $u_n = v_n$ $(n \geq 1)$ is then defined by

$$\mathbf{V}[[u_n = v_n\ (n \geq 1)]] = \bigcup_{m \geq 1} \mathbf{W}_m. \tag{7.2.9}$$

To exemplify this notion, let us return to the case of finite groups. A finite monoid is a group if and only if every element x has the property that $x^n = 1$ for some n. If G has *exponent* m (i.e., if m is the smallest positive integer such that $g^m = 1$ for every g in G) then certainly G satisfies the equations

$$x^{m!} = 1, \ x^{(m+1)!} = 1, \ldots.$$

(Here $m! = m(m-1)\ldots 2.1$.) The F-variety **FGp** is ultimately defined by the equations

$$x^{n!} = 1 \ (n = 1, 2, \ldots).$$

It is fairly clear that a class $\mathbf{V}[[u_n = v_n]]$ of finite monoids ultimately defined by a sequence of equations $u_n = v_n \ (n \geq 1)$ is an F-variety. To see this, let A be a monoid in the class $\mathbf{V}[[u_n = v_n]]$. Then, in the notation used above, A is in the F-variety $\mathbf{W}_m$ for all sufficiently large m. It follows that every submonoid and every quotient monoid of A is in $\mathbf{W}_m$ and so in $\mathbf{V}[[u_n = v_n]]$. To verify the property (7.2.4), consider monoids A and B in $\mathbf{V}[[u_n = v_n]]$. Then there exist m_1, m_2 such that $A \in \mathbf{W}_m$ for all $m \geq m_1$ and $B \in \mathbf{W}_m$ for all $m \geq m_2$. It follows that $A \times B \in \mathbf{W}_m$ for all $m \geq \max\{m_1, m_2\}$. Hence $A \times B \in \mathbf{V}[[u_n = v_n]]$, as required.

The converse result is much less obvious, and is due to Eilenberg and Schutzenberger (1976):

Theorem 7.2.10. *Every F-variety of monoids is ultimately defined by a sequence of equations.*

Proof. Let $\mathbf{V}$ be an F-variety of monoids. We can, in some arbitrary order, produce a list

$$M_1, \ M_2, \ M_3, \ldots \tag{7.2.11}$$

of the distinct (that is to say, non-isomorphic) monoids in $\mathbf{V}$. (One way would be to begin by listing the monoids of order 1 in $\mathbf{V}$, then the monoids of order 2 in $\mathbf{V}$, and so on. At each stage the number of monoids of order k in $\mathbf{V}$ is finite.) Having obtained the list, we now define

$$P_n = M_1 \times M_2 \times \cdots \times M_n,$$

the direct product of $M_1, M_2, \ldots, M_n$.

Let $X = \{x_1, x_2, \ldots\}$ be an infinite alphabet and for each $n \geq 1$ let

$$X_n = \{x_1, x_2, \ldots, x_n\}.$$

Let τ_n be the congruence on X_n^* defined by the rule that $(u, v) \in \tau_n$ if and only if $\varphi(u) = \varphi(v)$ for every morphism $\varphi : X_n^* \to P_n$.

In fact there are only finitely many morphisms from X_n^* into the finite semigroup P_n. To see this, notice that every such morphism is determined by its action on the generators $x_1, x_2, \ldots x_n$, and so the number of possible morphisms is at most $|P_n|^n$. Let us list the morphisms from X_n^* into P_n as

$$\varphi_1, \ \varphi_2, \ldots, \varphi_N.$$

Now consider the subset

$$Q = \left\{ (\varphi_1(w), \varphi_2(w), \ldots, \varphi_N(w)) : w \in X_n^* \right\}$$

of the monoid P_n^N. It is easy to check that Q is a submonoid. The map $\delta : Q \to X_n^*/\tau_n$ defined by

$$\delta\left((\varphi_1(w), \varphi_2(w), \ldots \varphi_N(w)) \right) = w\tau_n \quad (w \in X_n^*)$$

is an isomorphism, since the definition of τ_n implies that $u\tau_n = v\tau_n$ if and only if

$$(\varphi_1(u), \varphi_2(u), \ldots \varphi_N(u)) = (\varphi_1(v), \varphi_2(v), \ldots \varphi_N(v)).$$

We have shown that X_n^*/τ_n is a homomorphic image of a submonoid of P_n^N, i.e., that X_n^*/τ_n divides P_n^N. It follows in particular that X_n^*/τ_n is finite. Accordingly, we can appeal to Theorem 7.1.2 and deduce that $\tau_n = \mathbf{E}_n^\#$, where $\mathbf{E}_n$ is a finite set such that

$$\mathbf{E}_n \subset X_n^* \times X_n^* \subset X^* \times X^*.$$

Let

$$\mathbf{E} = \bigcup_{n \geq 1} \mathbf{E}_n.$$

We shall show that the set of pairs constituting $\mathbf{E}$ is in effect a set of equations ultimately defining the F-variety $\mathbf{V}$.

Every morphism $\varphi : X^* \to P_n$ is determined by the values of $\varphi(x_1), \varphi(x_2), \ldots$. The pairs (u, v) in $\mathbf{E}_n$ necessarily belong to $X_n^* \times X_n^* (\subset X^* \times X^*)$. So whether or not a morphism $\varphi : X^* \to P_n$ has the property (for a pair (u, v) in $X_n^* \times X_n^*$) that $\varphi(u) = \varphi(v)$ depends entirely on whether its restriction to X_n^* has the same property. But the restriction of φ to X_n^* is one of $\varphi_1, \varphi_2, \ldots, \varphi_N$, and by definition we have

$$\varphi_1(u) = \varphi_1(v), \ldots, \varphi_N(u) = \varphi_N(v).$$

Hence certainly for every (u, v) in $\mathbf{E}_n$ $(\subset \tau_n)$ we have that $\varphi(u) = \varphi(v)$ for every morphism $\varphi : X^* \to P_n$. That is, P_n satisfies the equations $u = v$ for all (u, v) in $\mathbf{E}_n$.

If $M \in \mathbf{V}$ then M appears in the list (7.2.11). Let us suppose that $M = M_k$. Then M divides P_n for all $n \geq k$ and so M satisfies the equations in $\mathbf{E}_n$ for all $n \geq k$. We have shown that $\mathbf{V}$ is contained in the class of monoids ultimately defined by the equations in $\mathbf{E}$. It remains to show that this class coincides with $\mathbf{V}$.

Accordingly, let M be a monoid satisfying the quations in $\mathbf{E}_n$ for all $n \geq k$. Let

$$n \geq \max\{k, |M|\}$$

and let $\varphi : X_n^* \to M$ be an arbitrarily chosen surjective morphism. By our assumption, $\varphi(u) = \varphi(v)$ for all (u, v) in $\mathbf{E}_n$, and hence also for all (u, v) in $\mathbf{E}_n^\# = \tau_n$. Hence by Theorem 1.5.7 there is a surjective morphism $\theta : X_n^*/\tau_n \to M$ such that the diagram

$$
\begin{array}{ccc}
X_n^* & \xrightarrow{\varphi} & M \\
\downarrow{\scriptstyle \tau_n^\natural} & \theta & \\
X_n^*/\tau_n & &
\end{array}
$$

commutes. Certainly M divides X_n^*/τ_n. Now, as we saw before, X_n^*/τ_n divides the direct product P_n^N of N copies of P_n. Since $P_n \in \mathbf{V}$ it now follows that $M \in \mathbf{V}$, as required. ∎

7.3 The variety theorem

Let $\mathbf{V}$ be an F-variety of monoids and let A be an alphabet. We define $\mathcal{L}_\mathbf{V}(A)$ to be the set of languages in A^* whose syntactic monoids are in the F-variety $\mathbf{V}$. Notice that the languages in $\mathcal{L}_\mathbf{V}(A)$ are all rational, since their syntactic monoids are finite.

It is convenient at this stage to give another criterion for membership of $\mathcal{L}_\mathbf{V}(A)$ that can sometimes be easier to apply. Recall from Section 3.1 that a language in A^* is said to be *recognized* by a finite monoid M if there is a morphism $\varphi : A^* \to M$ and a subset P of M such that $L = \varphi^{-1}(P)$.

Theorem 7.3.1. *With the above definitions, $L \in \mathcal{L}_\mathbf{V}(A)$ if and only if L is recognized by a monoid in $\mathbf{V}$.*

Proof. One way round this is obvious, since if $L \in \mathcal{L}_\mathbf{V}(A)$ then L is recognized by the monoid $\mathrm{Syn}(L)$. (See Theorem 3.1.6.) Conversely, if L is recognized by a monoid M in $\mathbf{V}$, then, again by Theorem 3.1.6, $\mathrm{Syn}(L)$ divides M. Hence $\mathrm{Syn}(L) \in \mathbf{V}$, by (7.2.7). ∎

Now we have:

Theorem 7.3.2. *Let* **V**, **W** *be F-varieties of monoids. Then* **V** $\subseteq$ **W** *if and only if* $\mathcal{L}_\mathbf{V}(A) \subseteq \mathcal{L}_\mathbf{W}(A)$ *for every finite alphabet* A.

Proof. If **V** $\subseteq$ **W** and $L \in \operatorname{Rat} A^*$, where A is an arbitrary finite alphabet, then

$$L \in \mathcal{L}_\mathbf{V}(A) \Rightarrow \operatorname{Syn}(L) \in \mathbf{V} \subseteq \mathbf{W}$$
$$\Rightarrow L \in \mathcal{L}_\mathbf{W}(A)$$

and so $\mathcal{L}_\mathbf{V}(A) \subseteq \mathcal{L}_\mathbf{W}(A)$.

Conversely, suppose that $\mathcal{L}_\mathbf{V}(A) \subseteq \mathcal{L}_\mathbf{W}(A)$ for every finite alphabet A. Let $M \in \mathbf{V}$. By Theorem 1.7.4 we can find a finite alphabet A and a surjective morphism $\varphi : A^* \to M$. For each z in M the language $\varphi^{-1}(z)$ is recognized by M and so

$$\varphi^{-1}(z) \in \mathcal{L}_\mathbf{V}(A) \subseteq \mathcal{L}_\mathbf{W}(A).$$

Denote the language $\varphi^{-1}(z)$ by L_z. Then $\operatorname{Syn}(L_z) \in \mathbf{W}$ for every z and so

$$P = \prod_{z \in M} \operatorname{Syn}(L_z)$$

is also in **W**. We shall show that M divides P, from which it follows that $M \in \mathbf{W}$.

Denote the elements of M by $z_1, z_2, \ldots, z_n$, and for the sake of notational simplicity denote the syntactic morphism associated with L_{z_i} by $\zeta_i : A^* \to \operatorname{Syn}(L_{z_i})$. Let Q be the submonoid of P given by

$$Q = \left\{ \left(\zeta_1(w), \zeta_2(w), \ldots, \zeta_n(w) \right) : w \in A^* \right\},$$

and let $\delta : Q \to M$ be defined by

$$\delta \Big(\left(\zeta_1(w), \zeta_2(w), \ldots, \zeta_n(w) \right) \Big) = \varphi(w).$$

This is a well-defined map. For suppose that $\zeta_i(w) = \zeta_i(w')$ for $i = 1, 2, \ldots, n$. Then $(w, w') \in \sigma_{\varphi^{-1}(z)}$ for all z in M and so, for all x, y in A^* and all z in M,

$$\varphi(xwy) = z \text{ if and only if } \varphi(xw'y) = z.$$

In particular, for all z in M, we have

$$\varphi(w) = z \text{ if and only if } \varphi(w') = z,$$

and this is just a complicated way of saying that $\varphi(w) = \varphi(w')$. The map δ is then easily seen to be a surjective morphism and so M divides P as required. ∎

As an obvious consequence we have:

Corollary 7.3.3. *Let* **V**, **W** *be F-varieties of monoids. Then* **V** = **W** *if and only if* $\mathcal{L}_{\mathbf{V}}(A) = \mathcal{L}_{\mathbf{W}}(A)$ *for every finite alphabet* A. ∎

We say that $\mathcal{L}$ is a *rational language map* (or an *RL-map*) if it associates with every finite alphabet A a subset $\mathcal{L}(A)$ of Rat A^*. The map $\mathcal{L}_{\mathbf{V}}$ we have been examining is an example of an RL-map. An RL-map $\mathcal{L}$ is said to be *varietal*, or to be a *VRL-map*, if:

(7.3.4) for every A the set $\mathcal{L}(A)$ of languages in A^* is closed under the operations of union and complementation;

(7.3.5) whenever A, B are finite alphabets and $\varphi : A^* \to B^*$ is a morphism, then $L \in \mathcal{L}(B)$ implies that $\varphi^{-1}(L) \in \mathcal{L}(A)$;

(7.3.6) for every finite alphabet A, every L in $\mathcal{L}(A)$ and every a in A, the subsets $a^{-1}L$ and La^{-1} are in $\mathcal{L}(A)$.

Remark 7.3.7. Condition (7.3.4) implies that $\mathcal{L}(A)$ is closed under all the 'Boolean' operations of union, intersection and complementation. Condition (7.3.6) implies (by an easy inductive proof) that if $L \in \mathcal{L}(A)$ and $u \in A^*$ then $u^{-1}L$ and Lu^{-1} are in $\mathcal{L}(A)$.

It is now fairly easy to show:

Theorem 7.3.8. *If* **V** *is an F-variety of monoids, then the map* $\mathcal{L}_{\mathbf{V}}$ *defined above is a VRL-map.*

Proof. We first show that $\mathcal{L}_{\mathbf{V}}(A)$ is closed under the operations \ and ∪. First, let $L \in \mathcal{L}_{\mathbf{V}}(A)$. Then

$$A^* \backslash L = (\sigma_L^\natural)^{-1}(P),$$

where

$$P = \mathrm{Syn}(L) \backslash \sigma_L^\natural(L).$$

Thus $A^* \backslash L$ is recognized by $\mathrm{Syn}(L)$ ($\in$ **V**) and so is in $\mathcal{L}_{\mathbf{V}}(A)$.

Next, let L_1, $L_2 \in \mathcal{L}_{\mathbf{V}}(A)$. If we write φ for the morphism from A^* into $\mathrm{Syn}(L_1) \times \mathrm{Syn}(L_2)$ defined by

$$\phi(w) = \left(\sigma_{L_1}^\natural(w),\ \sigma_{L_2}^\natural(w)\right) \quad (w \in A^*)$$

we see that $L_1 \cup L_2 = \varphi^{-1}(P)$, where

$$P = \left[\sigma_{L_1}^\natural(L_1) \times \mathrm{Syn}(L_2)\right] \cup \left[\mathrm{Syn}(L_1) \times \sigma_{L_2}^\natural(L_2)\right].$$

Thus $L_1 \cup L_2$ is recognized by the monoid $\mathrm{Syn}(L_1) \times \mathrm{Syn}(L_2)$ from **V**. We have now verified condition (7.3.4).

Next, let $\varphi : A^* \to B^*$ be a morphism and suppose that $L \in \mathcal{L}_{\mathbf{V}}(B)$. Then $\mathrm{Syn}(L)$ recognizes L. Let

$$\psi = \sigma_L^\natural \circ \varphi : A^* \to \mathrm{Syn}(L)$$

and let $P = \sigma_L^\natural(L) \ (\subseteq \mathrm{Syn}(L))$. Then

$$\phi^{-1}(L) = \psi^{-1}(P)$$

and so $\mathrm{Syn}(L)$ recognizes $\varphi^{-1}(L)$. We have now verified condition (7.3.5).

Finally, let $L \in \mathcal{L}_{\mathbf{V}}(A)$ and let $a \in A$. Let

$$P = \{z\sigma_L \in \mathrm{Syn}(L) : (a\sigma_L)(z\sigma_L) \in \sigma_L^\natural(L)\}.$$

Then

$$z \in a^{-1}L \text{ if and only if } az \in L,$$
$$\text{i.e., if and only if } (a\sigma_L)(z\sigma_L) \in \sigma_L^\natural(L),$$
$$\text{i.e., if and only if } \sigma_L^\natural(z) \in P,$$
$$\text{i.e., if and only if } z \in (\sigma_L^\natural)^{-1}(P).$$

Thus

$$a^{-1}L = (\sigma_L^\natural)^{-1}(P),$$

and so $\mathrm{Syn}(L)$ recognizes $a^{-1}L$. A similar argument applies to La^{-1}. This completes the proof of Theorem 7.3.8. ∎

Much less immediate is the converse result, due to Eilenberg (1975):

Theorem 7.3.9. *For every VRL-map $\mathcal{L}$ there exists an F-variety $\mathbf{V}$ of monoids such that $\mathcal{L} = \mathcal{L}_{\mathbf{V}}$.*

Proof. For $n \geq 1$ let X_n be the finite alphabet $\{x_1, x_2, \ldots, x_n\}$. The image of X_n under the VRL-map $\mathcal{L}$ is a set $\mathcal{L}(X_n)$ of rational subsets of X_n^*. Each L in $\mathcal{L}(X_n)$ has a finite syntactic monoid $\mathrm{Syn}(L)$. Let $\mathbf{V}$ be the F-variety generated by

$$\{\mathrm{Syn}(L) : L \in \mathcal{L}(X_n), \ n \geq 1\}.$$

We shall show that $\mathcal{L} = \mathcal{L}_{\mathbf{V}}$.

One way round this is clear. Every finite alphabet A in effect coincides with X_n for some n. Since $\mathrm{Syn}(L) \in \mathbf{V}$ for all L in $\mathcal{L}(A)$ it is clear that $\mathcal{L}(A) \subseteq \mathcal{L}_{\mathbf{V}}(A)$.

It remains to show that $\mathcal{L}_{\mathbf{V}}(A) \subseteq \mathcal{L}(A)$ for every finite alphabet A. So suppose that $L \in \mathcal{L}_{\mathbf{V}}(A)$. Then $\mathrm{Syn}(L) \in \mathbf{V}$ and so by definition of $\mathbf{V}$

there exist $k \geq 1$, finite alphabets $A_1, A_2, \ldots, A_k$ and rational languages $L_i \in \mathcal{L}(A_i)$ $(i = 1, 2, \ldots, k)$ such that

$$\mathrm{Syn}(L) \text{ divides } \mathrm{Syn}(L_1) \times \mathrm{Syn}(L_2) \times \cdots \times \mathrm{Syn}(L_k).$$

Let us denote $\mathrm{Syn}(L_1) \times \mathrm{Syn}(L_2) \times \cdots \times \mathrm{Syn}(L_k)$ by M, and for $i = 1, 2, \ldots, k$ let $\pi_i : M \to \mathrm{Syn}(L_i)$ be the projection map given by

$$\pi_i(z_1, z_2, \ldots, z_k) = z_i \quad \left((z_1, z_2, \ldots, z_k) \in M \right).$$

Since $\mathrm{Syn}(L)$ divides M it follows from Theorem 3.1.6 that M recognizes L. That is, there exists a morphism $\varphi : A^* \to M$ and a subset P of M such that

$$L = \varphi^{-1}(P).$$

We wish to show that $L \in \mathcal{L}(A)$. Since

$$L = \bigcup_{z \in P} \varphi^{-1}(z)$$

and since $\mathcal{L}(A)$ is by assumption closed under the taking of finite unions, it will be sufficient to prove that $\varphi^{-1}(z) \in \mathcal{L}(A)$ for every z in M.

We can reduce the problem still further. To see this, let us write φ_i for the map

$$\pi_i \circ \varphi : A^* \to \mathrm{Syn}(L_i) \quad (i = 1, 2, \ldots, k).$$

Let $z = (z_1, z_2, \ldots, z_k) \in M$. Then

$$w \in \varphi^{-1}(z) \text{ if and only if } \varphi(w) = z = (z_1, z_2, \ldots, z_k),$$
$$\text{i.e., if and only if } \varphi_i(w) = z_i \quad (i = 1, 2, \ldots, k),$$
$$\text{i.e., if and only if } w \in \varphi_i^{-1}(z_i) \quad (i = 1, 2, \ldots, k).$$

Thus

$$\varphi^{-1}(z) = \bigcap_{i=1}^{k} \varphi_i^{-1}(z_i).$$

Now by Remark 7.3.7 the set $\mathcal{L}(A)$ is closed under the taking of finite intersections. Hence it will be enough to show that $\varphi_i^{-1}(z_i) \in \mathcal{L}(A)$ for $i = 1, 2, \ldots, k$ and for every z_i in $\mathrm{Syn}(L_i)$.

We have a diagram

$$A^* \xrightarrow{\varphi_i} \mathrm{Syn}(L_i)$$
$$\uparrow{\scriptstyle \tau_i}$$
$$A_i^*$$

where for the sake of notational simplicity we are writing τ_i for the syntactic morphism $\sigma^{\natural}_{L_i}$.

Certainly τ_i is surjective, and so for each a in A we can choose u_i in A_i^* such that $\tau_i(u_i) = \varphi_i(a)$. We can regard this choosing process as defining a map ψ_i from A into A_i^* (taking u_i as $\psi_i(a)$), and by Theorem 1.7.3 this extends to a morphism $\psi_i : A^* \to A_i^*$ such that the diagram

$$A^* \xrightarrow{\varphi_i} \mathrm{Syn}(L_i)$$

$$\psi_i \qquad \uparrow \tau_i$$

$$A_i^*$$

commutes.

This observation enables us to reduce the problem still further. For we now have

$$\varphi_i^{-1}(z_i) = \psi_i^{-1}\left(\tau_i^{-1}(z_i)\right) \quad (z_i \in \mathrm{Syn}(L_i)).$$

Hence by the property (7.3.6) of a VRL-map it will follow that $\varphi_i^{-1}(z_i) \in \mathcal{L}(A)$ if we can show that $\tau_i^{-1}(z_i) \in \mathcal{L}(A)$.

Now $\tau_i : A^* \to \mathrm{Syn}(L_i)$ is the syntactic morphism. From Exercise 3.4 we have

$$\tau_i^{-1}(z_i) = \left[\bigcap_{(u,v)\in K(w)} u^{-1}L_iv^{-1} \right] \Big\backslash \left[\bigcup_{(u,v)\notin K(w)} u^{-1}L_iv^{-1} \right], \qquad (7.3.10)$$

where $w \in \tau_i^{-1}(z_i)$ and where

$$K(w) = \{(u,v) \in A_i^* \times A_i^* : uwv \in L_i\}$$
$$= \{(u,v) \in A_i^* \times A_i^* : w \in u^{-1}L_iv^{-1}\}.$$

Moreover, since L is a rational language it follows from Exercise 3.3 that both the union and the intersection involved in (7.3.11) are finite. Hence we will have proved that $\tau_i^{-1}(z_i) \in \mathcal{L}(A)$ if we can show that $u^{-1}L_iv^{-1} \in \mathcal{L}(A)$ for all u, v in A^*. But this now follows fron Remark 7.3.7, since by assumption $L_i \in \mathcal{L}(A)$. This completes the proof of Theorem 7.3.9. ∎

We summarize the main result:

Theorem 7.3.11 (The Variety Theorem). *Every F-variety* **V** *of monoids determines a VRL-map* $\mathcal{L}_{\mathbf{V}}$ *given by*

$$\mathcal{L}_{\mathbf{V}}(A) = \{L \subseteq A^* : \mathrm{Syn}(L) \in \mathbf{V}\}.$$

Conversely, for every VRL-map $\mathcal{L}$ *there exists an F-variety* **V** *of monoids such that* $\mathcal{L} = \mathcal{L}_{\mathbf{V}}$. ∎

Example 7.3.12. If **Mon** is the F-variety of *all* finite monoids then the corresponding VRL-map associates with each A the set of *all* rational languages in A^*.

Example 7.3.13. At the other extreme, if **Triv** is the F-variety of trivial monoids, consisting of the single monoid $\{1\}$, then for each A the set $\mathcal{L}_{\mathbf{Triv}}(A)$ consists of the rational languages L for which $\mathrm{Syn}(L)$ is the trivial monoid $\{1\}$, that is, for which the syntactic congruence is the universal congruence. It follows that $\mathcal{L}_{\mathbf{Triv}}(A) = \{\emptyset, A^*\}$.

In practice it can be quite difficult to describe $\mathcal{L}_{\mathbf{V}}(A)$ for a given F-variety **V** of monoids. If **V** is an F-variety generated by a single monoid M then we obtain the following useful simplification:

Theorem 7.3.14. *Let* $\mathbf{V} = \mathbf{V}\langle M \rangle$, *the F-variety generated by a single monoid* M, *and let* A *be a finite alphabet. Then* $\mathcal{L}(A)$ *is the Boolean algebra generated by the languages* $\varphi^{-1}(z)$ *for all* z *in* M *and all morphisms* $\varphi : A^* \to M$.

Proof. Clearly M recognizes $\varphi^{-1}(z)$ and so $\varphi^{-1}(z) \in \mathcal{L}_{\mathbf{V}}(A)$. This holds for every z in M and for every morphism $\varphi : A^* \to M$. By the Variety Theorem $\mathcal{L}_{\mathbf{V}}(A)$ is closed with respect to the boolean algebra operations. Hence the Boolean algebra generated by all the $\varphi^{-1}(z)$ lies inside $\mathcal{L}_{\mathbf{V}}(A)$.

Conversely, let $L \in \mathcal{L}_{\mathbf{V}}(A)$. Then $\mathrm{Syn}(L) \in \mathbf{V}$ and so $\mathrm{Syn}(L)$ divides M^n for some $n \geq 1$. (Here M^n denotes $M \times M \times \cdots \times M$, with n factors.) Hence, by Theorem 3.1.6, M^n recognizes L and so there exists $\varphi : A^* \to M^n$ and $P \subseteq M^n$ such that

$$L = \varphi^{-1}(P) = \bigcup_{z \in P} \varphi^{-1}(z).$$

There are projections $\pi_i : M^n \to M$ given by

$$\pi_i\big((z_1, z_2, \ldots, z_n)\big) = z_i \quad (i = 1, 2, \ldots, n),$$

and so there are morphisms

$$\varphi_i = \pi_i \circ \varphi : A^* \to M.$$

Since $w \in \varphi^{-1}(z)$ if and only if $\varphi(w) = z$, i.e., if and only if $\varphi_i(w) = \pi_i(z)$, $(i = 1, 2, \ldots, n)$, we have

$$\varphi^{-1}(z) = \bigcap_{i=1}^{n} \varphi_i^{-1}\big(\pi_i(z)\big)$$

and hence

$$L = \bigcup_{z \in P} \left(\bigcap_{i=1}^{n} \varphi_i^{-1}\big(\pi_i(z)\big) \right).$$

This is exactly what we require.   ∎

To find an example of an F-variety generated by a single monoid let us consider the monoid $U_1 = \{0, 1\}$, with

$$1.0 = 0.1 = 0.0 = 0, \quad 1.1 = 1.$$

It is clear that U_1 satisfies the equations

$$x^2 = x, \quad xy = yx,$$

and so certainly the F-variety $\mathbf{V}\langle U_1 \rangle$ generated by U_1 is contained in the F-variety $\mathbf{V}_F[x^2 = x, \ xy = yx]$. In fact

$$\mathbf{V}\langle U_1 \rangle = \mathbf{V}_F[x^2 = x, \ xy = yx],$$

but this is less obvious. Suppose that $u = v$ is an equation satisfied by all the monoids in $\mathbf{V}\langle U_1 \rangle$. Then in particular it is satisfied by U_1 itself. Recall from Section 1.7 that if w is a word in a free monoid then $C(w)$, the *content* of w, is defined as the set of letters appearing in w. If there exists X in $C(u)\backslash C(v)$ then substituting 0 for x and 1 for all other letters gives $0 = 1$, a contradiction. Equally, there can be no x in $C(v)\backslash C(u)$. Thus $C(u) = C(v)$. It follows that v can be reached from u by finitely many applications of the equations $xy = yx$ and $x^2 = x$. Thus

$$\mathbf{V}\langle U_1 \rangle = \mathbf{V}_F[x^2 = x, \ xy = yx].$$

We denote $\mathbf{V}\langle U_1 \rangle$ by $\mathbf{SL}$. It is the F-variety of *semilattice monoids*. Using Theorem 7.3.14 we can now show

Theorem 7.3.15. *Let* $\mathbf{SL}$ *be the F-variety of semilattice monoids and let* A *be a finite alphabet. Then* $\mathcal{L}_{\mathbf{SL}}(A)$ *is the Boolean algebra generated by the languages* B^*, *where* $B \subseteq A$.

Proof. For each subset B of A let $\varphi_B : A \to U_1$ be given by

$$\varphi_B(a) = \begin{cases} 1 & \text{if } a \in B, \\ 0 & \text{if } a \notin B. \end{cases}$$

Then by Theorem 1.7.3 φ_B extends to a morphism $\varphi_B : A^* \to U_1$, and

$$\varphi_B^{-1}(1) = B^*.$$

Thus $\mathcal{L}_{\mathbf{SL}}(A)$ contains every B^*, and so contains the Boolean algebra generated by all the sets B^*.

Conversely, let $\varphi : A^* \to U_1$ be a morphism. If $B = \{a \in A : \varphi(a) = 1\}$ then

$$\varphi^{-1}(1) = B^*, \quad \varphi^{-1}(0) = A^* \backslash B^*.$$

If $L \in \mathcal{L}_{\mathrm{SL}}(A)$ then it is in the Boolean algebra generated by the sets $\varphi^{-1}(1)$, $\varphi^{-1}(0)$, where φ runs over all possible morphisms from A^* into U_1. Thus L is in the Boolean algebra generated by the sets B^* ($B \subseteq A$). ∎

We end this Section with a result, due to Pin, Straubing and Thérien (1984), which demonstrates that even for an F-variety generated by a single monoid the application of the Variety Theorem is not a trivial matter.

Let $Z = \{1, p, q\}$ be the monoid with multiplication given by the table

	1	p	q
1	1	p	q
p	p	p	q
q	q	p	q

Then we have:

Theorem 7.3.16. *With the above definition of Z:*

(i) $\mathbf{V}\langle Z \rangle = \mathbf{V}_F[xyx = yx]$;

(ii) $\mathcal{L}_{\mathbf{V}\langle Z \rangle}(A)$ *is the Boolean algebra generated by all languages of the form B^* ($B \subseteq A$) and $A^* a B^*$ ($a \in A$ and $B \subset A$).*

Proof. (i) Since Z satisfies the equation $xyx = yx$, it is clear that

$$\mathbf{V}\langle Z \rangle \subseteq \mathbf{V}_F[xyx = yx]. \tag{7.3.17}$$

Conversely, suppose that $u = v$ is an equation satisfied by all monoids in $\mathbf{V}\langle Z \rangle$ and *not* deducible from $xyz = yx$. In particular, $u = v$ is satisfied by the monoid Z itself.

We may suppose that the equation $u = v$ is chosen so that $|u| + |v|$ is as small as possible. If u contains any factor of the form wzw with $w \neq 1$ then by (7.3.17) all monoids in $\mathbf{V}\langle Z \rangle$ satisfy the equation $u' = v$, where u' is obtained from u by substituting zw for wzw. Since $|u'| + |v| < |u| + |v|$ the equation $u' = v$ is deducible from $xyx = yx$. But then $u = v$ also is deducible from $xyx = yx$ and we have a contradiction. The only possible conclusion is that u (and by the same argument also v) contains no factor of the form wzw. Hence neither u nor v can contain a repeated letter.

If there exists x in $C(u)\backslash C(v)$ then substituting p for x and 1 for every other letter gives $p = 1$ in Z, a contradiction. Equally there can be no x in $C(v)\backslash C(u)$, and so $C(u) = C(v)$. Thus u and v must be permutations of each other:

$$u = x_1 x_2 \ldots x_k, \quad v = x_{\sigma(1)} x_{\sigma(2)} \cdots x_{\sigma(k)},$$

where $\sigma : \{1, 2, \ldots, k\} \to \{1, 2, \ldots, k\}$ is a bijection. Our assumption is still that $u = v$ is not deducible from $xyx = yx$, and so assuredly u and v are not identical. Hence there exist i, j in $\{1, 2, \ldots, k\}$ with $i < j$, $\sigma^{-1}(i) > \sigma^{-1}(j)$. Substituting p for x_i, q for x_j and 1 for all other letters now gives $pq = qp$ in Z, a contradiction. Hence

$$\mathbf{V}\langle Z \rangle = \mathbf{V}_F[xyx = yx],$$

as required.

(ii) We show first that the monoid $Z = \{1, a, b\}$ recognizes every B^* and every $A^* a B^*$. First, let $\varphi_1 : A^* \to Z$ be the morphism defined by

$$\varphi_1(b) = 1 \ (b \in B), \quad \varphi_1(c) = p \ (c \in A\backslash B).$$

Then $\varphi_1^{-1}(1) = B^*$. Secondly, for a given a in $A\backslash B$ let $\varphi_2 : A^* \to Z$ be the morphism defined by

$$\varphi_2(b) = 1 \ (b \in B), \quad \varphi_2(a) = p, \quad \varphi_2(c) = q \ (c \in A\backslash B, \ c \neq a).$$

Then clearly $A^* a B^* \subseteq \varphi_2^{-1}(p)$, since for every $z = uav$ with $u \in A^*$ and $v \in B^*$ we have

$$\varphi_2(z) = \varphi_2(u).p.1 = p.$$

Conversely, if $z = x_1 x_2 \ldots x_k \in \varphi_2^{-1}(p)$ then $\varphi_2(z) = p$ and so the letter a appears somewhere in z. We consider x_j, the *last* such appearance, after which we must have $\varphi_2(x_i) = 1$, since otherwise $\varphi(z) = q$. So there exists j in $\{1, 2, \ldots, k\}$ such that

$$x_j = a, \quad x_{j+1}, \ldots, x_k \in B;$$

hence

$$z = (x_1 \ldots x_{j-1})a(x_{j+1} \ldots x_k) \in A^* a B^*.$$

We have now shown that Z recognizes languages of the forms

$$B^* \ (B \subseteq A) \quad \text{and} \quad A^* a B^* \ (B \subset A, \ a \in A\backslash B).$$

Hence all such languages are in $\mathcal{L}_{\mathbf{V}\langle Z \rangle}(A)$, and consequently the Boolean algebra generated by languages of this type is contained in $\mathcal{L}_{\mathbf{V}\langle Z \rangle}(A)$.

Conversely, suppose that $L \in \mathcal{L}_{\mathbf{V}\langle Z \rangle}(A)$. Then by Theorem 7.3.15 L is in the Boolean algebra generated by the sets $\varphi^{-1}(z)$ for all elements z of Z and all morphisms $\varphi : A^* \to Z$. We must show that this Boolean algebra is contained in the one generated by the sets of the type

$$B^* \ (B \subseteq A) \text{ and } A^*aB^* \ (B \subset A, \ a \in A\backslash B).$$

Accordingly, let $\varphi : A^* \to Z$ be a morphism. Then φ is determined by its effect on the elements of A. Let

$$\varphi^{-1}(1) = B, \quad \varphi^{-1}(p) = P, \quad \varphi^{-1}(q) = Q;$$

thus $B \cup P \cup Q = A^*$. If P and Q are both empty then

$$\varphi^{-1}(1) = A^*, \quad \varphi^{-1}(p) = \varphi^{-1}(q) = \emptyset. \tag{7.3.18}$$

If exactly one of P and Q is non-empty (say P), then

$$\varphi^{-1}(1) = B^*, \quad \varphi^{-1}(p) = B^*\backslash A^*, \quad \varphi^{-1}(q) = \emptyset. \tag{7.3.19}$$

If P and Q are both non-empty then the image of an element $w = x_1 x_2 \ldots x_k$ of A^* is 1 if and only if all x_i are in B, is p if and only if the last x_i not in B is in P, and is q if and only if the last x_i not in B is in Q. That is,

$$\varphi^{-1}(1) = B^*, \quad \varphi^{-1}(p) = A^*a_1B^*, \quad \varphi^{-1}(q) = A^*a_2B^*, \tag{7.3.20}$$

where $a_1 \in P \subseteq A\backslash B$, $a_2 \in Q \subseteq A\backslash B$.

In all cases ((7.3.18), (7.3.19), (7.3.20)) $\varphi^{-1}(z)$ is in the Boolean algebra generated by the languages

$$B^* \ (B \subseteq A) \text{ and } A^*aB^* \ (B \subset A, \ a \in A\backslash B). \qquad \blacksquare$$

7.4 Star-free languages and aperiodic monoids

Our final illustration of the correspondence given by the Variety Theorem (Theorem 7.3.11) is significant enough to merit a new section. Historically it predates the Variety Theorem, being due to Schutzenberger (1965).

We shall write $\mathbf{Ap}$ for the F-variety of *aperiodic* monoids. This is the F-variety $\mathbf{V}[\![x^n = x^{n+1}]\!]$ ultimately defined by the sequence of equations $x^n = x^{n+1}$ $(n \geq 1)$. It is not hard to obtain other characterizations of the monoids in $\mathbf{Ap}$. We confine ourselves to two. (For the definition of $W(z)$ see Section 7.1.)

Theorem 7.4.1. *Let M be a finite monoid. Then the following statements are equivalent:*
  (i) $M \in \mathbf{Ap}$;
  (ii) M *has no non-trivial subgroups;*
  (iii) *for all z in M, $(zM \cap Mz)\backslash W(z) = \{z\}$.*

Proof. (i) $\Rightarrow$ (ii). Let $M \in \mathbf{Ap}$ and let z be an element of a subgroup H of M. Denote the identity of H by e. For some sufficiently large n we have $z^{n+1} = z^n$, from which it follows by cancellation within the group H that $z = e$. Thus $H = \{e\}$.

  (ii) $\Rightarrow$ (i). From (1.4.7) we know that for every element x in a finite monoid M there exist integers $m \geq 0$, $r \geq 1$ such that $x^{m+r} = x^m$. Moreover,
$$\langle x \rangle = \{x, x^2, \ldots, x^{m+r-1}\},$$

and $\{x^m, \ldots, x^{m+r-1}\}$ is a subgroup of M of order r. We are assuming that all subgroups are trivial, and so we conclude that $r = 1$ for every x. That is, $M \in \mathbf{Ap}$.

  (i) $\Rightarrow$ (iii). Let $v \in (zM \cap Mz)\backslash W(z)$. Then there exist a, b in M such that
$$v = za = bz. \tag{7.4.2}$$

Also $v \notin W(z)$ and so by definition $z \in MvM$; that is, there exist c, d in M such that
$$z = cvd. \tag{7.4.3}$$

From (7.4.2) and (7.4.3) we have
$$v = cvda = c^2 v(da)^2 = \cdots = c^n v(da)^n = \cdots,$$

and, choosing n so that $c^{n+1} = c^n$, we deduce that $cv = v$. A similar argument using $v = bcvd$ gives $vd = v$, and we deduce that $v = cvd = z$. Thus
$$(zM \cap Mz)\backslash W(z) = \{z\}.$$

  (iii) $\Rightarrow$ (ii) Let H be a subgroup of M with identity e and let $v \in H$. Then $v \notin W(e)$ since $e = v^{-1}ve \in MvM$. Also
$$v = ev = ve \in eM \cap Me.$$

Thus
$$v \in (eM \cap Me)\backslash W(e) = \{e\};$$

that is, $v = e$. We deduce that every subgroup of M is trivial. ∎

A further property of aperiodic monoids is given by the next result:

Theorem 7.4.4. *If M is an aperiodic monoid then $W(1) = M \backslash \{1\}$.*

Proof. Suppose that $z \notin W(1)$, so that $1 \in MzM$. Thus there exists a, b in M such that $azb = 1$. Hence $az(bz) = z$, and more generally

$$a^n z (bz)^n = z \quad (n = 1, 2, \ldots).$$

If we choose n so that $a^n = a^{n+1}$ then we obtain

$$z = a^{n+1} z (bz)^n = az.$$

Similarly $z = zb$ and so $z = azb = 1$. Since $1 \notin W(1)$ the result is now clear. ∎

In proving the main theorem of this section we shall require to make use of a construction known as the *Schutzenberger product* (or the *Boolean product*). To describe this construction, let us consider two monoids A, B. For every $X \subseteq A \times B$ and every $a \in A$, $b \in B$ we define

$$aX = \{(ax, y) : (x, y) \in X\}, \quad Xb = \{(x, yb) : (x, y) \in X\}. \quad (7.4.5)$$

The *Schutzenberger product* $A \Diamond B$ is defined to be the set

$$A \Diamond B = \{(a, X, b) : a \in A, \, b \in B, \, X \subseteq A \times B\} \quad (7.4.6)$$

endowed with the multiplication

$$(a_1, X_1, b_1)(a_2, X_2, b_2) = (a_1 a_2, X_1 b_2 \cup a_1 X_2, b_1 b_2). \quad (7.4.7)$$

That this is an associative operation follows from the easily verified set-theoretic identity

$$(X_1 b_2 \cup a_1 X_2)b_3 \cup a_1(a_2 X_3) = (X_1 b_2)b_3 \cup a_1(X_2 b_3 \cup a_2 X_3).$$

Indeed $A \Diamond B$ is even a monoid, the identity element being $(1_A, \emptyset, 1_B)$.

The importance of this construction in the context of this section lies in the following result:

Theorem 7.4.8. *If $A, B \in \mathbf{Ap}$ then so does $A \Diamond B$.*

Proof. Let $A, B \in \mathbf{Ap}$, and let H be a subgroup of $A \Diamond B$ with identity (e, E, f). This identity must be an idempotent of $A \Diamond B$ and so

$$e^2 = e, \; f^2 = f, \; Ef \cup eE = E. \quad (7.4.9)$$

Now consider the projection morphism $\pi : A \Diamond B \to A \times B$ given by

$$\pi(a, X, b) = (a, b).$$

The image of H under π is a subgroup of $A \times B$ with identity (e, f). Since $A \times B \in \mathbf{Ap}$ the image is in fact the trivial subgroup $\{(e, f)\}$, and so every element of H is of the form (e, X, f) for some $X \subseteq A \times B$.

Since H is a subgroup it follows that for every (e, X, f) in H there exists (e, X', f) in H such that

$$(e, X', f)(e, X, f) = (e, E, f).$$

Also,

$$(e, E, f)(e, X, f) = (e, X, f) = (e, X, f)(e, E, f),$$
$$(e, E, f)(e, X', f) = (e, X', f) = (e, X', f)(e, E, f).$$

In set-theoretic terms these equations give

$$X'f \cup eX = E, \tag{7.4.10}$$
$$Ef \cup eX = X = Xf \cup eE, \tag{7.4.11}$$
$$Ef \cup eX' = X' = X'f \cup eE.$$

From (7.4.11) and (7.4.9) we obtain

$$X = (Ef \cup eX) \cup Ef = (Xf \cup eE) \cup Ef$$
$$= Xf \cup (Ef \cup eE) = Xf \cup E,$$

and so $E \subseteq X$. The same argument also gives us that $E \subseteq X'$. Then from (7.4.10) and (7.4.11) we have

$$X = Ef \cup eX \subseteq X'f \cup eX = E.$$

Thus $X = E$ and so $H = \{(e, E, f)\}$, a trivial subgroup. By Theorem 7.4.1 it now follows that $A \Diamond B \in \mathbf{Ap}$. ∎

Recall now that every rational language in a free monoid A^* can be obtained from finite languages by finitely many applications of the operations

$$\cup \text{ (union)}, \quad . \text{ (multiplication)}, \quad \backslash \text{ (complementation)}$$

$$\text{and } (\ )^* \text{ (Kleene's star operation)}.$$

(The complementation operation is not strictly necessary, but it is clear from Exercise 2.7 that it does no harm.) A rational language that can be obtained from finite languages without the use of $(\ )^*$ (that is to say, using $\cup$, . and $\backslash$ only) is called *star-free*.

We can define an RL-map $\mathcal{S}$ by associating with every alphabet A the set $\mathcal{S}(A)$ of all star-free languages in A^*. We aim to prove the following theorem:

Theorem 7.4.12 (Schutzenberger's Theorem). *Let $\mathcal{L}_{\mathbf{Ap}}$ be the VRL-map associated with the F-variety $\mathbf{Ap}$ of aperiodic monoids. Then $\mathcal{L}_{\mathbf{Ap}} = \mathcal{S}$.*

Proof. We show first that $\mathcal{S}(A) \subseteq \mathcal{L}_{\mathbf{Ap}}(A)$ for every alphabet A. From Exercise 3.6 it follows that for every finite language L in A^* the syntactic monoid $\mathrm{Syn}(L)$ is aperiodic and so $L \in \mathcal{L}_{\mathbf{Ap}}(A)$. By Theorem 7.3.8 we have that if $L, L_1, L_2 \in \mathcal{L}_{\mathbf{Ap}}(A)$ then $L_1 \cup L_2$, $L_1 \cap L_2$ and $A^* \backslash L$ all belong to $\mathcal{L}_{\mathbf{Ap}}(A)$. To show that $\mathcal{S}(A) \subseteq \mathcal{L}_{\mathbf{Ap}}(A)$ it remains to prove the implication

$$L_1, L_2 \in \mathcal{L}_{\mathbf{Ap}}(A) \;\Rightarrow\; L_1 L_2 \in \mathcal{L}_{\mathbf{Ap}}(A). \tag{7.4.13}$$

So suppose that $L_1, L_2 \in \mathcal{L}_{\mathbf{Ap}}(A)$. By Theorem 3.1.6 there are finite monoids M_1, M_2 in $\mathbf{Ap}$, morphisms $\varphi_1 A^* \to M_1$, $\varphi_2 : A^* \to M_2$ and subsets $P_1 \subseteq M_1$, $P_2 \subseteq M_2$ such that

$$L_1 = \varphi_1^{-1}(P_1), \quad L_2 = \varphi_2^{-1}(P_2).$$

Let $M = M_1 \Diamond M_2$, the Schutzenberger product of M_1 and M_2, given by (7.4.6) and (7.4.7). By Theorem 7.4.8 we have $M \in \mathbf{Ap}$.

Define $\varphi : A^* \to M$ by the rule that

$$\varphi(w) = \big(\varphi_1(w), X(w), \varphi_2(w)\big),$$

where

$$X(w) = \big\{ \big(\varphi_1(u), \varphi_2(v)\big) : u, v \in A^*, \; uv = w \big\}. \tag{7.4.14}$$

It is convenient to have an alternative expression for $X(w)$. Let us denote the set of left factors of w by $LF(w)$, and if $u \in LF(w)$ let us denote the unique v in A^* such that $uv = w$ by $u^{-1}w$. Then

$$X(w) = \big\{ \big(\varphi_1(u), \varphi_2(u^{-1}w)\big) : u \in LF(w) \big\}.$$

To show that φ is a morphism we must show that for all w, w' in A^*

$$X(ww') \;=\; \varphi_1(w)X(w') \cup X(w)\varphi_2(w').$$

The key to this identity is the observation that

$$LF(ww') \;=\; LF(w) \cup wLF(w').$$

Then we have

$$\begin{aligned}
X(ww') &= \big\{ \big(\varphi_1(z), \varphi_2(z^{-1}ww')\big) : z \in LF(ww') \big\} \\
&= \big\{ \big(\varphi_1(u), \varphi_2(u^{-1}w)\big)\varphi_2(w') : u \in LF(w) \big\} \\
&\quad \cup \big\{ \varphi_1(w)\big(\varphi_1(v), \varphi_2(v^{-1}w')\big) : v \in LF(w') \big\} \\
&= \varphi_1(w)X(w') \cup X(w)\varphi_2(w'),
\end{aligned}$$

exactly as required.

Now let $Q = \varphi(L_1 L_2)$. Then certainly

$$L_1 L_2 \subseteq \varphi^{-1}(Q).$$

To show the reverse inclusion, let $w \in \varphi^{-1}(Q)$. Then $\varphi(w) \in Q$ and so there exist $l_1 \in L_1$, $l_2 \in L_2$ such that $\varphi(w) = \varphi(l_1 l_2)$, i.e., such that

$$\big(\varphi_1(w), X(w), \varphi_2(w)\big) = \big(\varphi_1(l_1 l_2), X(l_1 l_2), \varphi_2(l_1 l_2)\big).$$

Now by the definition (7.4.14) it is clear that $X(l_1 l_2)$ contains the pair $\big(\varphi_1(l_1), \varphi_2(l_2)\big)$. Hence so does $X(w)$ and so $l_1 l_2$ is a factorization of w. Thus $w \in L_1 L_2$ as required. This completes the proof of the implication (7.4.10) and so we have now established that

$$\mathcal{S}(A) \subseteq \mathcal{L}_{\mathbf{Ap}}(A).$$

To establish the reverse inclusion, that

$$\mathcal{L}_{\mathbf{Ap}}(A) \subseteq \mathcal{S}(A),$$

let us consider a language L in $\mathcal{L}_{\mathbf{Ap}}(A)$. Then we have a monoid M in **Ap**, a morphism $\varphi : A^* \to M$ and a subset P of M such that $L = \varphi^{-1}(P)$. Since $\operatorname{im} \varphi$ is a submonoid of M and hence still in **Ap**, there will be no loss of generality if we assume that φ maps A^* onto M. Before embarking on the main argument, it is helpful to establish two fairly technical preliminary lemmas. These make reference to the ideal $W(z)$ discussed in Section 7.1.

Lemma 7.4.15. *With the above definitions, suppose that M has a zero element, and let*

$$A_0 = \{a \in A : \varphi(a) = 0\}.$$

Then

$$\varphi^{-1}(0) = A^* A_0 A^* \cup A^* \Big[\bigcup_{p \in P} Z_p \Big] A^*,$$

where $P = \{p \in M : W(p) \geq 2\}$ and where each Z_p is of the form $a_1 \varphi^{-1}(p) a_2$ with a_1, a_2 in A and is such that $\varphi(Z_p) = 0$.

Proof. It is routine to verify that

$$A^* A_0 A^* \cup A^* \Big[\bigcup_{p \in P} Z_p \Big] A^* \subseteq \varphi^{-1}(0).$$

To prove the reverse inclusion, let $w \in \varphi^{-1}(0)$ and suppose that $w \notin A^* A_0 A^*$. Let u be a factor of w of smallest possible length such that

$\varphi(u) = 0$. Since $w \notin A^* A_0 A^*$ we must have $|u| \geq 2$ and we may write $u = a_1 v a_2$ with a_1, a_2 in A and v in A^*. Since v is a factor of w with $|v| < |u|$ we may assume that $\varphi(v) \neq 0$; let us write $\varphi(v) = p$. Thus $w \in A^* Z_p A^*$, where $Z_p = a_1 \varphi^{-1}(p) a_2$ and where $\varphi(Z_p) = \varphi(u) = 0$.

It remains to show that $W(p) \geq 2$. Since M contains a zero element we may assume (by virtue of the remarks immediately preceding Theorem 7.1.13) that $W(p) \neq \emptyset$. Suppose by way of contradiction that $|W(p)| = 1$. Then, by Theorem 7.1.13, $W(p) = \{0\}$. Hence $z \notin W(p)$ for all $z \neq 0$ in M and so p belongs to the ideal I, where

$$I = \bigcap_{z \in M \setminus \{0\}} M z M.$$

In particular,

$$p \in M \varphi(a_1) p M, \quad p \in M p \varphi(a_2) M.$$

From the first of these we have that

$$p = c \varphi(a_1) p d$$

for some c, d in M. From Theorem 7.1.10 we then deduce that $p \mathcal{L} \varphi(a_1) p$, i.e., that there exists g in M such that

$$p = g \varphi(a_1) p.$$

In a similar manner we deduce that

$$p = p \varphi(a_2) h$$

for some h in M, and then we have

$$p = g \varphi(a_1) p \varphi(a_2) h = g \varphi(u) h = 0,$$

a contradiction. Hence $W(p) \geq 2$. ■

Next, we have:

Lemma 7.4.16. *Let $z \in M$, where $M \in \mathbf{Ap}$, and let $\varphi : A^* \to M$ be a surjective morphism. Then:*

(i) $\varphi^{-1}(z) = \left[\varphi^{-1}(zM) \cap \varphi^{-1}(Mz)\right] \setminus \varphi^{-1}(W(z))$;
(ii) *If $z \neq 1$, then*

$$\varphi^{-1}(zM) = \varphi^{-1}(zM \cap W(z)) \cup \left[\bigcup_{p \in P} \varphi^{-1}(p) A_p\right] A^*,$$

where

$$P \subseteq \{p \in M : W(p) \supset W(z)\}, \quad A_p = \{a \in A : p \varphi(a) \in zM\}.$$

Proof. (i) This follows directly from Theorem 7.4.1 (iii).

(ii) It is routine to verify that

$$\varphi^{-1}(zM \cap W(z)) \cup \left[\bigcup_{p \in P} \varphi^{-1}(p)A_p\right] A^* \subseteq \varphi^{-1}(zM).$$

To show the reverse inclusion, consider an element w in $\varphi^{-1}(zM)$, where $z \neq 1$. If $\varphi(w) \in W(z)$ then $w \in \varphi^{-1}(zM \cap W(z))$. So suppose that $\varphi(w) \notin W(z)$, so that $z \in M\varphi(w)M$. That is, there exist c, d, g in M such that

$$z = c\varphi(w)d, \quad \varphi(w) = zg,$$

and from Theorem 7.1.9 it follows that $z \mathcal{R} \varphi(w)$.

Let u be the shortest left factor of w with the property that $z \mathcal{R} \varphi(u)$. If $u = 1$ (the empty word) then $\varphi(u) = 1$ and it follows from Theorem 7.1.11 that $z \in H_1$. Hence, since M is aperiodic, $z = 1$, a contradiction. So we may assume that $|u| \geq 1$ and write $u = va$, where $v \in A^*$, $a \in A$. Let $p = \varphi(v)$. Then

$$p\varphi(a) = \varphi(u) \in zM$$

and so $a \in A_p$. We have shown that

$$w \in vaA^* \subseteq \left[\varphi^{-1}(p)A_p\right]A^*.$$

To complete the proof we need to show that $W(z) \subset W(p)$. First, to show that $W(z) \subseteq W(p)$, suppose that $y \in W(z)$, so that $z \notin MyM$. Suppose by way of contradiction that $y \notin W(p)$, so that $p \in MyM$; then certainly $\varphi(u) \left(= p\varphi(a)\right) \in MyM$. But $\varphi(u) \mathcal{R} z$, and so

$$z \in \varphi(u)M \subseteq MyM,$$

a contradiction. Hence $y \in W(p)$. We have shown that $W(z) \subseteq W(p)$.

To show that the containment is in fact proper, suppose by way of contradiction that $W(z) = W(p)$. Since $z \notin W(z)$ by Theorem 7.1.13, we must have $z \notin W(p)$. Thus $p \in MzM$; that is, there exist c, d in M such that

$$p = czd.$$

On the other hand,

$$z \in \varphi(u)M = p\varphi(a)M$$

and so $z = pg$ for some g in M. By Theorem 7.1.9 it now follows that $z \mathcal{R} p$. Thus $z \mathcal{R} \varphi(v)$, where v is a factor of w and is *shorter* than u. This contradicts the definition of u. Hence $W(z) \subset W(p)$ as required. ∎

We now make use of these lemmas in carrying out an inductive proof of Schutzenberger's Theorem. Consider the proposition

$\mathbf{P}(n)$: *If $\varphi : A^* \to M$ is a morphism onto a finite aperiodic monoid M of order n, then $\varphi^{-1}(z) \in \mathcal{S}(A)$ for all z in M.*

If we can prove this for all $n \geq 1$, then this will be sufficient, since every L in $\mathcal{L}_{\mathbf{Ap}}(A)$ is expressible as

$$L = \bigcup_{z \in P} \varphi^{-1}(z)$$

for some morphism φ from A^* onto a finite aperiodic monoid M and some subset P of M.

It is easy to establish $\mathbf{P}(1)$, since there is only one morphism $\varphi : A^* \to \{1\}$, and $\varphi^{-1}(1) = A^*$, wich certainly belongs to $\mathcal{S}(A)$.

As regards $\mathbf{P}(2)$, notice that the only aperiodic monoid of order 2 is $U_1 = \{0, 1\}$, and we saw earlier (Theorem 7.3.15) that if $\varphi : A^* \to U_1$ is defined by

$$\varphi(a) = 1 \ (a \in B), \quad \varphi(a) = 0 \ (a \in A \backslash B)$$

then $\varphi^{-1}(1) = B^*$ and $\varphi^{-1}(0) = A^* \backslash B^*$. Now on the face of it B^* does not appear to be a star-free language, but in fact

$$A^* \backslash B^* = \bigcup_{a \in A \backslash B} A^* a A^* \in \mathcal{S}(A)$$

and hence

$$B^* = A^* \backslash (A^* \backslash B^*) \in \mathcal{S}(A).$$

This establishes $\mathbf{P}(2)$.

Suppose now that we have a surjective morphism $\varphi : A^* \to M$, where $M \in \mathbf{Ap}$, $|M| \geq 3$, and suppose inductively that $\mathbf{P}(n)$ is true for all $n < k$. Let $z \in M$.

We now distinguish three cases: (1) $|W(z)| \geq 2$; (2) $|W(z)| = 1$; (3) $W(z) = \emptyset$.

In case (1) we consider the natural morphism κ of M onto the Rees quotient $M/W(z)$. (See Section 7.1.) Since $z \notin W(z)$ we may identify $\kappa(z)$ with z. We have morphisms

$$A^* \xrightarrow{\varphi} M \xrightarrow{\kappa} M/W(z),$$

and $\varphi^{-1}(z) = (\kappa \circ \varphi)^{-1}(z)$. Certainly $M/W(z)$ is aperiodic and has order $n - |W(z)| + 1$. Hence by the induction hypothesis we may conclude that $(\kappa \circ \varphi)^{-1}(z) \in \mathcal{S}(A)$, i.e., that $\varphi^{-1}(z) \in \mathcal{S}(A)$.

In case (2) we apply Theorem 7.1.13 (ii) and deduce that M has a zero and $W(z) = \{0\}$. It is easy to see that $W(0) = \emptyset$; hence if we have

$|W(z)| = 1$ then we must have $z \neq 0$. Suppose first that $z \neq 1$. Then from Lemma 7.4.16 (ii)

$$\varphi^{-1}(zM) = \varphi^{-1}(zM \cap W(z)) \cup \left[\bigcup_{p \in P} \varphi^{-1}(p) A_p \right] A^*$$

$$= \varphi^{-1}(0) \cup \left[\bigcup_{p \in P} \varphi^{-1}(p) A_p \right] A^*.$$

Now, from Lemma 7.4.15 and from Case (1) above we may deduce that $\varphi^{-1}(0) \in \mathcal{S}(A)$. Also, for all p in P in the above formula we have $|W(p)| \geq 2$; hence, by Case (1), we have $\varphi^{-1}(p) \in \mathcal{S}(A)$. It follows that $\varphi^{-1}(zM) \in \mathcal{S}(A)$. By the dual version of Lemma 7.4.16 (ii) we can equally well show that $\varphi^{-1}(Mz) \in \mathcal{S}(A)$. Then finally, by Lemma 7.4.16 (i), we have

$$\varphi^{-1}(z) = \left[\varphi^{-1}(zM) \cup \varphi^{-1}(Mz) \right] \backslash \varphi^{-1}(0),$$

and so $\varphi^{-1}(z) \in \mathcal{S}(A)$.

To complete the proof of Case (2) we must consider the case where $z = 1$. Here $W(z) = M \backslash \{1\}$, by Theorem 7.4.4. Since $|M| \geq 3$ this gives $|W(z)| \geq 2$ and so this case cannot in fact arise.

In Case (3) Lemma 7.4.16 (ii) gives

$$\varphi^{-1}(zM) = \left[\bigcup_{p \in P} \varphi^{-1}(p) A_p \right] A^*,$$

and since $|W(p)| \geq 1$ for all p in P we may, by Cases (1) and (2), deduce that $\varphi^{-1}(p) \in \mathcal{S}(A)$. Hence $\varphi^{-1}(zM)$ (and dually $\varphi^{-1}(Mz)$) is contained in $\mathcal{S}(A)$. Since by Lemma 7.4.16 (i) we have

$$\varphi^{-1}(z) = \varphi^{-1}(zM) \cap \varphi^{-1}(Mz),$$

we deduce that $\varphi^{-1}(z) \in \mathcal{S}(A)$. This completes the proof of Theorem 7.4.12. ∎

Exercises 7

7.1. Show that an F-variety generated by a finite collection

$$M_1, M_2, \ldots, M_n$$

of monoids is generated by the single monoid $M_1 \times M_2 \times \cdots \times M_n$.

7.2. Show that an F-variety generated by a single monoid is defined (as opposed to ultimately defined) by a sequence of equations.

7.3. Show that the class of all monoids having no non-trivial subgroups is *not* a variety in the sense of Birkhoff. [Hint: consider the element

$$(a_1, a_2, a_3, \ldots) \in \langle a_1 \rangle \times \langle a_2 \rangle \times \langle a_3 \rangle \times \cdots,$$

where a_n satisfies $a_n^n = a_n^{n+1}$ $(n = 1, 2, 3, \ldots)$.]

7.4. Show that the F-varieties **FGp** (finite groups) and **Ap** (finite aperiodic monoids) are not finitely generated.

7.5. Let $\mathbf{V}_n$ be the F-variety of monoids generated by $\mathbf{Z}_n$, the cyclic group of order n. Show that $\mathcal{L}_{\mathbf{V}_n}(A)$ is the Boolean algebra generated by all the languages

$$L(a, k) = \{w \in A^* : |w|_a \equiv k \,(\mathrm{mod}\, n)\},$$

where $a \in A$ and $0 \le k < n$. [Hint: if

$$\varphi : A^* \to \mathbf{Z}_n = \{0, 1, \ldots, n-1\}$$

is a morphism, then show that

$$\varphi^{-1}(l) \in \bigcap_{a \in A} L(a, k_a),$$

where the integers k_a $(a \in A)$ are such that

$$\sum_{a \in A} \varphi(a) k_a \equiv l \,(\mathrm{mod}\, n).]$$

7.6. From the result that every finite abelian group is a direct product of cyclic groups, deduce that for the F-variety **FAb** of finite abelian groups $\mathcal{L}_{\mathbf{FAb}}(A)$ is the Boolean algebra generated by all languages

$$L(n, a, k) = \{w \in A^* : |w|_a \equiv k \,(\mathrm{mod}\, n)\},$$

where $a \in A$, $0 \le k < n$ and $n = 2, 3, \ldots$.

Many of the results concerning monoids in this chapter can be proved instead for semigroups. To do this we must redefine a language as a subset of the free semigroup A^+, and an RL-map $\mathcal{L}$ becomes a map from A into the set of all rational subsets of A^+. The definition of an F-variety of semigroups is obvious, and the Variety Theorem (Theorem 7.3.12) works for semigroups if we make the natural modifications. The next two exercises refer to F-varieties of semigroups.

7.7. Let k (≥ 2) be a fixed integer, and let $\mathbf{Nil}_k$ be the class of finite semigroups S with the property that S contains a zero element 0 and $S^k = \{0\}$.

(i) Show that $\mathbf{Nil}_k$ is an F-variety of semigroups.

(ii) Show that $\mathbf{Nil}_k$ is defined by the equation

$$x_1 x_2 \ldots x_k = y_1 y_2 \ldots y_k.$$

(iii) Show that $\mathcal{L}_{\mathbf{Nil}_k}(A)$ is the Boolean algebra generated by all languages $\{w\}$, where $1 \leq |w| < k$.

7.8. Let $\mathbf{Nil}$ be the class of finite semigroups S containing 0 with the property that $S^k = \{0\}$ for some $k \geq 2$.

(i) Show that $\mathbf{Nil}$ is an F-variety of semigroups.

(ii) Show that $\mathbf{Nil}$ is ultimately defined by the equations

$$x_1 x_2 \ldots x_n = y_1 y_2 \ldots y_n \quad (n \geq 1).$$

(iii) Show that $\mathcal{L}_{\mathbf{Nil}}(A)$ is the Boolean algebra consisting of all subsets of A^+ that are either finite or cofinite. (A set is said to be *cofinite* if its complement is finite.)

7.9. Use the Schutzenberger Theorem (Theorem 7.4.12) to deduce that the 12 element monoid encountered in Example 3.5.2 is aperiodic. Verify this directly.

7.10 Let $A = \{a, b\}$. Use the Schutzenberger Theorem to show that the language $(a^2)^*$ is not star-free, but that $(ab)^*$ is star-free. Find an expression for $(ab)^*$ that demonstrates the star-free property directly.

7.11. Let A be an arbitrary finite alphabet and let $n \geq 2$ be an integer. Show that the language

$$\{w \in A^* : |w| \equiv 0 \,(\mathrm{mod}\, n)\}$$

is not star-free.

Solutions to exercises

Exercises 1

1.1. (i)
$$A + B = (A \cap B') \cup (A' \cap B)$$
$$= [(A \cap B') \cup A'] \cap [(A \cap B') \cup B] \quad \text{by (4)},$$
$$= (A \cup A') \cap (B' \cup A') \cap (A \cup B) \cap (B' \cap B)$$
$$\text{by (4) and (1)},$$
$$= (A \cup B) \cap (B' \cup A') \quad \text{by (8), (5) and (2)},$$
$$= (A \cup B) \cap (A \cap B)' \quad \text{by (2) and (9)}.$$

(ii)

$$A \cap (B + C) = A \cap [(B \cap C') \cup (B' \cap C)]$$
$$= (A \cap B \cap C') \cup (A \cap B' \cap C) \quad \text{by (4) and (1)}.$$

$$(A \cap B) + (A \cap C) = [(A \cap B) \cap (A \cap C)'] \cup [(A \cap B)' \cap (A \cap C)]$$
$$= [(A \cap B) \cap (A' \cup C')] \cup [(A' \cup B') \cap (A \cap C)] \quad \text{by (9)}$$
$$= [(A \cap B \cap A') \cup (A \cap B \cap C')] \cup [(A' \cap A \cap C) \cup (B' \cap A \cap C)]$$
$$\text{by (4)}$$
$$= \emptyset \cup (A \cap B \cap C') \cup \emptyset \cup (A \cap B' \cap C) \quad \text{by (2), (8) and (6)}$$
$$= (A \cap B \cap C') \cup (A \cap B' \cap C) \quad \text{by (5)}.$$

(iii) $(A + B) + C = [(A + B) \cap C'] \cup [(A + B)' \cap C]$. Now,

$$(A + B) \cap C' = [(A \cap B') \cup (A' \cap B)] \cap C'$$
$$= (A \cap B' \cap C') \cup (A' \cap B \cap C'),$$

229

and

$$(A + B)' \cap C = [(A \cup B) \cap (A \cap B)']' \quad \text{by part (i) above}$$
$$= [(A \cup B)' \cup (A \cap B)] \cap C \quad \text{by (9) and (7)}$$
$$= [(A' \cap B') \cup (A \cap B)] \cap C \quad \text{by (9)}$$
$$= (A' \cap B' \cap C) \cup (A \cap B \cap C) \quad \text{by (4) and (1)}.$$

Thus

$$(A + B) + C$$
$$= (A \cap B \cap C) \cup (A \cap B' \cap C') \cup (A' \cap B \cap C') \cup (A' \cap B' \cap C).$$

Similarly,

$$A + (B + C) = [A \cap (B + C)'] \cup [A' \cap (B + C)],$$

and

$$A \cap (B + C)' = (A \cap B' \cap C') \cup (A \cap B \cap C),$$
$$A' \cap (B + C) = (A' \cap B \cap C') \cup (A' \cap B' \cap C).$$

Thus

$$A + (B + C)$$
$$= (A \cap B \cap C) \cup (A \cap B' \cap C') \cup (A' \cap B \cap C') \cup (A' \cap B' \cap C),$$

and the result follows.

1.2. (i) Begin by writing the list

$$1, \quad \tfrac{1}{2}, \quad \tfrac{1}{3}, \quad \tfrac{2}{3}, \quad \tfrac{1}{4}, \quad \tfrac{2}{4}, \quad \tfrac{3}{4}, \quad \tfrac{1}{5}, \quad \tfrac{2}{5}, \quad \tfrac{3}{5}, \quad \tfrac{4}{5}, \quad \tfrac{1}{6}, \quad \tfrac{2}{6}, \quad \cdots$$

Then cross off repetitions and obtain the list

$$1, \quad \tfrac{1}{2}, \quad \tfrac{1}{3}, \quad \tfrac{2}{3}, \quad \tfrac{1}{4}, \quad \tfrac{3}{4}, \quad \tfrac{1}{5}, \quad \tfrac{2}{5}, \quad \tfrac{3}{5}, \quad \tfrac{4}{5}, \quad \tfrac{1}{6}, \quad \tfrac{5}{6}, \quad \tfrac{1}{7}, \quad \cdots .$$

(ii) For $n = 10$ (for example) we have the sets

$$\{10\}, \ \{1, 9\}, \ \{2, 8\}, \ \{3, 7\}, \ \{4, 6\},$$

$$\{1, 2, 7\}, \ \{1, 3, 6\}, \ \{1, 4, 5\}, \ \{2, 3, 5\},$$

$$\{1, 2, 3, 4\}.$$

Then we make the single list

$$\{1\}; \ \{2\}; \ \{3\}, \ \{1, 2\}; \ \{4\}, \{1, 3\}; \ \{5\}, \ \{1, 4\}, \ \{2, 3\};$$

$$\{6\},\ \{1,5\},\ \{2,4\},\ \{1,2,3\};$$
$$\{7\},\ \{1,6\},\ \{2,5\},\ \{3,4\},\ \{1,2,4\};$$
$$\{8\},\ \{1,7\},\ \{2,6\},\ \{3,5\},\ \{1,2,5\},\ \{1,3,4\};$$
$$\{9\},\ \{1,8\},\ \{2,7\},\ \{3,6\},\ \{4,5\},\ \{1,2,6\},\{1,3,5\},\{2,3,4\};$$
$$\{10\},\ \{1,9\},\ \{2,8\},\ \{3,7\},\ \{4,6\},$$
$$\{1,2,7\},\ \{1,3,6\},\ \{1,4,5\},\ \{2,3,5\},\ \{1,2,3,4\},\dots.$$

1.3. (i) Let $\rho \subseteq \sigma$ and let $a\rho$ be a ρ-class. Then

$$x \in a\rho \ \Rightarrow\ x\,\rho\,a \ \Rightarrow\ x\,\sigma\,a \ \Rightarrow\ x \in a\sigma.$$

Thus $a\rho \subseteq a\sigma$ and so every σ-class is a union of one or more ρ-classes. Conversely, if every σ-class is a union of ρ-classes, suppose that $a\,\rho\,b$. Then a and b belong to the same ρ-class and hence certainly to the same σ-class. Thus $a\,\sigma\,b$, and we deduce that $\rho \subseteq \sigma$.

(ii) Each σ-class is a union of two ρ-classes:

$$\{0, \pm3, \pm6, \pm9, \dots\} = \{0 \pm 6, \pm12, \dots\} \cup \{\pm3, \pm9, \pm15, \dots\};$$

and

$$\{\dots, -5, -2, 1, 4, 7, 10, \dots\}$$
$$= \{\dots, -5, 1, 7, 13, \dots\} \cup \{\dots, -8, -2, 4, 10, \dots\},$$
$$\{\dots, -7, -4, -1, 2, 5, 8, \dots\}$$
$$= \{\dots, -7, -1, 5, 11, \dots\} \cup \{\dots, -4, 2, 8, 14, \dots\}.$$

(iii) Suppose that $|X/\rho| = k$ and that the ρ-classes are

$$A_1, A_2, \dots, A_k.$$

The σ-classes are unions of ρ-classes, and $\rho \subset \sigma$ (properly) if and only if at least two ρ-classes coalesce to form a σ-class, that is, if and only if $|X/\sigma| < k$.

1.4. The required Cayley table is

	e	a	f	b
e	e	a	f	b
a	a	e	b	f
f	f	b	f	b
b	b	f	b	f

S is a monoid with identity element e. Also $H_f = \{f, b\}$ is a subgroup which is not a submonoid.

1.5. The required Cayley table is

	0	e	f	a	b
0	0	0	0	0	0
e	0	e	0	a	0
f	0	0	f	0	b
a	0	0	a	0	e
b	0	b	0	f	0

The idempotents are e and f, and $H_e = \{e\}$, $H_f = \{f\}$.

1.6. (i) We obtain the effect of $(a_1 a_k) \circ (a_1 a_{k-1}) \circ \cdots \circ (a_1 a_2)$ on a_1 by reading from the right:

$$a_1 \rightarrow a_2 \rightarrow a_2 \rightarrow \cdots \rightarrow a_2.$$

In general, the effect on a_i (for $i = 2, \ldots, k - 1$) is

$$a_i \rightarrow \cdots \rightarrow a_i \rightarrow a_1 \rightarrow a_{i+1} \rightarrow \cdots \rightarrow a_{i+1},$$

and the effect on a_k is

$$a_k \rightarrow \cdots \rightarrow a_k \rightarrow a_1.$$

Since neither $(a_1 a_k) \circ (a_1 a_{k-1}) \circ \cdots \circ (a_1 a_2)$ nor $(a_1 a_2 \ldots a_k)$ has any effect on elements outside $\{a_1, a_2, \ldots, a_k\}$, we deduce that

$$(a_1 a_k) \circ (a_1 a_{k-1}) \circ \cdots \circ (a_1 a_2) = (a_1 a_2 \ldots a_k).$$

Since every element of S_n is a product (that is, a composition) of cycles, and every cycle is a product of transpositions, S_n is generated by all the transpositions (ij) $(i < j)$.

(ii) It is clear that ζ^{n-i} maps i to n and that ζ^i then maps n to i. Thus, for every i, ζ^n maps i to i. Hence $\zeta^n = \mathrm{id}$ and hence $\zeta^{-1} = \zeta^{n-1}$.

The effect of $\zeta \circ \tau \circ \zeta^{-1}$ on 2 is given by

$$2 \rightarrow 1 \rightarrow 2 \rightarrow 3,$$

and on 3 is given by

$$3 \to 2 \to 1 \to 2.$$

The effect on 1 is $1 \to n \to n \to 1$ and on $i \, (\geq 3)$ is

$$i \to i - 1 \to i - 1 \to i.$$

Thus $\zeta \circ \tau \circ \zeta^{-1} = (23)$. Now suppose inductively that

$$\zeta^{i-2} \circ \tau \circ \zeta^{-i+2} = (i - 1 \, i).$$

Then if $i \leq n - 1$ the effect of

$$\zeta^{i-1} \circ \tau \circ \zeta^{-i+1} = \zeta \circ \left(\zeta^{i-2} \circ \tau \circ \zeta^{-i+2} \right) \circ \zeta^{-1}$$

is given by

$$
\begin{aligned}
i &\to i - 1 \to i \to i + 1, \\
i + 1 &\to i \to i - 1 \to i, \\
j &\to j - 1 \to j - 1 \to j \quad (j \notin \{i, i+1, 1\}), \\
1 &\to n \to n \to 1.
\end{aligned}
$$

So

$$\zeta^{i-1} \circ \tau \circ \zeta^{-i+1} = (i \, i + 1). \tag{1}$$

The effect of

$$(j \, j + 1) \circ \cdots \circ (2 \, 3) \circ (1 \, 2) \circ (2 \, 3) \circ \cdots (j \, j + 1)$$

is given by

$$
\begin{aligned}
1 &\to 1 \to \cdots \to 1 \to 2 \to 3 \to 4 \to \cdots \to j + 1, \\
j + 1 &\to j \to j - 1 \to \cdots \to 2 \to 1 \to 1 \to 1 \to \cdots \to 1, \\
i &\to i \to \cdots \to i + 1 \to \cdots \to i + 1 \to i \to \cdots \to i, \\
&\qquad (2 \leq i \leq j), \\
i &\to i \to \cdots \to i \quad (j + 2 \leq i \leq n).
\end{aligned}
$$

Hence

$$(j \, j + 1) \circ \cdots \circ (2 \, 3) \circ (1 \, 2) \circ (2 \, 3) \circ \cdots (j \, j + 1) = (1 \, j + 1). \tag{2}$$

The effect of $\zeta^{i-1} \circ (1\ j+1) \circ \zeta^{-i+1}$ is given by

$$
\begin{aligned}
& i \to 1 \to j+1 \to i+j, \\
& i+j \to j+1 \to 1 \to i, \\
& k \to k-i+1 \to k-i+1 \to k \quad (k \ge i+1,\ k \ne i+j), \\
& k \to k-i+n+1 \to k-i+n-1 \to k \quad (k \le i-1).
\end{aligned}
$$

Thus

$$
\zeta^{i-1} \circ (1\ j+1) \circ \zeta^{-i+1} = (i\ i+j). \tag{3}
$$

(iii) We know that the transpositions $(i\ i+j)$ (where $1 \le i \le n-1$, $1 \le j \le n-i$) generate S_n. Hence, by (3), S_n is generated by ζ together with the transpositions $(1\ j+1)$. By (2) it then follows that S_n is generated by ζ and τ together with the transpositions $(2\ 3)$, $(3\ 4)\ \ldots$, $(n-1\ n)$. Finally, by (1), we deduce that S_n is generated by ζ and τ.

1.7. (i) Under $(1\ i) \circ \|1\ 2\| \circ (1\ i)$ we have

$$
\begin{aligned}
& i \to 1 \to 2 \to 2, \quad 2 \to 2 \to 2 \to 2, \\
& 1 \to i \to i \to 1, \quad j \to j \to j \to j\ (j \notin \{1,2,i\}).
\end{aligned}
$$

Thus
$$
(1\ i) \circ \|1\ 2\| \circ (1\ i) = \|i\ 2\|.
$$

Similarly, the effect of $(2\ j) \circ \|1\ 2\| \circ (2\ j)$ is given by

$$
\begin{aligned}
& 1 \to 1 \to 2 \to j, \quad 2 \to j \to j \to 2, \\
& j \to 2 \to 2 \to j, \quad i \to i \to i \to i\ (i \notin \{1,2,j\}),
\end{aligned}
$$

and so
$$
(2\ j) \circ \|1\ 2\| \circ (2\ j) = \|1\ j\|.
$$

The effect of

$$
(1\ i) \circ (2\ j) \circ \|1\ 2\| \circ (2\ j) \circ (1\ i) = (1\ i) \circ \|1\ j\| \circ (1\ i)
$$

is given by

$$
\begin{aligned}
& 1 \to i \to i \to 1, \quad i \to 1 \to j \to j, \\
& j \to j \to j \to j, \quad k \to k \to k \to k\ (k \notin \{1,i,j\}).
\end{aligned}
$$

Thus
$$
(1\ i) \circ \|1\ j\| \circ (1\ i) = \|i\ j\|.
$$

Finally, the effect of $(i\,j) \circ \|i\,j\| \circ (i\,j)$ is given by

$$i \to j \to j \to i, \quad j \to i \to j \to i,$$

$$k \to k \to k \to k \quad (k \notin \{i,j\}).$$

Thus

$$(i\,j) \circ \|i\,j\| \circ (i\,j) = \|j\,i\|.$$

The effect of these identities is to establish that the submonoid $\langle \zeta, \tau, \pi \rangle$ contains every $\|i\,j\|$.

(ii) We have

$$\big(\hat{\varphi} \circ \|i\,j\|\big)(i) = \hat{\varphi}(j) = \varphi(j) = \varphi(i),$$
$$\big(\hat{\varphi} \circ \|i\,j\|\big)(k) = \hat{\varphi}(k) = \varphi(k) \quad (k \neq i).$$

(iii) Notice that $\operatorname{im} \hat{\varphi} = \operatorname{im} \varphi \cup \{x\}$, and so

$$|\operatorname{im} \hat{\varphi}| = r + 1.$$

Applying this repeatedly, we find that a map φ in T_n such that $|\operatorname{im} \varphi| = r$ is expressible as

$$\theta \circ \|i_1 j_1\| \circ \cdots \circ \|i_{n-r} j_{n-r}\|,$$

where $\theta \in S_n$. It now follows from (ii) and from Exercise 1.6 that

$$\langle \zeta, \tau, \pi \rangle.$$

1.8. Let a be an element of M, where M is a submonoid of a finite group G. Denote the identity of the group by 1. By (1.4.7) there exist $m, r \geq 1$ such that $a^{m+r} = a^m$. It follows that

$$(a^m)^{-1} a^{m+r} = (a^m)^{-1} a^m,$$

i.e., that $a^r = 1$. The element a certainly has an inverse a^{-1} in G, and it now follows that $a^{-1} = a^{r-1} \in M$. Thus M is a group.

1.9. Suppose first that ρ satisfies (1.5.1), i.e., that

$$(a,b) \in \rho \text{ and } c \in S \ \Rightarrow \ (ca, cb) \in \rho, \ (ac, bc) \in \rho.$$

Suppose now that $(a,b), (c,d) \in \rho$. Then $(ac, bc) \in \rho$ and $(bc, bd) \in \rho$. Hence by transitivity we have $(ac, bd) \in \rho$, as required.

Suppose conversely that ρ satisfies (1.5.2), i.e., that

$$(a, b),\ (c, d)\ \in \rho \ \Rightarrow\ (ac, bd) \in \rho.$$

Let $(a, b)\ \in \rho$ and $c \in S$. Then $(c, c)\ \in \rho$ by reflexivity, and so $(ca, cb) \in \rho$ and $(ac, bc) \in \rho$.

1.10. In U we have the following factorizations of a^6:

$$a^6 = a^2(a^2 a^2) = (a^3)(a^3).$$

Since it is not possible to find u in U such that $a^3 = a^2 u$ or such that $a^2 = a^3 u$, the semigroup U is not equidivisible. Hence U is not free.

1.11. (i) It is clear that $S^2 = \{w \in A^+ : |w| \geq 2\}$. Thus

$$S \backslash S^2 = \{w \in A^+ : |w| = 1\} = A.$$

(ii) The set of generators $\{ab, ba, aba, bab\}$ is irreducible, in the sense that none is expressible as a product of the others. All other elements of U are finite products of these four elements, and so are in U^2. Thus

$$U \backslash U^2 = \{ab, ba, aba, bab\}.$$

If U is a free semigroup, then its generators are $x_1 = ab$, $x_2 = ba$, $x_3 = aba$, $x_4 = bab$, and two words in these generators are equal if and only if they are identical. However, it is easy to see that $x_1^3 = x_3 x_4$, and so U is *not* a free semigroup.

Alternatively, we see that U is not equidivisible, since

$$(ab)\big((ab)(ab)\big) = (aba)(bab),$$

and there is no u in U for which $abu = aba$ or for which $ab = abau$.

1.12. Let $l = \min\{|w| : w \in S\}$. Then $l \geq 1$. Evidently,

$$w \in S^n \ \Rightarrow\ |w| \geq nl.$$

For all w in A^+ we may choose n to be greater than $|w|/l$, and deduce that $w \notin S^n$. Hence

$$\bigcap_{n=1}^{\infty} S^n = \emptyset.$$

1.13. It is easy to see that ρ is an equivalence. Suppose now that $w\,\rho\,z$ and that $u \in S$. If $w = z$ then certainly $uw = uz$; if $|w|, |z| \geq 3$ then certainly $|uw|, |uz| \geq 3$. Thus $uw\,\rho\,uz$, and similarly $wu\,\rho\,zu$. Thus ρ is a congruence, with classes

$$\{1\}, \{a\}, \{b\}, \{a^2\}, \{ab\}, \{ba\}, \{b^2\}, \; Z(= \{w : |w| \geq 3\}).$$

The Cayley table is

	1	a	b	a^2	ab	ba	b^2	Z
1	1	a	b	a^2	ab	ba	b^2	Z
a	a	a^2	ab	Z	Z	Z	Z	Z
b	b	ba	b^2	Z	Z	Z	Z	Z
a^2	a^2	Z	Z	Z	Z	Z	Z	Z
ab	ab	Z	Z	Z	Z	Z	Z	Z
ba	ba	Z	Z	Z	Z	Z	Z	Z
b^2	b^2	Z	Z	Z	Z	Z	Z	Z
Z	Z	Z	Z	Z	Z	Z	Z	Z

1.14. We have, for all w, z in A^*,

$$\lambda(wz) = |wz| = |w| + |z|,$$
$$\varphi_a(wz) = a^{|wz|_a} = a^{|w|_a + |z|_a} = (a^{|w|_a})(a^{|z|_a}) = \varphi_a(w)\varphi_a(z).$$

Also $\lambda(1) = 0$, $\varphi(1) = a^0 = 1$. Thus λ and φ_a are morphisms. For all n in $\mathbf{N}^0$, $\lambda^{-1}(n)$ contains a^n, and so λ is surjective. For all a^n in a^*, $\varphi_a^{-1}(a^n)$ contains a^n, and so φ_a is surjective.

Exercises 2

2.1. (i) $L(\mathcal{A}) = (ab)^*$; (ii) $L(\mathcal{A}) = a^* \cup b^*$; (iii) $L(\mathcal{A}) = A^* abA^*$.

2.2. (i) $L(\mathcal{A}) = (ab)^* a^2 \cup (ab)^+$.

(ii) $\{1\}a = \{2,4\}, \{1\}b = \emptyset$; $\{2,4\}a = \{3\}, \{2,4\}b = \{1,5\}$; $\{3\}a = \emptyset, \{3\}b = \emptyset$; $\{1,5\}a = \{2,4\}, \{1,5\}b = \emptyset$; $\emptyset a = \emptyset b = \emptyset$. The state diagram is

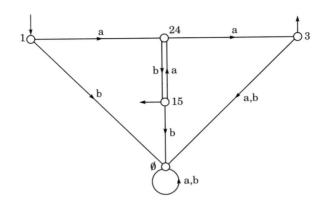

This is complete, deterministic and accessible. We trim it by removing all states that are not coaccessible; that is, we remove $\emptyset$ and all edges leading to it. We obtain

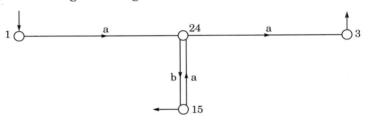

This is trim and deterministic, but not complete.

2.3. (i) $L(\mathcal{A}) = A^*a^2A^* \cup A^*b^2A^*$ (with $A = \{a, b\}$).

(ii)

$$\{1\}a = \{1, 2\}, \{1\}b = \{1, 3\};$$

$$\{1, 2\}a = \{1, 2, 4\}, \{1, 2\}b = \{1, 3\};$$

$$\{1, 3\}a = \{1, 2\}, \{1, 3\}b = \{1, 3, 4\};$$

$$\{1, 2, 4\}a = \{1, 2, 4\}, \{1, 2, 4\}b = \{1, 3, 4\};$$

$$\{1, 3, 4\}a = \{1, 2, 4\}, \{1, 3, 4\}b = \{1, 3, 4\}.$$

The state diagram is

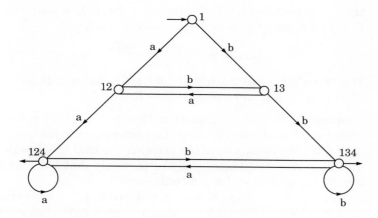

and A is complete, deterministic and trim.

2.4. Suppose first that L has the given property. Construct an automaton A by taking Q (the set of states) as $\{q_0, q_1, \ldots, q_{l+k-1}\}$, and defining the action of A on Q by the rules

$$q_i a = q_{i+1} \ (i = 0, 1, \ldots, l + k - 2), \quad q_{l+k-1} a = q_l.$$

Let q_0 be the initial state of A, and let

$$T = \{q_j : 0 \le j \le l + k - 1, a^j \in L\}.$$

It is thus clear that, for all $n \le l + k - 1$, $a^n \in L(A)$ if and only if $a^n \in L$.

Now let $n \ge l + k$, and suppose that $a^n \in L$. We can write $n = l + pk + r$ for some $p \ge 1$ and some r such that $0 \le r \le k - 1$. By the given condition we have $a^{l+r} \in L$, and as we have already seen this implies that $a^{l+r} \in L(A)$, i.e., that $q_0 a^{l+r} \in T$. Since by definition of A we have $q_0 a^n = q_0 a^{l+r}$ it thus follows that $q_0 a^n \in T$, i.e., that $a^n \in L(A)$.

Conversely, suppose that $a^n \in L(A)$ (with $n \ge l + k$). Then we may write n as $l + pk + r$, with $p \ge 1$ and $0 \le r \le k - 1$. Since $q_0 a^{l+r} = q_0 a^n \in T$ and since $l + r \le l + k - 1$ it now follows that $a^{l+r} \in L$, and hence (by the given condition) that $a^n \in L$.

Suppose now that L is recognizable, so that $L = L(A)$ for some automaton

$$A = (Q, A, i, T).$$

Write q_j for the state ia^j $(j = 1, 2, \ldots)$. The states $q_0, q_1, \ldots$ are not all distinct, since Q is finite. So let q_l (with $l \ge 0$) be the first state to repeat and let q_{l+k} (with $k \geq 1$) be its first reappearance. If $n \geq l$ we can write it as $l + m$, with $m \geq 0$. Then

$$q_{n+k} = q_0 a^{l+k+m} = q_0 a^{l+m} = q_n.$$

Thus for all $n \geq l$ we have that $q_0 a^n \in T$ if and only if $q_0 a^{n+k} \in T$. That is, $a^n \in L$ if and only if $a^{n+k} \in L$.

2.5. Suppose, by way of contradiction, that $\{a^{n^2} : n \geq 1\}$ is recognizable. Let l, k be defined as in the last exercise and choose n so that $n^2 \geq l$. It then follows from the last exercise that $n^2 + k = (n+1)^2$, $n^2 + 2k = (n + 2)^2$. These imply respectively that $k = 2n + 1$ and $k = 2n + 2$, and so we have a contradiction.

In the same way, if we suppose that $\{a^p : p \text{ is prime}\}$ is recognizable and choose $p \geq l$ then we must have that $p + k, p + 2k, \ldots$ are all prime. In particular $p + pk = p(k+1)$ is prime — and this is obviously false. So $\{a^p : p \text{ is prime}\}$ is not recognizable.

2.6. Denote $\{a^n b^{2n} : n \geq 1\}$ by L. Let N be as defined in the statement of the Pumping Lemma (2.2.7), and choose $n > N$. The factorization of $a^n b^{2n}$ into uvw predicted by the lemma must be of the form
$$u = a^p, \; v = a^q, \; w = a^r b^{2n},$$

where $p, r \geq 0$, $q > 0$ and $p + q + r = n$. But then the lemma implies that $uv^m w = a^{n+mq} b^{2n} \in L$ for all $m \geq 0$, and this is a contradiction.

The arguments for parts (ii) and (iii) are closely similar.

(iv) Denote $\{z^2 : z \in \{a, b\}^*\}$ by L and let N be as described in the Pumping Lemma. Let $z = a^n b$, where $n > N$. Then z^2 has a factorization uvw with $|uv| \leq N$ such that $uv^m w \in L$ for all $m \geq 0$. Now because $|uv| \leq N < n$ we must have

$$u = a^p, \; v = a^q, \; w = a^r b a^n b,$$

with $p, r \geq 0$, $q > 0$ and $p + q + r = n$. Hence by the Pumping Lemma $a^{n+2q} b a^n b \in L$. Now $|b a^n b| = n + 2 \leq n + 2q$. So if $a^{n+2q} b a^n b$ is to be a square then the first half of it consists entirely of a's. Since the second half contains b's we have a contradiction.

2.7. (i) We suppose that A is complete and deterministic, so that iw is uniquely defined for all w in A^*. If

$$A' = (Q, A, i, Q \backslash T)$$

then $w \in L(A')$ if and only if $iw \notin T$, i.e., if and only if $w \notin L$. Thus $L(A') = A^* \backslash L$.

(ii) Let
$$\mathcal{A} = (Q_1 \times Q_2, A, \psi, (i_1, i_2), T_1 \times T_2),$$
where
$$\psi\big((q_1, q_2), a\big) = \big(\varphi_1(q_1, a), \varphi_2(q_2, a)\big) \quad (a \in A)$$
and more generally
$$\psi\big((q_1, q_2), w\big) = \big(\varphi_1(q_1, w), \varphi_2(q_2, w)\big) \quad (w \in A^*).$$
Then
$$\psi\big((i_1, i_2), w\big) \in T_1 \times T_2$$
if and only if $\varphi_1(i_1, w) \in T_1$ and $\varphi_2(i_2, w) \in T_2$,

i.e., if and only if $w \in L_1$ and $w \in L_2$,

i.e., if and only if $w \in L_1 \cap L_2$.

So $L(\mathcal{A}) = L_1 \cap L_2$ as required.

2.8. By Exercise 2.7 and Kleene's Theorem (2.5.2) $A^* \backslash M \in \operatorname{Rat} A^*$ and so $L \cap (A^* \backslash M) \in \operatorname{Rat} A^*$.

2.9. Let $w \in A^*$. Then $w \in (L_1 \cup L_2)^R$ if and only if $w = z^R$, where $z \in L_1 \cup L_2$, i.e., if and only if $w \in L_1^R$ or $w \in L_2^R$, i.e., if and only if $w \in L_1^R \cup L_2^R$. Equally, $w \in (L_1 L_2)^R$ if and only if $w = (u_1 u_2)^R$, where $u_1 \in L_1$, $u_2 \in L_2$, i.e., if and only if $w = u_2^R u_1^R \in L_2^R L_1^R$. Finally, $w \in (L^*)^R$ if and only if $w = (u_1 u_2 \ldots u_n)^R$ for some $n \geq 0$ and some $u_1, u_2, \ldots, u_n$ in L, i.e., if and only if $w^R = u_n^R u_{n-1}^R \ldots u_1^R \in (L^R)^*$.

If $L \in \operatorname{Rat} A^*$ then L has a finite expression involving finite sets $L_1, L_2, \ldots, L_k$ and the operations $\cup$, . and *. By the above results, L^R has a very similar expression involving the finite sets $L_1^R, L_2^R, \ldots, L_k^R$ and the operations $\cup$, . and *, and so $L^R \in \operatorname{Rat} A^*$. Conversely, if $L^R \in \operatorname{Rat} A^*$ then by the above argument $L = (L^R)^R$ is in $\operatorname{Rat} A^*$.

2.10. If L is recognizable then by Exercise 2.9 so is $L^R = \{b^n a^m : m \geq n \geq 0\}$. Hence, since the rôles of a and b are interchangeable, so is $L' = \{a^n b^m : m \geq n \geq 0\}$. Then by Exercise 2.7 $L \cap L' = \{a^n b^n : n \geq 0\}$ is recognizable. This contradicts Theorem 2.2.6.

2.11. (i) Use the Pumping Lemma as in Example 2.2.8 but with $z = a^n b^2 a^n$.

(ii) Let $L = \{wcw^R : w \in \{a, b\}^*\}$, and suppose by way of contradiction that L is recognizable. By the Pumping Lemma there exists N such that every $z = wcw^R$ of sufficient length can be factorized into uvt, with $|uv| \leq N$, $v \geq 1$, in such a way that $uv^*t \subseteq L$. If we choose $|z| \geq 2N + 1$, so that $|w| \geq N$, then u and v must be subwords of the first w in z. More precisely,

$$w = uvx, \text{ with } x \in \{a, b\}^*.$$

By the Pumping Lemma we deduce that

$$uv^0 t = uxcw^R \in L,$$

and this is a contradiction.

2.12. The set a^*b^* is recognizable, being obviously rational. So if $L = \{w : |w|_a = |w|_b\}$ is recognizable then so is

$$L \cap a^*b^* = \{a^n b^n : n \geq 0\},$$

contrary to Theorem 2.2.6. Similarly, if $L' = \{w : |w|_a \geq |w|_b\}$ is recognizable then so is

$$L' \cap a^*b^* = \{a^m b^n : m \geq n\},$$

contrary to Exercise 2.10.

2.13. Evidently $\operatorname{Rec} A^* \subseteq \operatorname{Polyrec} A^*$. Conversely, suppose that L is recognized by a polyautomaton $\mathcal{A} = (Q, A, \varphi, I, T)$, with I as (non-empty) set of initial states. Thus $w \in L(\mathcal{A}) = L$ if and only if $\varphi(i, w) \in T$ for some i in I. Let

$$\mathcal{A}' = \big(\mathcal{P}(Q), A, \psi, I, T'\big),$$

where for each subset P of Q and each a in A,

$$\psi(P, a) = \{\varphi(p, a) : p \in P\},$$

and

$$T' = \{P \in \mathcal{P}(Q) : P \cap T \neq \emptyset\}.$$

Then $\mathcal{A}'$ is an ordinary automaton with single initial state I. Also

$$w \in L(\mathcal{A}')$$
if and only if $\psi(I, w) \in T'$,
i.e., if and only if $\{\varphi(i, w) : i \in I\} \cap T \neq \emptyset$,
i.e., if and only if $\varphi(i, w) \in T$ for some i in I,
i.e., if and only if $w \in L$.

2.14. (i) We have $i1 = i$, by definition. So if $1 \in L(\mathcal{A})$ then $i \in T$. That is, $i = t$.

(ii) A successful path is a path from i to i and is composed of a succession of paths each of which goes from i to i without visiting i in the way. So

$$L(\mathcal{A}) = \Big(Z\big(i, Q\backslash\{i\}, i\big)\Big)^*.$$

Certainly $a^*b^* \in \operatorname{Rec} A^*$. Since $1 \in a^*b^*$ any monoautomaton recognizing a^*b^* must be of the form $\mathcal{A} = (Q, A, i, i)$. It then follows that $a^*b^* = L^*$, where $L = Z\big(i, Q\backslash\{i\}, i\big)$. To see that this is not possible, notice that $a^*b^* = L^*$ implies that

$$a \in a^*b^* = L^*, \quad b \in a^*b^* = L^*,$$

and so $a, b \in L$. But then $ba \in L^*$, and we have a contradiction.

Exercises 3

3.1. The statements

$$(\forall x, y \in M)\, xay \in P \text{ if and only if } xby \in P,$$

$$(\forall x, y \in M)\, xay \notin P \text{ if and only if } xby \notin P$$

are equivalent. So $\sigma_P = \sigma_{M\backslash P}$.

3.2. Suppose first that $x\,\sigma_P\,y$. Let $u \in M$. Then, for all v in M,

$$v \in x^{-1}u^{-1}P \text{ if and only if } uxv \in P,$$
$$\text{i.e., if and only if } uyv \in P,$$
$$\text{i.e., if and only if } v \in y^{-1}u^{-1}P.$$

Hence $x^{-1}u^{-1}P = y^{-1}u^{-1}P$.

Conversely, suppose that (for all u in M) $x^{-1}u^{-1}P = y^{-1}u^{-1}P$. Then for all $u, v \in M$

$$uxv \in P \text{ if and only if } v \in x^{-1}u^{-1}P,$$
$$\text{i.e., if and only if } v \in y^{-1}u^{-1}P,$$
$$\text{i.e., if and only if } uyv \in P.$$

Thus $x\,\sigma_P\,y$.

3.3. By Theorem 3.2.8 the set $\{u^{-1}L : u \in A^*\}$ is finite: that is, every set $u^{-1}L$ coincides with one of a finite collection

$$u_1^{-1}L, \ u_2^{-1}L, \ \ldots, \ u_m^{-1}L.$$

Dually, every set Lv^{-1} coincides with one of a finite collection

$$Lv_1^{-1}, \ Lv_2^{-1}, \ldots, \ Lv_n^{-1}.$$

So every set $u^{-1}Lv^{-1}$ coincides with $u_i^{-1}Lv_j^{-1}$ for some i between 1 and m and some j between1 and n.

3.4. It is clear that $w \ \sigma_L \ w'$ if and only if $K(w) = K(w')$. Hence

$$w' \in w\sigma_L \text{ if and only if } K(w) = K(w'),$$
$$\text{i.e., if and only if } \{(u, v) : w' \in u^{-1}Lv^{-1}\} = K(w),$$
$$\text{i.e., if and only if } w' \in u^{-1}Lv^{-1} \text{ for all } (u, v) \in K(w)$$
$$\text{and } w' \notin u^{-1}Lv^{-1} \text{ for all } (u, v) \notin K(w),$$
$$\text{i.e., if and only if}$$

$$w' \in \left[\bigcap_{(u,v)\in K(w)} u^{-1}Lv^{-1} \right] \setminus \left[\bigcup_{(u,v)\notin K(w)} u^{-1}Lv^{-1} \right].$$

3.5. If neither x nor y is a segment of w then (in the notation of Exercise 3.3)

$$K(x) = K(y) = \emptyset$$

and so $x \ \sigma_{\{w\}} \ y$. If x is a segment of w then there exist u, v in A^* such that $uxv = w$. If $x \ \sigma_{\{w\}} \ y$ then we also have $uyv = w$. Thus $uxv = uyv$ in the free monoid A^*, and so by cancellation $x = y$.

If x is not a segment of w then for every y in A^* neither xy nor yx is a segment of w. So, writing L for $\{w\}$, we have

$$(x\sigma_L)(y\sigma_L) = (y\sigma_L)(x\sigma_L) = x\sigma_L,$$

and so $x\sigma_L$ acts as 0 in the syntactic monoid. If x and y are segments of w then

$$\{x\}\{y\} = (x\sigma_L)(y\sigma_L) = (xy)\sigma_L$$
$$= \begin{cases} \{xy\} & \text{if } xy \text{ is a segment of } w, \\ 0 & \text{otherwise.} \end{cases}$$

If $x_1, x_2, \ldots, x_{l+1} \in A^+$ then

$$(x_1\sigma_L)(x_2\sigma_L)\ldots(x_{l+1}\sigma_L) = (x_1x_2\ldots x_{l+1})\sigma_L.$$

Since $|x_1x_2\ldots x_{l+1}| \geq l+1$ the word $x_1x_2\ldots x_{l+1}$ cannot be a segment of w, and so

$$(x_1\sigma_L)(x_2\sigma_L)\ldots(x_{l+1}\sigma_L) = 0.$$

3.6. $K(x) = \{(u, v) \in A^* : uxv \in \{w_1, w_2, \ldots, w_n\}\}$, and this is clearly empty if $|x| \geq l+1$. If $|x| \geq l+1$ then $|xy|, |yx| \geq l+1$ for all y in A^* and so $K(xy) = K(yx) = K(x) = \emptyset$. Thus

$$(x\sigma_L)(y\sigma_L) = (y\sigma_L)(x\sigma_L) = x\sigma_L,$$

which is the required zero property. If $x_1, x_2, \ldots, x_{l+1} \in A^+$ then

$$|x_1x_2\ldots x_{l+1}| \geq l+1$$

and so $(x_1\sigma_L)(x_2\sigma_L)\ldots(x_{l+1}\sigma_L) = 0$.

3.7. The elements of $\mathrm{Syn}(\{aba, bab\})$ are

$$1, \ a, \ b, \ ab, \ ba, \ aba, \ bab, \ 0,$$

and the Cayley table is

	1	a	b	ab	ba	aba	bab	0
1	1	a	b	ab	ba	aba	bab	0
a	a	0	ab	0	aba	0	0	0
b	b	ba	0	bab	0	0	0	0
ab	ab	aba	0	0	0	0	0	0
ba	ba	0	bab	0	0	0	0	0
aba	aba	0	0	0	0	0	0	0
bab	bab	0	0	0	0	0	0	0
0	0	0	0	0	0	0	0	0

3.8. The transformation monoid of the given automaton is generated by

$$a = (12), \ b = (12\ldots n), \ c = \|12\|,$$

and so is the full transformation semigroup, by Exercise 1.7.

3.9. We have

$$\{1\}a = \{1,2\}, \qquad\qquad \{1\}b = \{1,4\},$$
$$\{1,2\}a = \{1,2,3\}, \qquad\qquad \{1,2\}b = \{1,4\},$$
$$\{1,4\}a = \{1,2\}, \qquad\qquad \{1,4\}b = \{1,4,5\},$$
$$\{1,2,3\}a = \{1,2,3\}, \qquad\qquad \{1,2,3\}b = \{1,3,4\},$$
$$\{1,4,5\}a = \{1,2,5\}, \qquad\qquad \{1,4,5\}b = \{1,4,5\},$$
$$\{1,3,4\}a = \{1,2,3\}, \qquad\qquad \{1,3,4\}b = \{1,3,4,5\},$$
$$\{1,2,5\}a = \{1,2,3,5\}, \qquad\qquad \{1,2,5\}b = \{1,4,5\},$$
$$\{1,3,4,5\}a = \{1,2,3,5\}, \qquad\qquad \{1,3,4,5\}b = \{1,3,4,5\},$$
$$\{1,2,3,5\}a = \{1,2,3,5\}, \qquad\qquad \{1,2,3,5\}b = \{1,3,4,5\}.$$

We obtain a deterministic, accessible automaton with 9 states. The initial state is $\{1\}$ and the terminal states are

$$\{1,2,3\}, \ \{1,4,5\}, \ \{1,2,5\}, \ \{1,3,4\},$$
$$\{1,2,3,5\}, \ \{1,3,4,5\}.$$

The equivalence ρ defined by (3.3.1) identifies all these terminal states (since $q^{-1}T = \{a,b\}^*$ for all of them) but otherwise brings about no identifications, since

$$\{1\}^{-1}T = A^* a^2 A^* \cup A^* b^2 A^*,$$
$$\{1,2\}^{-1}T = aA^* \cup A^* a^2 A^* \cup A^* b^2 A^*,$$
$$\{1,4\}^{-1}T = bA^* \cup A^* a^2 A^* \cup A^* b^2 A^*.$$

So the reduced automaton has 4 states and state diagram

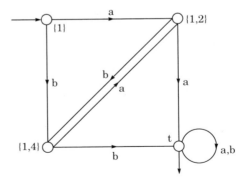

The algebraic method gives

$$L.a = aA^* \cup L = L_1,$$

$$L.b = bA^* \cup L = L_2,$$
$$L_1.a = A^*,$$
$$L_1.b = \{w : abw \in L\} = L_2,$$
$$L_2.a = \{w : baw \in L\} = L_1,$$
$$L_2.b = A^*,$$
$$A^*.a = A^*.b = A^*,$$

and a state diagram effectively identical to the one produced by the other method.

To find the syntactic monoid of L, note that (with a simplified notation) the transformation monoid of the minimal automaton is generated by

$$a = 2424, \quad b = 3344.$$

(We are writing 1 for $\{1\}$, 2 for $\{1, 2\}$, 3 for $\{1, 4\}$ and 4 for t; and $ijkl$ denotes the map that sends 1 to i, 2 to j, 3 to k and 4 to l.) Then

$$a^2 = b^2 = 4444 = 0,$$
$$ab = 3434, \ ba = 2244,$$
$$aba = a, \ bab = b.$$

Thus the syntactic monoid of L has 6 elements 0, 1, a, b, e $(= ab)$ and f $(= ba)$, with Cayley table

	1	a	b	e	f	0
1	1	a	b	e	f	0
a	a	0	e	0	a	0
b	b	f	0	b	0	0
e	e	a	0	e	0	0
f	f	0	b	0	f	0
0	0	0	0	0	0	0

3.10. (i) We have

$$\{1\}a = \{1, 2\}, \qquad \{1\}b = \emptyset,$$
$$\{1, 2\}a = \{1, 2\}, \qquad \{1, 2\}b = \{3\},$$
$$\{3\}a = \emptyset, \qquad \{3\}b = \{3\}.$$

The state diagram is

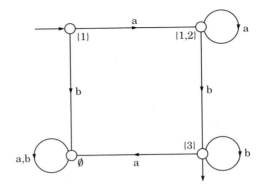

and the automaton is already reduced.

The algebraic method gives $i_L = L$ and

$$L.a = \{w : aw \in a^+b^+\} = a^*b^+ = L_1,$$
$$L.b = \{w : bw \in a^+b^+\} = \emptyset,$$
$$L_1.a = \{w : a^2w \in a^+b^+\} = L_1,$$
$$L_1.b = \{w : abw \in a^+b^+\} = b^* = L_2,$$
$$L_2.a = \{w : abaw \in a^+b^+\} = \emptyset,$$
$$L_2.b = \{w : ab^2w \in a^+b^+\} = L_2,$$

and the state diagram is in effect identical to the one previously obtained. Writing 1 for $\{1\}$, 2 for $\{1, 2\}$, 3 for $\{3\}$ and 4 for $\emptyset$, and using the notational convention of the previous exercise, we see that the minimal automaton is generated by

$$a = 2244, \quad b = 4334.$$

Then

$$a^2 = a, \qquad b^2 = b$$

$$ab = 3344, \quad ba = 4444.$$

The syntactic monoid has 5 elements 1, a, b, c $(= ab)$ and 0, and the Cayley table is

	1	a	b	c	0
1	1	a	b	c	0
a	a	a	c	c	0
b	b	0	b	0	0
c	c	0	c	0	0
0	0	0	0	0	0

(ii) This automaton is deterministic but not complete. We obtain a complete deterministic accessible automaton by adding a new 'sink' state 0 and defining

$$1b = 2a = 3b = 0a = 0b = 0.$$

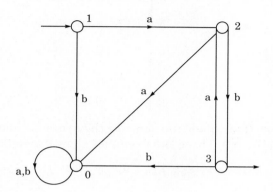

Algebraically, we have $i_L = L$ and

$$
\begin{aligned}
L.a &= b(ab)^* = L_1, \\
L.b &= \emptyset, \\
L_1.a &= \{w : a^2w \in (ab)^+\} = \emptyset, \\
L_1.b &= \{w : abw \in (ab)^+\} = (ab)^* = L_2, \\
L_2.a &= \{w : aw \in (ab)^*\} = b(ab)^* = L_1, \\
L_2.b &= \{w : bw \in (ab)^*\} = \emptyset, \\
\emptyset.a &= \emptyset.b = \emptyset,
\end{aligned}
$$

and the state diagram is identical to the one already found. The syntactic monoid of L is generated by

$$a = 2020, \quad b = 0300.$$

Then

$$a^2 = b^2 = 0000 \text{ (which we denote simply by 0)}$$
$$ab = 3030, \ ba = 0200, \ aba = a, \ bab = b.$$

The monoid has 6 elements 0, 1, a, b, e (= ab) and f (= ba), and the Cayley table is

	1	a	b	e	f	0
1	1	a	b	e	f	0
a	a	0	e	0	a	0
b	b	f	0	b	0	0
e	e	a	0	e	0	0
f	f	0	b	0	f	0
0	0	0	0	0	0	0

Notice that this is the same as the syntactic monoid obtained in Exercise 3.9. These two examples show incidentally that

$$\mathrm{Syn}(L_1) \simeq \mathrm{Syn}(L_2) \not\Rightarrow L_1 = L_2.$$

3.11. Let $a, b \in L$. Then $uav \in L$ for *all* u, v in A^*, and $ubv \in L$ for *all* u, v in A^*; so we certainly have (for all $u, v \in A^*$)

$$uav \in L \text{ if and only if } ubv \in L.$$

Hence $a\, \sigma_L\, b$. It follows that L is a single σ_L-class.

If $b\sigma_L$ is an arbitrary σ_L-class and $a\sigma_L = L$ (which happens precisely when $a \in L$, then

$$(a\sigma_L)(b\sigma_L) = (ab)\sigma_L = a\sigma_L\ (= L)$$

since $ab \in L$, and similarly

$$(b\sigma_L)(a\sigma_L) = a\sigma_L = L.$$

Thus L acts as a zero in $\mathrm{Syn}(L)$.

3.12. $P.a^i = \{w \in A^* : a^i w \in P\}$ contains ba^i. On the other hand $ba^i \notin P.a^j$ if $j \neq i$. So $P.a^i \neq P.a^j$. Thus

$$P,\ P.a,\ P.a^2, \ldots$$

are all distinct.

3.13. (i) Suppose that $(q_1, q_2) \in \tau_m$. Then

$$q_1^{-1} T \cap A_l = q_2^{-1} T \cap A_l$$

for all $l \leq m - 1$, and hence $(q_1, q_2) \in \tau_{m-1}$. We have shown that $\tau_m \subseteq \tau_{m-1}$.

Let $(q_1, q_2) \in \tau$. Then $q_1^{-1}T = q_2^{-1}T$ and so certainly

$$q_1^{-1}T \cap A_l = q_2^{-1}T \cap A_l$$

for all $l \geq 0$. Thus $(q_1, q_2) \in \tau_m$ for all $m \geq 0$. We have shown that

$$\tau \subseteq \bigcap_{m \geq 0} \tau_m.$$

To show the reverse inclusion, suppose that $(q_1, q_2) \in \tau_m$ for all $m \geq 0$. Thus

$$q_1^{-1}T \cap A_l = q_2^{-1}T \cap A_l$$

for all $l \geq 0$. If $w \in q_1^{-1}T$ then

$$w \in q_1^{-1}T \cap A_l = q_2^{-1}T \cap A_l \subseteq q_2^{-1}T,$$

where $m = |w|$. Thus $q_1^{-1}T \subseteq q_2^{-1}T$, and similarly $q_2^{-1}T \subseteq q_1^{-1}T$. Thus $(q_1, q_2) \in \tau$, and so

$$\tau = \bigcap_{m \geq 0} \tau_m$$

as required.

(ii) Suppose first that $(q_1, q_2) \in \tau_{m+1}$, so that

$$q_1^{-1}T \cap A_l = q_2^{-1}T \cap A_l \tag{1}$$

for all $l \leq m + 1$. Certainly $(q_1, q_2) \in \tau_0$. Let $a \in A$ and let $k \leq m$. Then

$$w \in (q_1 a)^{-1}T \cap A_k$$
$$\text{if and only if } |w| = k \text{ and } (q_1 a)w \in T,$$
$$\text{i.e., if and only if } |aw| = k + 1 \text{ and } q_1(aw) \in T,$$
$$\text{i.e., if and only if } aw \in q_1^{-1}T \cap A_{k+1},$$
$$\text{i.e., if and only if } aw \in q_2^{-1}T \cap A_{k+1}$$
$$\text{by (1), since } k + 1 \leq m + 1)$$
$$\text{i.e., if and only if } w \in (q_2 a)^{-1}T \cap A_k.$$

Thus

$$(q_1 a)^{-1}T \cap A_k = (q_2 a)^{-1}T \cap A_k$$

for all $k \leq m$, and so $(q_1 a, q_2 a) \in \tau_m$ as required.

Conversely, suppose that $q_1 \tau_0 q_2$ and that (for all a in A) $q_1 a \tau_m q_2 a$. Then

$$(q_1 a)^{-1} T \cap A_k = (q_2 a)^{-1} T \cap A_k \tag{2}$$

for all $k \leq m$. Let $w \in A^*$ with $|w| = l \geq 1$. Then $w = az$ for some a in A, z in A^*, with $|z| = l - 1$. For all l in $\{1, 2, \ldots, m + 1\}$ we have that

$$w \in q_1^{-1} T \cap A_l \text{ if and only if } |w| = l \text{ and } q_1 w \in T,$$
$$\text{i.e., if and only if } w = az, \ |z| = l - 1, \ q_1(az) \in T,$$
$$\text{i.e., if and only if } w = az, \ |z| = l - 1, \ (q_1 a)z \in T,$$
$$\text{i.e., if and only if } w = az, \ z \in (q_1 a)^{-1} T \cap A_{l-1},$$
$$\text{i.e., if and only if } w = az, \ z \in (q_2 a)^{-1} T \cap A_{l-1},$$
$$\text{(by (2), since } l - 1 \leq m)$$
$$\text{i.e., if and only if } w \in q_2^{-1} T \cap A_l.$$

Since the condition that $q_1 \tau_0 q_2$ gives the property that

$$q_1^{-1} T \cap A_0 = q_2^{-1} T \cap A_0,$$

we thus have

$$q_1^{-1} T \cap A_l = q_2^{-1} T \cap A_l$$

for all $l \leq m + 1$, and so $(q_1, q_2) \in \tau_{m+1}$, as required.

(iii) Suppose that $\tau_m = \tau_{m+1}$. Since $\tau_{m+2} \subseteq \tau_{m+1}$ in any event, we require only to prove that $\tau_{m+1} \subseteq \tau_{m+2}$. Accordingly, suppose that $q_1 \tau_{m+1} q_2$. The by part (ii) above $q_1 \tau_0 q_2$ and (for all a in A) $q_1 a \tau_m q_2 a$. Hence $q_1 \tau_0 q_2$ and (for all a in A) $q_1 a \tau_{m+1} q_2 a$. Hence (again by part (ii)) $q_1 \tau_{m+2} q_2$.

In this case the descent stabilizes:

$$\tau_0 \supseteq \tau_1 \supseteq \cdots \supseteq \tau_m = \tau_{m+1} = \tau_{m+2} = \cdots,$$

and

$$\tau = \bigcap_{j \geq 0} \tau_j = \tau_0 \cap \tau_1 \cap \ldots \cap \tau_m = \tau_m.$$

(iv) If P is a finite set and if ρ, σ are equivalences on P such that $\rho \subseteq \sigma$, then $|P/\sigma| < |P/\rho|$ if and only if $\rho \subset \sigma$. So if

$$\tau_0 \supset \tau_1 \supset \tau_2 \supset \cdots \text{ (properly)}$$

then
$$|Q/\tau_0| < |Q/\tau_1| < |Q/\tau_2| < \cdots \text{ (strictly)}.$$

An infinite ascent in a finite set is not possible, and so we eventually reach τ_m such that
$$\tau_m = \tau_{m+1} = \tau_{m+2} = \cdots$$

and $\tau = \tau_m$. Moreover, since $|Q/\tau_j| \geq j + 1$, we must have $m \leq |Q| - 1$.

3.14. It is clear that ρ_L is an equivalence. To show that it is a right congruence, suppose that $w_1 \, \rho_L \, w_2$ and $z \in A^*$. Then $u \in (w_1 z)^{-1}L$ if and only if $w_1 z u \in L$, i.e., if and only if $z u \in w_1^{-1}L = w_2^{-1}L$, i.e., if and only if $u \in (w_2 z)^{-1}L$. Thus $w_1 z \, \rho_L \, w_2 z$, as required.

If $w_1 \in L$ and $w_1 \rho_L w_2$, then $1 \in w_1^{-1}L = w_2^{-1}L$ and so $w_2 \in L$. Thus ρ_L respects L. Now let σ be a right congruence on A^* respecting L, and suppose that $w_1 \, \sigma \, w_2$. Let $z \in A^*$. Then $w_1 z \, \sigma \, w_2 z$, and it follows that

$$z \in w_1^{-1}L \text{ if and only if } w_1 z \in L,$$
$$\text{i.e., if and only if } w_2 z \in L,$$
$$\text{i.e., if and only if } z \in w_2^{-1}L.$$

Notice now that A^*/ρ_L is finite if and only if $Q_L = \{u^{-1}L : u \in A^*\}$ is finite, that is, if and only if L is rational.

Exercises 4

4.1. (i) ab^*; (ii) $a\{a,b\}^*b$; (iii) $\{a,b\}^*aba\{a,b\}^*$.

4.2. (i) The minimal automaton has state diagram

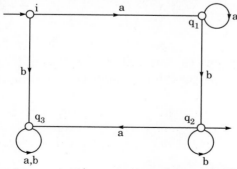

where $i = L$, $q_1 = a^*b^+$, $q_2 = b^*$ and $q_3 = \emptyset$. The corresponding regular grammar has productions

$$i \to aq_1, \; q_1 \to aq_1, \; q_1 \to bq_2, \; q_2 \to bq_2, \; q_2 \to 1.$$

(ii) The minimal automaton has state diagram

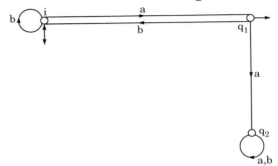

where $i = L$, $q_1 = L \cup aA^*$, $q_2 = A^*$. The corresponding regular grammar has productions

$$i \to aq_1, \; i \to bi, \; q_1 \to aq_2, \; q_1 \to bi,$$

$$q_2 \to aq_2, \; q_2 \to bq_2, \; q_2 \to 1.$$

(iii) The minimal automaton has state diagram

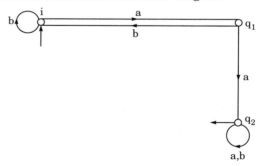

(i.e., as for part (ii) but with a complementary set of terminal states). The corresponding regular grammar has productions

$$i \to aq_1, \; i \to bi, \; q_1 \to aq_2, \; q_1 \to bi,$$

$$q_2 \to aq_2, \; q_2 \to bq_2, \; i \to 1, \; q_1 \to 1.$$

(iv) The minimal automaton has state diagram

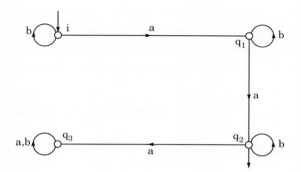

where $i = L$, $q_1 = \{w : |w|_a = 1\}$, $q_2 = b^*$, $q_3 = \emptyset$. The corresponding regular grammar has productions

$$i \to aq_1, \; i \to bi, \; q_1 \to aq_2, \; q_1 \to bq_1,$$

$$q_2 \to bq_2, \; q_2 \to 1.$$

4.3. Suppose that $L = L(\Gamma)$, where Γ is left regular. By the left-right dual of Lemma 4.2.3 we may assume that every production of Γ is of the form $\alpha \to \beta a$ (with $\alpha, \beta \in V \backslash A$, $a \in A$) or of the form $\alpha \to 1$, and the derivation of a typical word $w = a_1 a_2 \ldots a_n$ in L is of the form

$$\sigma \Rightarrow \beta_{n-1} a_n \Rightarrow \beta_{n-2} a_{n-1} a_n \Rightarrow \cdots$$

$$\cdots \Rightarrow \beta_0 a_1 a_2 \ldots a_n \Rightarrow a_1 a_2 \cdots a_n,$$

where

$$\sigma \to \beta_{n-1} a_n, \; \beta_{n-1} \to \beta_{n-2} a_{n-1}, \ldots, \beta_1 \to \beta_0 a_1, \; \beta_0 \to 1$$

are productions in Γ.

Form a right linear grammar Γ' having a production $\alpha \to a\beta$ whenever $\alpha \to \beta_a$ is a production in Γ, and having a production $\alpha \to 1$ whenever Γ has a production $\alpha \to 1$. Then

$$\sigma \Rightarrow a_n \beta_{n-1} \Rightarrow a_n a_{n-1} \beta_{n-2} \Rightarrow \cdots$$

$$\cdots \Rightarrow a_n a_{n-1} \ldots a_1 \beta_0 \Rightarrow a_n a_{n-1} \ldots a_1$$

is a derivation of w^R in Γ'. In fact it is easy to see that $L(\Gamma') = \left(L(\Gamma)\right)^R$. Thus $\left(L(\Gamma)\right)^R$ is a rational set, and by Exercise 2.9 it follows that $L(\Gamma)$ is rational.

Conversely, if $L(\Gamma)$ is rational then so is $\left(L(\Gamma)\right)^R$ and so $\left(L(\Gamma)\right)^R$ can be generated by a right linear grammar. By an argument dual to the argument in the last paragraph we find that $L(\Gamma) = \left(\left(L(\Gamma)\right)^R\right)^R$ is generated by a left linear grammar.

4.4. (i) $\{a^n b^{2n} : n \geq 0\}$. (ii) $\{ww^R : w \in \{a, b\}^+\}$, the set of palindromes of even length. Neither language is regular, by Exercises 2.6 and 2.11.

4.5. (i) $\sigma \to a\lambda b^2$, $\lambda \to a\lambda b^2$, $\lambda \to 1$.
(ii) $\sigma \to \alpha\beta$, $\alpha \to a\lambda b$, $\lambda \to a\lambda b$, $\beta \to b\mu a$, $\mu \to b\mu a$, $\lambda \to 1$, $\mu \to 1$.
(iii)

$$\sigma \to a\lambda b, \ \lambda \to a\mu, \ \mu \to a\mu, \ \mu \to a\mu b,$$

$$\lambda \to \nu b, \nu \to \nu b, \nu \to a\nu b, \mu \to 1, \nu \to 1.$$

(iv) Modify the grammar studied in the proof of Theorem 4.3.7, by interchanging σ and λ:

$$\sigma \to a, \ \sigma \to a\lambda, \ \sigma \to b\sigma^2, \ \lambda \to a\mu,$$

$$\lambda \to b\sigma, \ \mu \to b, \ \mu \to b\lambda, \ \mu \to a\mu^2.$$

(v) $\sigma \to a\sigma a$, $\sigma \to b\sigma b$, $\sigma \to c$.

4.6. The required grammar has productions

$$\sigma \to a\sigma a, \ \sigma \to b\sigma b, \ \sigma \to a, \ \sigma \to b.$$

In Chomsky Normal Form this is

$$\sigma \to \beta_a\gamma, \ \gamma \to \sigma\beta_a, \ \sigma \to \beta_b\delta, \ \delta \to \sigma\beta_b,$$

$$\sigma \to a, \ \sigma \to b, \ \beta_a \to a, \ \beta_b \to b,$$

and a leftmost derivation of $a^2 baba^2$ is

$$\sigma \Rightarrow \beta_a\gamma \Rightarrow a\gamma \Rightarrow a\sigma\beta_a \Rightarrow a\beta_a\gamma\beta_a$$
$$\Rightarrow a^2\gamma\beta_a \Rightarrow a^2\sigma\beta_a^2 \Rightarrow a^2\beta_b\delta\beta_a^2 \Rightarrow a^2 b\delta\beta_a^2$$
$$\Rightarrow a^2 b\sigma\beta_b\beta_a^2 \Rightarrow a^2 ba\beta_b\beta_a^2 \Rightarrow a^2 bab\beta_a^2$$
$$\Rightarrow a^2 baba\beta_a \Rightarrow a^2 baba^2.$$

4.7. (i) A typical derivation is

$$\sigma \Rightarrow a\lambda a \Rightarrow ab\lambda a^2 \Rightarrow aba\lambda a^3$$
$$\Rightarrow (ab)^2\lambda a^4 \Rightarrow (ab)^2 a^4,$$

and $L(\Gamma) = \{wa^{|w|} : w \in \{a, b\}^+\}$.

(ii) If $L = L(\Gamma)$ were regular then $L \cap b^* a^*$ would be regular also, i.e., $\{b^m a^n : n \geq m\}$ would be regular. But from Exercise 2.9 we know that this is not so.

(iii) In Chomsky Normal Form the grammar is

$$\sigma \to \beta_a \gamma, \; \gamma \to \lambda \beta_a, \; \sigma \to \beta_b \gamma,$$

$$\lambda \to \beta_a \gamma, \; \lambda \to \beta_b \gamma, \; \sigma \to \beta_a^2, \; \sigma \to \beta_b \beta_a,$$

$$\lambda \to \beta_a^2, \; \lambda \to \beta_b \beta_a, \; \beta_a \to a, \; \beta_b \to b,$$

and a leftmost derivation of $ab^2 a^5$ is

$$\sigma \Rightarrow \beta_a \gamma \Rightarrow a\gamma \Rightarrow a\lambda\beta_a \Rightarrow a\beta_b\gamma\beta_a$$

$$\Rightarrow ab\gamma\beta_a \Rightarrow ab\lambda\beta_a^2 \Rightarrow ab\beta_b\gamma\beta_a^2 \Rightarrow ab^2\gamma\beta_a^2$$

$$\Rightarrow ab^2\lambda\beta_a^3 \Rightarrow ab^2\beta_a^5 \Rightarrow ab^2 a\beta_a^4 \Rightarrow ab^2 a^2 \beta_a^3$$

$$\Rightarrow ab^2 a^3 \beta_a^2 \Rightarrow ab^2 a^4 \beta_a \Rightarrow ab^2 a^5.$$

4.8. A suitable grammar is described by the productions

$$\sigma \to \lambda\mu, \; \lambda \to a\lambda b, \; \mu \to b^2 \mu a,$$

$$\lambda \to 1, \mu \to 1.$$

Stage 1 of the reduction gives productions

$$\sigma \to \lambda\mu, \; \lambda \to a\lambda b, \; \mu \to b^2 \mu a,$$

$$\lambda \to ab, \; \mu \to b^2 a$$

and gives a grammar generating $L \backslash \{1\}$. A grammar in Chomsky Normal Form generating $L \backslash \{1\}$ is then decribed by the productions

$$\sigma \to \lambda\mu, \; \lambda \to \beta_a \gamma, \; \gamma \to \lambda\beta_b, \; \mu \to \beta_b \delta_1,$$

$$\delta_1 \to \beta_b \delta_2, \; \delta_2 \to \mu\beta_a, \; \lambda \to \beta_a \beta_b, \; \mu \to \beta_b \zeta,$$

$$\zeta \to \beta_b \beta_a, \; \beta_a \to a, \; \beta_b \to b.$$

Then to this list of productions add

$$\sigma \to \alpha\beta, \; \alpha \to 1, \; \beta \to 1.$$

Exercises 5

5.1 (i) Let $Q = \{i, q, t\}$, $M = \{\zeta, \alpha\}$, $T = \{t\}$, $\mu = \emptyset$ and let

$$\rho(i, \zeta, a) = (i, \alpha^2\zeta), \quad \rho(q, \alpha, a) = (i, \alpha^3),$$
$$\rho(i, \alpha, b) = (q, 1), \quad \rho(q, \alpha, b) = (q, 1),$$

$$\rho(q, \zeta, 1) = (t, \zeta).$$

Then a typical successful computation is

$$(i, \zeta, a^n b^{2n}) \overset{*}{\to} (i, \alpha^{2n}\zeta, b^{2n}) \to (q, \alpha^{2n-1}\zeta, b^{2n-1})$$
$$\overset{*}{\to} (q, \zeta, 1) \to (t, \zeta, 1).$$

Moreover, if a word w in $\{a, b\}^+$ begins with b, or if it is in a^+ or in $a^+ b^+ a^+ \{a, b\}^*$, or if it is of the form $a^n b^p$ with $p \neq 2n$, then the computation fails. So $L(\mathcal{P}) = \{a^n b^{2n} : n \geq 1\}$.

(ii) Let $Q = \{i, q, q', t\}$, $M = \{\zeta, \alpha, \beta\}$, $T = \{t\}$, $\mu = \emptyset$, and let

$$\rho(i, \zeta, a) = (i, \alpha\zeta), \quad \rho(i, \alpha, a) = (i, \alpha^2),$$
$$\rho(i, \alpha, b) = (q, 1), \quad \rho(q, \alpha, b) = (q, 1),$$
$$\rho(q, \zeta, b) = (q, \beta\zeta), \quad \rho(q, \beta, b) = (q, \beta^2),$$
$$\rho(q, \beta, a) = (q', 1), \quad \rho(q', \beta, a) = (q', 1),$$

$$\rho(q', \zeta, 1) = (t, \zeta).$$

Then a typical successful computation is

$$(i, \zeta, a^p b^{p+q} a^q) \overset{*}{\to} (i, \alpha^p\zeta, b^{p+q} a^q)$$
$$\to (q, \alpha^{p-1}\zeta, b^{p+q-1} a^q) \overset{*}{\to} (q, \zeta, b^q a^q)$$
$$\overset{*}{\to} (q, \beta^q\zeta, a^q) \to (q', \beta^{q-1}\zeta, a^{q-1})$$
$$\overset{*}{\to} (q', \zeta, 1) \to (t, \zeta, 1),$$

and a computation beginning with (i, ζ, w) fails unless $w \in a^+ b^+ a^+$ and unless $w = a^p b^r a^q$ with $r = p + q$.

(iii) Let $Q = \{i, q, t\}$, $M = \{\zeta, \alpha\}$, $T = \{t\}$, $\mu = \emptyset$, and let

$$\rho(i, \zeta, a) = (i, \alpha\zeta), \quad \rho(i, \alpha, a) = (i, \alpha^2),$$
$$\rho(i, \alpha, b) = (q, 1), \quad \rho(q, \alpha, b) = (q, 1),$$

$$\rho(q, \alpha, 1) = (t, \alpha).$$

Then a typical successful computation is

$$(i, \zeta, a^m b^n) \overset{*}{\to} (i, \alpha^m \zeta, b^n) \to (q, \alpha^{m-1} \zeta, b^{n-1})$$
$$\overset{*}{\to} (q, \alpha^{m-n} \zeta, 1) \quad \text{(provided } m > n\text{)}$$
$$\to (t, \alpha^{m-n}, 1).$$

The computation fails if $w \notin a^+ b^+$ or if $w = a^m b^n$ with $m \leq n$.

(iv) Let $Q = \{i, q, t\}$, $M = \{\zeta, \alpha, \beta\}$, $T = \{t\}$, $\mu = \emptyset$, and let

$$\rho(i, \zeta, a) = (i, \alpha\zeta), \quad \rho(i, \alpha, a) = (i, \alpha^2),$$
$$\rho(i, \beta, a) = (i, 1), \quad \rho(i, \zeta, b) = (i, \beta, \zeta),$$
$$\rho(i, \beta, b) = (i, \beta^2), \quad \rho(i, \alpha, b) = (i, 1),$$
$$\rho(i, \alpha, 1) = (q, 1), \quad \rho(q, \zeta, 1) = (t, \zeta).$$

Then a typical successful computation is

$$(i, \zeta, ab^2 aba^2) \to (i, \alpha\zeta, b^2 aba^2) \to (i, \zeta, baba^2)$$
$$\to (i, \beta\zeta, aba^2) \to (i, \zeta, ba^2) \to (i, \beta\zeta, a^2)$$
$$\to (i, \zeta, a) \to (i, \alpha\zeta, 1) \to (q, \zeta, 1) \to (t, \zeta, 1).$$

In general, if $w = uv$ then the computation beginning with (i, ζ, w) arrives at $(i, \psi(u)\zeta, v)$, where

$$\psi(u) = \begin{cases} \alpha^{|u|_a - |u|_b} & \text{if } |u|_a \geq |u|_b, \\ \beta^{|u|_b - |u|_a} & \text{if } |u|_b > |u|_a. \end{cases}$$

Eventually it reaches $(i, \psi(w)\zeta, 1)$, and the computation fails at this stage unless $|w|_a > |w|_b$. If $|w|_a > |w|_b$ then the next stage is

$$(q, \alpha^{|w|_a - |w|_b - 1} \zeta, 1),$$

and now the computation fails unless $|w|_a - |w|_b - 1 = 0$.

(v) Let $Q = \{i, q, t\}$, $M = \{\zeta, \alpha\}$, $T = \{t\}$, $\mu = \emptyset$, and let

$$\rho(i, \zeta, a) = \{(i, \alpha\zeta)\}, \quad \rho(i, \alpha, a) = \{(i, \alpha^2), (q, 1)\},$$
$$\rho(i, \zeta, b) = \{(i, \alpha\zeta)\}, \quad \rho(i, \alpha, b) = \{(i, \alpha^2)\},$$
$$\rho(q, \alpha, a) = \{(q, 1)\}, \quad \rho(q, \zeta, 1) = \{(t, \zeta)\}.$$

Then a typical successful computation is

$$(i, \zeta, a^3 b^2 a^7) \overset{*}{\to} (i, \alpha^6 \zeta, a^6) \to (q, \alpha^5 \zeta, a^5)$$
$$\overset{*}{\to} (q, \zeta, 1) \to (t, \zeta, 1).$$

If $w = uv$ then $(i, \alpha^{|u|}\zeta, v)$. If the computation is to succeed then it must at some stage invoke the second choice for $\rho(i, \alpha, a)$. This is available if $v = av'$ and the next step then takes us to

$$(q, \alpha^{|u|-1}\zeta, v').$$

The computation thus succeeds if and only if $v = a^{|u|}$ for some factorization uv of w.

5.2. Look first at the successful computation

$$(i, \zeta, a^3 b^4 a^{10}) \overset{*}{\to} (i, a^6 \zeta, b^4 a^{10}) \to (q, \alpha^7 \zeta, b^3 a^{10})$$
$$\overset{*}{\to} (q, \alpha^{10}\zeta, a^{10}) \to (q', \alpha^9 \zeta, a^9) \overset{*}{\to} (q', \zeta, 1) \to (t, \zeta, 1).$$

The PDA can enter state q only if w has a left factor $a^m b$ and can enter state q' only if w has a factor $a^m b^n a$. At this point the memory stack contains $\alpha^{2m+n-1}\zeta$ and the computation fails unless the third entry of the ID is a^{2m+n-1}. So the language recognized by the PDA is $\{a^m b^n a^{2m+n} : m, n \geq 1\}$.

5.3. Look first at computations

$$(i, \zeta, a^2 b^2 a^2) \overset{*}{\to} (i, \alpha^2 \zeta, b^2 a^2) \to (q, \alpha\zeta, ba^2)$$
$$\to (i, \zeta, ba^2) \to (i, \beta^2 \zeta, a^2) \overset{*}{\to} (i, \zeta, 1) \to (t, \zeta, 1),$$
$$(i, \zeta, ab^2 a^3) \to (i, \alpha\zeta, b^2 a^3) \to (q, \zeta, ba^3)$$
$$\to (i, \beta\zeta, ba^3) \to (i, \beta^3 \zeta, a^3) \overset{*}{\to} (i, \zeta, 1) \to (t, \zeta, 1).$$

Notice that in both cases $|w|_a = 2|w|_b$.

Next, we show by induction on the length of w that there is a computation

$$(i, \zeta, w) \overset{*}{\to} (i, \psi(w)\zeta, 1),$$

where

$$\psi(w) = \begin{cases} \alpha^{|w|_a - 2|w|_b} & \text{if } |w|_a - 2|w|_b \geq 0, \\ \beta^{2|w|_b - |w|_a} & \text{if } |w|_a - 2|w|_b < 0. \end{cases}$$

If $|w| = 1$ then either $w = a$ (and $\psi(w) = \alpha$) or $w = b$ (and $\psi(w) = \beta^2$). We have computations

$$(i, \zeta, a) \to (i, \alpha\zeta, 1), \quad (i, \zeta, b) \to (i, \beta^2\zeta, 1)$$

as required.

Now suppose first that $w = za$, with $z \in A^*$, $|z| = n - 1$, $|w| = n$. We may assume that there is a computation

$$(i, \zeta, z) \overset{*}{\to} (i, \psi(z)\zeta, 1),$$

and hence we have a computation

$$(i, \zeta, w) \overset{*}{\to} (i, \psi(z)\zeta, a).$$

If $|z|_a - 2|z|_b \geq 0$ then the next step in the computation is

$$(i, \alpha^{|z|_a - 2|z|_b}\zeta, a) \to (i, \alpha^{|z|_a - 2|z|_b + 1}\zeta, 1) = (i, \alpha^{|w|_a - 2|w|_b}\zeta, 1);$$

if $|z|_a - 2|z|_b < 0$ then the next step is

$$(i, \beta^{2|z|_b - |z|_a}\zeta, a) \to (i, \beta^{2|z|_b - |z|_a - 1}\zeta, 1) = (i, \beta^{2|w|_b - |w|_a}\zeta, 1).$$

In both cases we have a computation

$$(i, \zeta, w) \overset{*}{\to} (i, \psi(w)\zeta, 1).$$

Now suppose that $w = zb$. As in the previous case we may assume that there is a computation

$$(i, \zeta, w) \overset{*}{\to} (i, \psi(z)\zeta, b).$$

We now distinguish three cases: (i) $|z|_a - 2|z|_b \geq 2$; (ii) $|z|_a - 2|z|_b = 1$; (iii) $|z|_a - 2|z|_b \leq 0$.
In case (i) we have

$$|w|_a - 2|w|_b = |z|_a - 2|z|_b - 2 \geq 0,$$

and the next steps in the computation are

$$(i, \alpha^{|z|_a - 2|z|_b}\zeta, b) \to (q, \alpha^{|z|_a - 2|z|_b - 1}\zeta, 1)$$
$$\to (i, \alpha^{|z|_a - 2|z|_b - 2}\zeta, 1) = (i, \psi(w)\zeta, 1).$$

In case (ii) $|w|_a - 2|w|_b = -1$, and the next steps in the computation are

$$(i, \alpha\zeta, b) \to (q, \zeta, 1) \to (i, \beta\zeta, 1) = (i, \psi(w)\zeta, 1).$$

In case (iii) the next step in the computation is

$$(i, \beta^{2|z|_b - |z|_a}\zeta, b) \to (i, \beta^{2|z|_b - |z|_a + 2}\zeta, 1) = (i, \psi(w)\zeta, 1).$$

In all cases there is a computation

$$(i, \zeta, w) \overset{*}{\to} (i, \psi(w)\zeta, 1)$$

as required.

At this point the computation fails unless $\psi(w) = 1$. So the language recognized by the given PDA is $\{w \in \{a,b\}^* : |w|_a = 2|w|_b\}$.

5.4. (i) A simple modification of the PDA of 5.1(iv) is what we require. Let $Q = \{i, t\}$, $M = \{\zeta, \alpha, \beta\}$, $\mu = \emptyset$, and let

$$\begin{aligned}
\rho(i, \alpha, a) &= (i, \alpha^2), & \rho(i, \beta, a) &= (i, 1), \\
\rho(i, \zeta, b) &= (i, \beta\zeta), & \rho(i, \beta, b) &= (i, 1), \\
\rho(i, \alpha, b) &= (i, 1), & \rho(i, \alpha, 1) &= (t, \alpha).
\end{aligned}$$

A typical successful computation is

$$(i, \zeta, b^2 a^5 b) \xrightarrow{*} (i, \beta^2\zeta, a^5 b) \xrightarrow{*} (i, \zeta, a^3 b)$$
$$\xrightarrow{*} (i, \alpha^3\zeta, b) \rightarrow (i, \alpha^2\zeta, 1) \rightarrow (t, \alpha^2\zeta, 1).$$

(ii) We can easily modify this specification so that recognition is by empty memory stack. We change the last specification to

$$\rho(i, \alpha, 1) = (t, 1)$$

and add

$$\rho(t, \alpha, 1) = (t, 1), \quad \rho(t, \zeta, 1) = (t, 1),$$

so that whenever the computation enters the terminal phase the memory stack is eaten up.

Using the algorithm described in the proof of Theorem 5.4.1, we take the set of non-terminal symbols as

$$(\{i, t\} \times \{\zeta, \alpha, \beta\} \times \{i, t\}) \cup \{\sigma\}$$

—13 symbols in all. Write

$$\begin{aligned}
[i, \zeta, i] &= \lambda_1, & [i, \alpha, i] &= \mu_1, & [i, \beta, i] &= \nu_1, \\
[i, \zeta, t] &= \lambda_2, & [i, \alpha, t] &= \mu_2, & [i, \beta, t] &= \nu_2, \\
[t, \zeta, i] &= \lambda_3, & [t, \alpha, i] &= \mu_3, & [t, \beta, i] &= \nu_3, \\
[t, \zeta, t] &= \lambda_4, & [t, \alpha, t] &= \mu_4, & [t, \beta, t] &= \nu_4.
\end{aligned}$$

Then we have productions

$$\sigma \to \lambda_1, \qquad\qquad \sigma \to \lambda_2$$

$$
\begin{array}{ll}
\lambda_1 \to a\mu_1\lambda_1 & \lambda_1 \to b\nu_1\lambda_1 \\
[\lambda_1 \to a\mu_2\lambda_3] & [\lambda_1 \to b\nu_2\lambda_3] \\
\lambda_2 \to a\mu_1\lambda_2 & \lambda_2 \to b\nu_1\lambda_2 \\
\lambda_2 \to a\mu_2\lambda_4 & \lambda_2 \to b\nu_2\lambda_4
\end{array}
$$

$$
\begin{array}{ll}
\mu_1 \to a\mu_1^2 & \nu_1 \to b\nu_1^2 \\
[\mu_1 \to a\mu_2\mu_3] & [\nu_1 \to b\nu_2\nu_3] \\
\mu_2 \to a\mu_1\mu_2 & \nu_2 \to b\nu_1\nu_2 \\
\mu_2 \to a\mu_2\mu_4 & \nu_2 \to b\nu_2\nu_4
\end{array}
$$

and

$$\nu_1 \to a, \ \mu_1 \to b, \ \mu_2 \to 1, \ \mu_4 \to 1, \ \lambda_4 \to 1.$$

(Since $\lambda_3, \mu_3, \nu_3, \nu_4$ never appear on the left of a production, no derivation of a word in A^* can involve them. So we may remove the productions inside square brackets.)

A derivation of (e.g.) b^2a^5b is

$$
\begin{aligned}
\sigma &\Rightarrow \lambda_2 \Rightarrow b\nu_1\lambda_2 \Rightarrow b^2\nu_1^2\lambda_2 \overset{*}{\Rightarrow} b^2a^2\lambda_2 \\
&\Rightarrow b^2a^3\mu_2\lambda_4 \Rightarrow b^2a^4\mu_2\mu_4\lambda_4 \Rightarrow b^2a^5\mu_1\mu_2\mu_4\lambda_4 \\
&\Rightarrow b^2a^5b\mu_2\mu_4\lambda_4 \overset{*}{\Rightarrow} b^2a^5b.
\end{aligned}
$$

5.5 (i) Let $L = \{a^{n^2} : n \geq 1\}$, and suppose that L is context-free. By the Pumping Lemma, for some sufficiently large n we can factorize a^{n^2} into $uvwxy$ in such a way that $|vx| \geq 1$ and $uv^mwx^my \in L$ for all $m \geq 0$. Now if $|vx| = k$ then

$$|uv^mwx^my| = n^2 + (m-1)k;$$

so L contains

$$a^{n^2}, \ a^{n^2+k}, \ a^{n^2+2k}, \ \ldots.$$

In particular, if we put $m - 1 = k$, we have that $a^{n^2+k^2} \in L$; thus $n^2 + k^2 = p^2$ for some $p > k$. But then putting $m - 1 = k + 1$ gives

$$n^2 + k(k+1) = q^2$$

for some $q > p$, and we deduce that

$$k = [n^2 + k(k+1)] - [n^2 + k^2] = q^2 - p^2$$
$$\geq (p+1)^2 - p^2 = 2p + 1 > 2k + 1,$$

a contradiction.

(ii) Suppose that $L = \{a^p : p \text{ is prime}\}$ is context-free. Then by the same Pumping Lemma technique as above there is a prime p and an integer $k \geq 1$ such that

$$p, \ p + k, \ p + 2k, \ \ldots$$

are all prime. In particular $p + pk = p(k+1)$ is prime, and this is a contradiction.

5.6. (i) Suppose that $L = \{a^n b^n c^n : n \geq 1\}$ is context-free. Then for some sufficiently large n we have

$$a^n b^n c^n = uvwxy,$$

with $|vx| \geq 1$ and $uv^m wx^m y \in L$ for all $m \geq 0$. Using the 'boundaries argument' of Example 5.5.6 we may deduce that each of v, x is a power of a single letter from $\{a, b, c\}$.

All words z in L are 'balanced', in the sense that they satisfy $|z|_a = |z|_b = |z|_c$. If this property is to hold good in every $uv^m wx^m y$ then every imbalance in v must be corrected by an opposite imbalance in x. So

$$|v|_a - |v|_b = |x|_b - |x|_a, \quad |v|_a - |v|_c = |x|_c - |x|_a. \tag{1}$$

Since v and x cannot both be 1 there is no real loss of generality in supposing that $|v| \geq 1$. So suppose now that $v = a^k$, with $k \geq 1$. Then from (1) we must have

$$|x|_b - |x|_a = |x|_c - |x|_a = k$$

and so x must involve both b's and c's, contrary to the conclusion of the last paragraph. Since $v = b^k$ and $v = c^k$ are similarly impossible, we must conclude that L is not context-free.

(ii) A very similar argument works in this case.

(iii) Let $L = \{a^n b^{2n} a^n : n \geq 1\}$, and suppose that L is context-free. For some sufficiently large n we must have

$$a^n b^n c^n = uvwxy,$$

with $|vx| \geq 1$ and $uv^m wx^m y \in L$ for all $m \geq 0$. By the 'boundaries' argument of Example 5.5.6 each of v, x is a power of a single letter and so lies in a single 'section' of $a^n b^{2n} a^n$ (i.e., in the initial block of a's, or in the central block of b's, or in the final block of a's). Conceivably both v and x lie in the same section. At any rate, v and x can cover at most two sections, which as m increases must grow while the third stays constant. Thus it is not possible to preserve the property that the middle section is precisely twice as long as each of the outer sections. Hence L is not context-free.

5.7. In Exercise 5.1(iii) we saw that $L_1 = \{a^m b^n : m > n\}$ is context-free. A similar argument shows that $L_2 = \{a^m b^n : m < n\}$ is context-free. So $L = L_1 \cup L_2$ is context-free. Now,

$$A^* \backslash L = (A^* \backslash a^+ b^+) \cup \{a^n b^n : n \geq 1\},$$

being the union of a regular language and a context-free language, is context-free. If either L or its complement were regular, then both would be regular, and it would then follow that

$$(A^* \backslash L) \backslash (A^* \backslash a^+ b^+) = \{a^n b^n : n \geq 1\}$$

is regular. Since this is not the case, we deduce that L is not regular.

5.8. (i) If $L = \{w \in \{a, b, c\}^+ : |w|_a = |w|_b = |w|_c\}$ is context-free, then so is its intersection with the regular language $a^* b^* c^*$, by Theorem 5.6.5. That is, $\{a^n b^n c^n : n \geq 1\}$ is context-free, and we have seen in Exercise 5.6(i) that this is not the case.

(ii) If $L = \{w w^R w : w \in \{a, b\}^*\}$ is context-free, then so is its intersection with the regular language $a^* b^* a^*$, i.e., so is

$$L' = \{a^m b^n . b^n a^m . a^m b^n : m, n \geq 0\} = \{a^m b^{2n} a^{2m} b^n : m, n \geq 0\}.$$

Each sufficiently long word $a^m b^{2n} a^{2m} b^n$ in L' must factorize into $uvwxy$ with $|vx| \geq 1$ in such a way that $uv^m wx^m y \in L'$ for all $m \geq 0$. By the 'boundaries' argument of Example 5.5.6 each of v and x is a power of a single letter. If (say) v is a positive power of a and x is a non-negative power of b then v is extracted from one or other of the two blocks of a's in $a^m b^{2n} a^{2m} b^n$ and so $uv^2 wx^2 y$ violates the condition that the second block of a's is exactly twice as long as the first.

There are in fact only two possibilities. Either

$$a^m b^{2n} a^{2m} b^n = a^p . a^q . a^r b^{2n} a^s . a^q . a^t b^n$$

(which is to say that

$$u = a^p, \ v = b^q, \ w = b^r a^{2m} b^s, \ x = b^q, \ y = b^t,$$

with

$$p, r, s, t \geq 0, \ q > 0, \ p + q + r = m, \ s + q + t = 2m);$$

or

$$a^m b^{2n} a^{2m} b^n = a^m b^p . b^q . b^r a^{2m} b^s . b^q . b^t$$

(which is to say that

$$u = a^m b^p, \ v = b^q, \ w = b^r a^{2m} b^s, \ x = b^q, \ y = b^t,$$

with

$$p, r, s, t \geq 0, \ q > 0, \ p + q + r = 2n, \ s + q + t = n).$$

But if we assume not only that $|a^m b^{2n} a^{2m} b^n| \ (= 3m + 3n) > N$ — see the statement (Theorem 5.5.1) of the Pumping Lemma — but also that $2m, 2n > N$ then neither solution satisfies the condition $|vwx| < N$. (In the first case $|vwx| \geq 2n$ and in the second case $|vwx| \geq 2m$.) So L' is not in fact context-free, and hence neither is L.

5.9. In Section 5.6 we showed that

$$A^* \backslash L = L_6 \cup L_7 \cup L_8,$$

where

$$L_6 = \{a^p b^q a^r : p, q, r \geq 1, \ p \neq q\},$$
$$L_7 = \{a^p b^q a^r : p, q, r \geq 1, \ q \neq r\},$$
$$L_8 = \{a, b\}^* \backslash a^+ b^+ a^+.$$

A context-free grammar generating L_6 has productions

$$
\begin{array}{lll}
\sigma_1 \to \rho_1 \tau_1, & \rho_1 \to a \lambda_1 b, & \lambda_1 \to a \mu_1, \\
\mu_1 \to a \mu_1, & \mu_1 \to a \mu_1 b, & \lambda_1 \to \nu_1 b, \\
\nu_1 \to \nu_1 b, & \nu_1 \to a \nu_1 b, & \mu_1 \to 1, \\
\nu_1 \to 1, & \tau_1 \to a \tau_1, & \tau_1 \to 1.
\end{array}
\tag{1}
$$

A context-free grammar generating L_7 is

$$
\begin{array}{lll}
\sigma \to \rho_2 \tau_2, & \rho_2 \to a \rho_2, & \rho_2 \to 1, \\
\tau_2 \to b \lambda_2 a, & \lambda_2 \to b \mu_2, & \mu_2 \to b \mu_2, \\
\mu_2 \to b \mu_2 a, & \lambda_2 \to \nu_2 a, & \nu_2 \to \nu_2 a, \\
\nu_2 \to b \nu_2 a, & \mu_2 \to 1, & \nu_2 \to 1.
\end{array}
\tag{2}
$$

A context-free (indeed regular) grammar generating L_8 has productions

$$\sigma_3 \to a\lambda_3, \qquad \sigma_3 \to b\zeta_3,$$
$$\lambda_3 \to a\lambda_3, \qquad \lambda_3 \to b\mu_3,$$
$$\mu_3 \to a\nu_3, \qquad \mu_3 \to b\mu_3, \tag{3}$$
$$\nu_3 \to a\zeta_3, \qquad \nu_3 \to b\zeta_3,$$
$$\zeta_3 \to a\zeta_3, \qquad \zeta_3 \to b\zeta_3,$$

together with

$$\sigma_3 \to 1, \; \lambda_3 \to 1, \; \mu_3 \to 1, \; \zeta_3 \to 1. \tag{4}$$

So a context-free grammar generating $L_6 \cup L_7 \cup L_8$ has all the symbols and productions of (1), (2), (3) and (4) together with a new sentence symbol σ and productions $\sigma \to \sigma_1$, $\sigma \to \sigma_2$, σ_3.

5.10. Let $\mathcal{D}$ be a DPDA, and suppose that $L = L(\mathcal{D})$ is not a prefix code. Then there exist $u, uv \in L$, with $v \in A^+$. There is a successful computation

$$(i, \zeta, u) \xrightarrow{*} (q, 1, 1)$$

for some q in Q and there is a successful computation

$$(i, \zeta, uv) \xrightarrow{*} (q', 1, 1)$$

for some q' in Q. But because of the deterministic property the latter computation must begin

$$(i, \zeta, uv) \xrightarrow{*} (q, 1, v)$$

and so fails at this point. It follows that L must be a prefix code.

The set $\{ww^R : w \in \{a, b\}^+\}$ contains ab^2a and also $ab^2a^2b^2a$, and so is not a prefix code. The set $\{a^m b^n : m \geq n\}$ contains a^3b and also a^3b^2, and so is not a prefix code.

5.11. A DPDA recognizing L_1 by empty memory stack is

$$\mathcal{D} = (\{i, q\}, \{a, b\}, \{\zeta, \alpha\}, \mu, \rho, i, \zeta),$$

where $\mu = \emptyset$ and

$$\rho(i, \zeta, a) = (i, \alpha\zeta), \quad \rho(i, \alpha, a) = (i, \alpha^2),$$
$$\rho(i, \alpha, b) = (q, 1), \quad \rho(q, \alpha, b) = (q, 1),$$
$$\rho(q, \zeta, 1) = (q, 1).$$

A DPDA recognizing L_2 is constructed simply by changing the first two specifications for ρ into

$$\rho(i, \zeta, a) = (i, \alpha^2 \zeta), \quad \rho(i, \alpha, a) = (i, \alpha^3).$$

On the other hand $L_1 \cup L_2$ is not a prefix code, since $ab, ab^2 \in L_1 \cup L_2$. Hence $L_1 \cup L_2$ cannot be recognized by a DPDA.

5.12. The language $L = \{a^n b^n a^n : n \geq 1\}$ is not context-free. (See Example 5.5.8.) On the other hand, L is a prefix code, for if $u = a^n b^n a^n$ and $v \in \{a, b\}^+$ then $uv \notin a^+ b^+ a^+$ unless $v \in a^+$. But then $uv = a^n b^n a^{n+k} \notin L$.

Exercises 6

6.1. (i) If we define

$$\theta(i, a) = (q_1, \bar{a}, R) \qquad\qquad \theta(q_4, \bar{b}) = (q_4, \bar{b}, L)$$
$$\theta(q_1, a) = (q_1, a, R) \qquad\qquad \theta(q_4, a) = (q_4, a, L)$$
$$\theta(q_1, \bar{b}) = (q_2, \bar{b}, R) \qquad\qquad \theta(q_4, \bar{a}) = (i, a, R)$$
$$\theta(q_1, b) = (q_3, \bar{b}, R)$$
$$\theta(q_2, b) = (q_3, \bar{b}, R)$$
$$\theta(q_2, \bar{b}) = (q_2, \bar{b}, R) \qquad\qquad \theta(i, \bar{b}) = (q_5, \bar{b}, R)$$
$$\theta(q_3, b) = (q_4, \bar{b}, L) \qquad\qquad \theta(q_5, \bar{b}) = (q_5, \bar{b}, R)$$
$$\theta(q_5, \Delta) = (t, \Delta, R)$$

then a typical successful computation is

$$ia^2 b^4 \to \bar{a} q_1 a b^4 \to \bar{a} a q_1 b^4 \to \bar{a} a \bar{b} q_3 b^3$$
$$\to \bar{a} a q_4 (\bar{b})^2 b^2 \xrightarrow{*} q_4 \bar{a} a (\bar{b})^2 b^2 \to a i a (\bar{b})^2 b^2$$
$$\to a \bar{a} q_1 (\bar{b})^2 b^2 \xrightarrow{*} a \bar{a} (\bar{b})^2 q_2 b^2 \to a \bar{a} (\bar{b})^3 q_3 b$$
$$\to a \bar{a} (\bar{b})^2 q_4 (\bar{b})^2 \xrightarrow{*} a q_4 \bar{a} (\bar{b})^4 \to a^2 i (\bar{b})^4$$
$$\to a^2 \bar{b} q_5 (\bar{b})^3 \xrightarrow{*} a^2 (\bar{b})^4 q_5 \Delta \to a^2 (\bar{b})^4 \Delta t \Delta.$$

If a word begins in b then the computation never gets started. If it begins $a^n b^m$ with $m < 2n$ then we reach a configuration $q_1 \Delta$ or $q_2 \Delta$ or $q_2 a$ and the computation fails. If it begins $a^n b^m$ with $m > 2n$ then we reach the configuration $q_5 b$ and the computation fails. If it is $a^n b^{2n} w$, with $w \in a\{a, b\}^*$, then we reach the configuration

$q_5 a$ and the computation fails. There is a successful computation if and only if the word is of the form $a^n b^{2n}$.

(ii) The strategy is to work inwards from the right and the left, checking that the same letter is encountered at each stage. The formal instructions are:

$$\theta(i, a) = (q_a, \#, R) \quad (a \in A)$$
$$\theta(q_a, b) = (q_a, b, R) \quad (a, b \in A)$$
$$\theta(q_a, \Delta) = (q'_a, \Delta, L) \quad (a \in A)$$
$$\theta(q'_a, a) = (q, \Delta, L) \quad (a \in A)$$
$$\theta(q, b) = (q, b, L) \quad (b \in A)$$
$$\theta(q, \#) = (i, \#, R)$$
$$\theta(i, \Delta) = (t, \Delta, R)$$
$$\theta(q'_a, \#) = (t, \#, R) \quad (a \in A).$$

Palindromes of length 1 are just single letters. Notice that for each a in A we have a successful computation

$$ia \rightarrow \#q_a\Delta \rightarrow q'_a\#\Delta \rightarrow \#t\Delta.$$

Palindromes of length 2 are squares. For each a in A we have a successful computation

$$ia^2 \rightarrow \#q_a a \rightarrow \#aq_a\Delta \rightarrow \#q'_a a\Delta \rightarrow q\#\Delta^2 \rightarrow \#i\Delta^3 \rightarrow \#\Delta t\Delta^2.$$

A palindrome of length $n \geq 3$ is of the form aza, where z is a palindrome of length $n - 2$. We have a computation

$$iaza \rightarrow \#q_a za \overset{*}{\rightarrow} \#zaq_a\Delta \rightarrow \#zq'_a a \overset{*}{\rightarrow} q\#z \rightarrow \#iz,$$

and the computation is ready to start again. So by induction all palindromes of length $n \geq 1$ are in $L(\mathcal{T})$. In fact, as is easy to see, $1 \in L(\mathcal{T})$ also.

If w is *not* a palindrome then at some stage we reach a configuration $q'_a b$ with $a \neq b$, and the computation fails.

6.2. (i) The strategy is to find and delete the first a, then the first b and then the first c, before returning to begin again. The formal instructions are

$$\theta(i, a) = (q_1, a, R)$$
$$\theta(q_1, a) = (q_1, a, R) \qquad \text{(Go right, changing}$$
$$\theta(q_1, \mu) = (q_1, \mu, R) \qquad \text{first } a \text{ to } \lambda,$$
$$\theta(q_1, b) = (q_2, \mu, R) \qquad \text{first } b \text{ to } \mu, \text{ and}$$

$$\theta(q_2, b) = (q_2, b, R) \qquad \text{first } c \text{ to } \nu;$$
$$\theta(q_2, \nu) = (q_2, \nu, R) \qquad \text{prepare to return.)}$$
$$\theta(q_2, c) = (q_3, \nu, L) \qquad (1)$$

For $x = \nu, b, \mu, a,$

$$\theta(q_3, x) = (q_3, x, L) \qquad \text{(Move left until } \lambda \text{ is met;}$$
$$\theta(q_3, \lambda) = (i, \lambda, R) \qquad \text{prepare to start again.)}$$

$$\theta(i, \mu) = (q_4, \mu, R) \qquad \text{(When all } a\text{'s have been deleted,}$$
$$\theta(q_4, \mu) = (q_4, \mu, R) \qquad \text{move right; enter terminal}$$
$$\theta(q_4, \nu) = (q_4, \nu, R) \qquad \text{state } t \text{ if no } a\text{'s, } b\text{'s or } c\text{'s}$$
$$\theta(q_4, \Delta) = (t, \Delta, R) \qquad \text{are encountered.)}$$

A typical successful computation is

$$ia^2b^2c^2 \rightarrow \lambda q_1 ab^2c^2 \rightarrow \lambda aq_1 b^2c^2 \rightarrow \lambda a\mu q_2 bc^2$$
$$\rightarrow \lambda a\mu bq_2 c^2 \rightarrow \lambda a\mu q_3 b\nu c \xrightarrow{*} q_3 \lambda a\mu b\nu c$$
$$\rightarrow \lambda ia\mu b\nu c \xrightarrow{*} \lambda^2 i\mu^2 \nu^2 \quad \text{(repeating the loop)}$$
$$\rightarrow \lambda^2 \mu q_4 \mu\nu^2 \xrightarrow{*} \lambda^2 \mu^2 \nu^2 q_4 \Delta \rightarrow \lambda^2 \mu^2 \nu^2 \Delta t \Delta.$$

It is clear that the computation fails on the first rightwards excursion unless the word being tested begins with $a^m b^n c^p$.

If $m < n$ we reach the configuration $q_4 b$ and the computation fails.

If $m > n$ we reach the configuration $q_1 \nu$ and the computation fails.

If $m = n > p$ we reach the configuration $q_2 \Delta$ and the computation fails.

If $m = n < p$ we reach the configuration $q_4 c$ and the computation fails.

If $w = a^n b^n c^n z$, where $z \in aA^* \cup bA^*$, we reach $\lambda^n \mu^n \nu^n q_4 z$ and the computation fails.

We deduce that $L(\mathcal{T}) = \{a^n b^n c^n : n \geq 1\}$.

(ii) It is fairly easy to modify the last TM by deleting the instruction marked (1), adding instructions

$$\theta(q_2, c) = (q', \nu, R)$$
$$\theta(q', c) = (q', c, R)$$
$$\theta(q', \rho) = (q', \rho, R)$$
$$\theta(q', d) = (q_3, \rho, L)$$
$$\theta(q_4, \rho) = (q_4, \rho, R)$$

and allowing x to take the values ρ, c, ν, b, μ, a in the instruction

$$\theta(q_3, x) = (q_3, x, L).$$

(iii) The same controlled deletion strategy will work here. The formal instructions are:

$$\theta(i, a) = (q_1, \lambda, R)$$
$$\theta(q_1, a) = (q_1, a, R)$$
$$\theta(q_1, b) = (q_2, \mu, R)$$
$$\theta(q_1, \mu) = (q_1, \mu, R)$$
$$\theta(q_2, b) = (q_3, \mu, R)$$
$$\theta(q_3, b) = (q_3, b, R)$$
$$\theta(q_3, \nu) = (q_3, \nu, R)$$
$$\theta(q_3, a) = (q_4, \nu, L)$$

(Move right, deleting one a, two b's and then one a; prepare to move left.)

For $x = \nu, b, \mu, a$,

$$\theta(q_4, x) = (q_4, x, L)$$
$$\theta(q_5, \lambda) = (i, \lambda, R)$$

(Move left until λ is met; prepare to start again.)

$$\theta(i, \mu) = (q_5, \mu, R)$$
$$\theta(q_5, \mu) = (q_5, \mu, R)$$
$$\theta(q_5, \nu) = (q_5, \nu, R)$$
$$\theta(q_5, \Delta) = (t, \Delta, R).$$

(When the initial block of a's is used up, move right; enter terminal state t if no a's or b's are encountered.)

A typical successful computation is

$$ia^2b^4a^2 \to \lambda q_1 ab^4a^2 \to \lambda a q_1 b^4 a^2 \to \lambda a \mu q_2 b^3 a^2$$
$$\to \lambda a \mu^2 q_3 b^2 a^2 \xrightarrow{*} \lambda a \mu^2 b^2 q_3 a^2 \to \lambda a \mu^2 b q_4 b \nu a$$
$$\xrightarrow{*} q_4 \lambda a \mu^2 b^2 \nu a \to \lambda i a \mu^2 b^2 \nu a$$
$$\xrightarrow{*} \lambda^2 i \mu^4 \nu^2 \quad \text{(repeating the loop)}$$
$$\to \lambda^2 \mu q_5 \mu^3 \nu^2 \xrightarrow{*} \lambda^2 \mu^4 \nu^2 q_5 \Delta \to \lambda^2 \mu^4 \nu^2 \Delta t \Delta.$$

We thus see that $\{a^n b^{2n} a^n : n \geq 1\} \subseteq L(\mathcal{T})$. To see the reverse inclusion, note first that if $w \in \{a, b\}^*$ does not begin with $a^m b^n a^p$ then the computation fails on the first rightwards excursion. Suppose that $w = a^m b^n a^p z$, where $z \in A^* \backslash a A^*$ and where $2m$, n and $2p$ are not all equal.

If $n = \min\{2m, n, 2p\}$ then the computation runs out of b's first and reaches the configuration $q_1 \nu$ or $q_2 \nu$.

If $2p = \min\{2m, n, 2p\}$ then the a's in the second block run out first and the computation reaches the configuration $q_3 \Delta$.

If $2m = \min\{2m, n, 2p\}$ then the a's in the first block run out first and the computation reaches the configuration $q_5 b$ or $q_5 a$.

Finally, if $2m = n = 2p$ and $z \neq 1$ then the computation reaches the configuration $q_5 b$.

The conclusion is that

$$L(\mathcal{T}) = \{a^n b^{2n} a^n : n \geq 1\}.$$

6.3. (i) The strategy is to scan the word, deleting an a, a b and a c on each sweep, and to enter a terminal state if and only if the a's, b's and c's run out simultaneously. The formal instructions are

$$\theta(i, x) = (q_x, \lambda, R) \qquad (x \in \{a, b, c\})$$
$$\theta(q_x, x) = (q_x, x, R)$$
$$\theta(q_x, y) = (q_{x,y}, \#, R) \qquad (x, y \in \{a, b, c\}, \ x \neq y)$$
$$\theta(q_{x,y}, x) = (q_{x,y}, x, R)$$
$$\theta(q_{x,y}, y) = (q_{x,y}, y, R)$$
$$\theta(q_{x,y}, z) = (q, \#, L) \quad (x, y, z \in \{a, b, c\}, \ x, y, z \text{ all distinct})$$

$$\theta(q, \zeta) = (q, \zeta, L) \qquad (\zeta = a, b, c, \#)$$
$$\theta(q, \lambda) = (i', \lambda, R)$$

$$\theta(i', x) = (q_x, \lambda, R) \qquad (x \in \{a, b, c\})$$
$$\theta(p, \#) = (p, \#, R) \qquad (p = q_x, q_{x,y}, i')$$
$$\theta(i', \Delta) = (t, \Delta, R).$$

A typical successful computation is

$$iac^2bab \rightarrow \lambda q_a c^2 bab \rightarrow \lambda\#q_{a,c}cbab \rightarrow \lambda\#cq_{a,c}bab$$
$$\rightarrow \lambda\#qc\#ab \overset{*}{\rightarrow} q\lambda\#c\#ab \rightarrow \lambda i'\#c\#ab$$
$$\rightarrow \lambda\#\lambda q_c\#ab \rightarrow \lambda\#\lambda\#\#q_{c,a}b \rightarrow \lambda\#\lambda\#q\#\#$$
$$\overset{*}{\rightarrow}\lambda\#q\lambda\#\#\# \rightarrow \lambda\#\lambda i'\#\#\#\overset{*}{\rightarrow}\lambda\#\lambda\#\#\#i'\Delta$$
$$\rightarrow \lambda\#\lambda\#\#\#\Delta t\Delta.$$

If the machine is applied to a word w in which $|w|_a$, $|w|_b$ and $|w|_c$ are not all equal, then we reach a configuration $q_x\Delta$ or $q_{x,y}\Delta$ and the computation fails. (Notice that the device of having i and i' is necessary to avoid having 1 in $L(\mathcal{T})$.)

(ii) A word ww^Rw has length divisible by 3. We begin by moving the cursor to a point $1/3$ of the way along the word:

$$\theta(i, x) = (q_1, x^*, R) \qquad (x = a, b)$$
$$\theta(q_1, x) = (q_1, x, R) \qquad (x = a, b)$$
$$\theta(q_1, \bar{x}) = (q_2, \bar{x}, L) \qquad (x = a, b)$$
$$\theta(q_1, \Delta) = (q_2, \Delta, L)$$
$$\theta(q_2, x) = (q_3, \bar{x}, L) \qquad (x = a, b)$$

$$\theta(q_3, x) = (q_4, \bar{x}, L) \qquad (x = a, b)$$
$$\theta(q_4, x) = (q_4, x, L) \qquad (x = a, b)$$
$$\theta(q_4, \bar{x}) = (i', \bar{x}, R) \qquad (x = a, b)$$
$$\theta(q_4, x^*) = (i', x^*, R) \qquad (x = a, b)$$
$$\theta(i', x) = (q_1, \bar{x}, R) \qquad (x = a, b).$$

For a word of length divisible by 3, this has the effect of changing the first symbol to a starred symbol, all others to barred symbols and moving the cursor (in state i') to the point $1/3$ of the way along. For example

$$iab^2a^2b \to a^*q_1b^2a^2b \overset{*}{\to} a^*b^2a^2bq_1\Delta \to a^*b^2a^2q_2b \to a^*b^2aq_3a\bar{b}$$
$$\to a^*b^2q_4a\bar{a}\bar{b} \overset{*}{\to} q_4a^*b^2a\bar{a}\bar{b} \to a^*i'b^2a\bar{a}\bar{b}$$
$$\to a^*\bar{b}q_1ba\bar{a}\bar{b} \overset{*}{\to} a^*\bar{b}baq_1\bar{a}\bar{b} \to a^*\bar{b}bq_2a\bar{a}\bar{b}$$
$$\to a^*\bar{b}q_3b\bar{a}\bar{a}\bar{b} \to a^*q_4(\bar{b})^2(\bar{a})^2\bar{b} \to a^*\bar{b}i'\bar{b}(\bar{a})^2\bar{b}.$$

Up to this point the palindromic property is irrelevant, but the computation fails during this first phase unless $|w|$ is divisible by 3. (If $|w| \equiv 1 \pmod 3$ we reach $q_2\bar{x}$, while if $|w| \equiv 2 \pmod 3$ we reach $q_3\bar{x}$.)

Next, we test the first $2/3$ of the word for the palindrome property:
$\theta(i', \bar{x}) = (q_x, x, L)\ (x = a, b)$ \qquad (Record and
$\theta(q_x, y) = (q_x, y, L)\ (x, y \in \{a, b\})$ \quad move left.)

$\theta(q_x, \bar{x}) = (q, x, R)$ \hfill (Test the first unbarred letter.)

$\theta(q, x) = (q, x, R)\ (x = a, b)$ \qquad (If successful, move right
$\theta(q, \bar{x}) = (q_x, x, L)\ (x = a, b)$ \qquad and repeat until the extreme
$\theta(q_x, x^*) = (p_x, x^*, R)\ (x = a, b)$ \quad left is reached.)

Finally, we test the first $1/3$ against the final $1/3$:
$\theta(p_x, y) = (p_x, y, R)\ (x, y \in \{a, b\})$ \quad (Record, move right, and
$\theta(p_x, \bar{x}) = (p, x, L)\ (x = a, b)$ \qquad test the first barred letter.
$\theta(p, x) = (p, x, L)\ (x = a, b)$ \qquad Move left to repeat.)

$\theta(p, x^*) = (p', x, R)\ (x = a, b)$ \qquad (Shift the starred marker
$\theta(p', x) = (p_x, x^*, R)\ (x = a, b)$ \qquad and move right to test.)

$\theta(p_x, \Delta) = (t, \Delta, R).$ \qquad (Enter terminal state.)

To understand the operation of this TM, consider a computation beginning with
$$ia^2b.ba^2.a^2b.$$

The first phase ends with the word $a^*\bar{a}b i'\bar{b}(\bar{a})^4\bar{b}$, and thereafter the computation proceeds as follows:

$$a^*\bar{a}b i'\bar{b}(\bar{a})^4\bar{b} \to a^*\bar{a}q_b\bar{b}b(\bar{a})^4\bar{b} \to a^*\bar{a}bqb(\bar{a})^4\bar{b}$$

$$\to a^*\bar{a}b^2 q(\bar{a})^4\bar{b} \to a^*\bar{a}bq_a ba(\bar{a})^3\bar{b} \xrightarrow{*} a^* q_a \bar{a}b^2 a(\bar{a})^3\bar{b}$$

$$\to a^* aqb^2 a(\bar{a})^3\bar{b} \xrightarrow{*} a^* ab^2 aq(\bar{a})^3\bar{b} \to a^* ab^2 q_a a^2(\bar{a})^2\bar{b}$$

$$\xrightarrow{*} q_a a^* ab^2 a^2(\bar{a})^2\bar{b} \to a^* p_a ab^2 a^2(\bar{a})^2\bar{b}$$

(end of second phase)

$$\xrightarrow{*} a^* ab^2 a^2 p_a(\bar{a})^2\bar{b} \to a^* ab^2 apa^2 \bar{a}\bar{b} \xrightarrow{*} pa^* ab^2 a^3 \bar{a}\bar{b}$$

$$\to ap' ab^2 a^3 \bar{a}\bar{b} \to aa^* p_a b^2 a^3 \bar{a}\bar{b} \xrightarrow{*} aa^* b^2 a^3 p_a \bar{a}\bar{b}$$

$$\to aa^* b^2 a^2 pa^2 \bar{b} \xrightarrow{*} apa^* b^2 a^4 \bar{b} \to a^2 p' b^2 a^4 \bar{b}$$

$$\to a^2 b^* p_b ba^4 \bar{b} \xrightarrow{*} a^2 b^* ba^4 p_b \bar{b} \to a^2 b^8 ba^3 pab$$

$$\xrightarrow{*} a^2 pb^* ba^4 b \to a^2 bp' ba^4 b \to a^2 bb^* p_b a^4 b \xrightarrow{*} a^2 bb^* a^4 bp_b \Delta$$

$$\to a^2 bb^* a^4 b \Delta t \Delta.$$

Clearly the computation fails unless the word is of the form $ww^R w$.

6.4. (i) Consider a word of the form $a^m b^n$, where $m > 0$, and suppose that $n = km + r$, where $0 \le r < m$. Then we have a computation

$$i a^m b^{km+r} \to a^* q_1 a^{m-1} b^{km+r} \qquad (1)$$

$$\xrightarrow{*} a^* a^{m-1} q_1 b^{km+r} \to a^* a^{m-2} q_3 a \bar{b} b^{km+r-1} \xrightarrow{*} q_3 a^* a^{m-1} \bar{b} b^{km+r-1}$$

$$\to a^* i' a^{m-1} \bar{b} b^{km+r-1} \to a^* \bar{a} q_1' a^{m-2} \bar{b} b^{km+r-1} \xrightarrow{*} a^* \bar{a} a^{m-2} \bar{b} q_2' b^{km+r-1}$$

$$\to a^* \bar{a} a^{m-2} q_3 (\bar{b})^2 b^{km+r-2} \xrightarrow{*} a^* q_3 \bar{a} a^{m-2} (\bar{b})^2 \to a^* a i' a^{m-2} (\bar{b})^2 b^{km+r-2}$$

$$\xrightarrow{*} \quad \text{(continuing to loop)}$$

$$\xrightarrow{*} a^* a^{m-1} i' (\bar{b})^m b^{(k-1)m+r} \to a^* a^{m-2} q_4 a(\bar{b})^m b^{(k-1)m+r}$$

$$\xrightarrow{*} q_4 a^* a^{m-1} (\bar{b})^m b^{(k-1)m+r} \to a^* q_1 a^{m-1} (\bar{b})^m b^{(k-1)m+r}$$

(essentially stage (1), but with m b's changed to $\bar{b}$)

$$\xrightarrow{*} \quad \text{(repeating the process)}$$

$$\xrightarrow{*} a^* q_1 a^{m-1} (\bar{b})^{2m} b^{(k-2)m+r}$$

$$\xrightarrow{*} \cdots \xrightarrow{*} a^* q_1 a^{m-1} (\bar{b})^{km} b^r.$$

If $r = 0$ the computation proceeds to a successful conclusion:

$$a^* q_1 a^{m-1} (\bar{b})^{km} \xrightarrow{*} a^* a^{m-1} (\bar{b})^{km} q_2 \Delta \to a^* a^{m-1} (\bar{b})^{km} \Delta t \Delta.$$

If $r > 0$ then the computation fails:

$$a^* q_1 a^{m-1} (\bar{b})^{km} b^r \xrightarrow{*} a^* a^{r-1} i' a^{m-r} (\bar{b})^{km} (\bar{b})^r$$

$$\to a^* a^r q_1' a^{m-r-1} (\bar{b})^{km+r} \xrightarrow{*} a^* a^{m-1} (\bar{b})^{km+r} q_2' \Delta.$$

For a word a^m $(= a^m b^0)$ we have a computation

$$i a^m \xrightarrow{*} a^* a^{n-1} q_1 \Delta \rightarrow a^* a^{n-1} \Delta t \Delta.$$

For a word w beginning with b a computation beginning with iw can never get off the ground, and a word $a^m b^k m a^p \ldots$ must lead to a configuration $q_2 a$ or $q_2' a$. Thus the set of words recognized by the TM is precisely

$$\{a^m b^n : m \text{ divides } n\}.$$

(ii) Leave out the instruction

$$\theta(q_1, \Delta) = (t, \Delta, R),$$

and replace the instruction $\theta(q_2, \Delta) = (t, \Delta, R)$ by

$$\theta(q_2', \Delta) = (t, \Delta, R).$$

Then the modified TM recognizes

$$\{a^m b^n : m \text{ does not divide } n\}.$$

6.5. Let

$$\mathcal{T} = (Q, A, M, \Delta, \theta, i, T)$$

be an LITM. Let $\mathcal{T}'$ be a TM with initial state i' and with an initial subroutine to insert a left marker λ:

$$i' a_1 a_2 \ldots a_n \xrightarrow{*} \lambda i a_1 a_2 \ldots a_n.$$

Suppose now that $\mathcal{T}'$ has all the instructions possessed by $\mathcal{T}$. Then whenever we have

$$i a_1 a_2 \ldots a_n \xrightarrow{*} q \alpha_1 \alpha_2 \ldots \alpha_k \rightarrow q' \Delta \beta \alpha_2 \ldots \alpha_k,$$

in $\mathcal{T}$ (this being the first time that the $\mathcal{T}$-computation goes off the left-hand edge), we have

$$\lambda i a_1 a_2 \ldots a_n \xrightarrow{*} \lambda q \alpha_1 \alpha_2 \ldots \alpha_k \rightarrow q' \lambda \beta \alpha_2 \ldots \alpha_k,$$

in $\mathcal{T}'$. We now suppose that any configuration involving $q\lambda$ triggers off a subroutine in $\mathcal{T}'$ leading to the insertion of Δ:

$$q' \lambda \beta \alpha_2 \ldots \alpha_k \xrightarrow{*} \lambda q' \Delta \beta \alpha_2 \ldots \alpha_k.$$

The conclusion is that we have a successful computation

$$ia_1 a_2 \ldots a_n \overset{*}{\to} \omega_1 t \omega_2$$

in $\mathcal{T}$ if and only if there is a successful computation

$$i' a_1 a_2 \ldots a_n \overset{*}{\to} \lambda i a_1 a_2 \ldots a_n \overset{*}{\to} \lambda \omega_1 t \omega_2$$

in $\mathcal{T}'$.

6.6. Suppose that in the TM

$$\mathcal{T} = (Q, A, M, \Delta, \theta, i, T)$$

we have $i \in T$. Let

$$\mathcal{T}' = \big(Q \cup \{i', q\}, A, M, \Delta, \theta', i', T\big),$$

where

$$\theta'(i', a) = (q, a, R), \quad \theta'(q, a) = (i, a, L) \quad (a \in A),$$

and where θ' is otherwise identical to θ. Then $L(\mathcal{T}') = L(\mathcal{T})$.

6.7. One possible subroutine is based on the instructions

$$\theta(i, a) = (q_1, a^*, R) \quad (a \in A)$$
$$\theta(q_1, a) = (q_1, a, R) \quad (a \in A)$$
$$\theta(q_1, \Delta) = (q_2, \Delta, L)$$

$$\theta(q_2, a) = (q_a, \bar{a}, R) \quad (a \in A)$$
$$\theta(q_a, \Delta) = (q_3, a, L) \quad (a \in A)$$
$$\theta(q_3, \#) = (q_3, \#, L)$$
$$\theta(q_3, a) = (q_3, a, L) \quad (a \in A)$$
$$\theta(q_3, \bar{a}) = (q_4, \#, L) \quad (a \in A)$$
$$\theta(q_4, a) = (q_a, \bar{a}, R) \quad (a \in A)$$
$$\theta(q_a, b) = (q_a, b, R) \quad (a, b \in A)$$
$$\theta(q_a, \#) = (q_a, \#, R)$$

$$\theta(q_4, a^*) = (q'_a, \#, R) \quad (a \in A)$$
$$\theta(q'_a, x) = (q'_a, x, R) \quad (x \in A \cup \{\#\})$$
$$\theta(q'_a, \Delta) = (t, a, R) \quad (a \in A).$$

Then, for example,

$$ia^2b \rightarrow a^*q_1ab \overset{*}{\rightarrow} a^*abq_1\Delta \rightarrow a^*aq_2b \rightarrow a^*a\bar{b}q_b\Delta$$
$$\rightarrow a^*aq_3\bar{b}bb \rightarrow a^*q_4a\#b \rightarrow a^*\bar{a}q_a\#b \overset{*}{\rightarrow} a^*\bar{a}\#bq_a\Delta$$
$$\rightarrow a^*\bar{a}\#q_3ba \overset{*}{\rightarrow} a^*q_3\bar{a}\#ba \rightarrow q_4a^*(\#)^2ba$$
$$\rightarrow \#q_a'(\#)^2ba \overset{*}{\rightarrow} (\#)^3baq_a'\Delta \rightarrow (\#)^3ba^2t\Delta.$$

6.8. One possible subroutine is based on the instructions

$$\theta(i,a) = (p_1,\bar{a},R)\,(a \in A)$$
$$\theta(p_1,a) = (p_1,a,R)\,(a \in A)$$
$$\theta(p_1,\#) = (p_2,\#,R)\,(a \in A)$$
$$\theta(p_2,a) = (p_a,\#,L)\,(a \in A)$$
$$\theta(p_a,b) = (p_a,b,L)\,(a,b \in A)$$
$$\theta(p_a,\#) = (p_a,\#,L)\,(a \in A)$$
$$\theta(p_a,\bar{b}) = (q_a,b,R)\,(a,b \in A)$$
$$\theta(q_a,b) = (r_b,a,R)\,(a,b \in A)$$
$$\theta(q_a,\#) = (t,a,R)\,(a \in A)$$
$$\theta(r_a,b) = (s_b,\bar{a},R)\,(a,b \in A)$$
$$\theta(r_a,\#) = (p_1,\bar{a},R)\,(a \in A)$$
$$\theta(s_a,b) = (s_b,a,R)\,(a \in A)$$
$$\theta(s_a,\#) = (p_1,a,R)\,(a \in A).$$

Then, for example,

$$iabc\#xyz \rightarrow \bar{a}p_1bc\#xyz \overset{*}{\rightarrow} \bar{a}bc\#p_2xyx \rightarrow \bar{a}bcp_x\#\#yz$$
$$\overset{*}{\rightarrow} p_x\bar{a}bc\#\#yz \rightarrow aq_xbc\#\#yz \rightarrow axr_bc\#\#yz$$
$$\rightarrow ax\bar{b}s_c\#\#yz \rightarrow ax\bar{b}cp_1\#yz \rightarrow ax\bar{b}c\#p_2yz$$
$$\rightarrow ax\bar{b}cp_y\#\#z \overset{*}{\rightarrow} axq_y\bar{b}c\#\#z \rightarrow axbq_yc\#\#z$$
$$\rightarrow axbyr_c\#\#z \rightarrow axby\bar{c}p_1\#z \rightarrow axby\bar{c}\#p_2z$$
$$\rightarrow axby\bar{c}q_z\#\# \rightarrow axbyp_z\bar{c}\#\# \rightarrow axbycq_z\#\# \rightarrow axbyczt\#.$$

6.9. This subroutine incorporates the duplication subroutine (6.2.6). We begin with an initial subroutine given by

$$\theta(i,\beta) = (q_1,\Delta,R)$$ (Delete the first two
$$\theta(q_1,\beta) = (q_2,\Delta,R)$$ β's, moving right
$$\theta(q_2,\beta) = (q_3,\beta,R)$$ past # and any $\bar{a}$'s
$$\theta(q_3,\bar{a}) = (q_3,\bar{a},R)$$ in place.)

This changes $i\beta^k \# a_1 a_2 \ldots a_n$ to

$$\Delta^2 \beta^{k-2} \# q_3 a_1 a_2 \ldots a_n.$$

Then the duplication subroutine comes into play and produces

$$\Delta^2 \beta^{k-2} \# q_4 \bar{a}_1 \bar{a}_2 \ldots \bar{a}_n a_1 a_2 \ldots a_n.$$

The next set of instructions is
$$\theta(q_4, \bar{a}) = (q_4, \bar{a}, L) \quad (a \in A)$$
$$\theta(q_4, \#) = (q_4, \#, L)$$
$$\theta(q_4, \beta) = (q_4, \beta, L)$$
$$\theta(q_4, \Delta) = (q_1, \Delta, R).$$
and essentially says 'Prepare to start again'. They lead into the initial subroutine, after which we obtain

$$\Delta^3 \beta^{k-3} \# \bar{a}_1 \bar{a}_2 \ldots \bar{a}_n q_3 a_1 a_2 a_n,$$

with another β deleted from the beginning. Eventually all the β's are removed and we reach

$$\Delta^k \# (\bar{a}_1 \bar{a}_2 \ldots \bar{a}_n)^{k-2} q_3 a_1 a_2 \ldots a_n.$$

The final application of the duplication subroutine gives

$$\Delta^k \# (\bar{a}_1 \bar{a}_2 \ldots \bar{a}_n)^{k-2} q_4 (\bar{a}_1 \bar{a}_2 \ldots \bar{a}_n) a_1 a_2 \ldots a_n.$$

This transforms to

$$\Delta^k q_1 \# (\bar{a}_1 \bar{a}_2 \ldots \bar{a}_n)^{k-1} a_1 a_2 \ldots a_n,$$

and now we require only a 'tidying up' subroutine based on
$$\theta(q_1, \#) = (q_5, \#, R)$$
$$\theta(q_5, \bar{a}) = (q_5, a, R) \quad (a \in A)$$
$$\theta(q_5, a) = (q_6, a, L) \quad (a \in A)$$
$$\theta(q_6, a) = (q_6, a, L) \quad (a \in A)$$
$$\theta(q_6, \#) = (t, \Delta, R)$$
to produce the final outcome $\Delta^{k+1} t (a_1 a_2 \ldots a_n)^k$.

6.10. It is enough to describe an NDTM. For each w in a^* we can describe a subroutine that will give a computation

$$iw \overset{*}{\to} w \# q\beta \# a \overset{*}{\to} w \# q\beta^2 \# a^2 \overset{*}{\to} w \# q\beta^3 \# a^3 \overset{*}{\to} \cdots.$$

At each stage $w \# q\beta^k \# a^k$ we can then use the routine of Exercise 6.9 to obtain

$$w \# q' a^{k^2}.$$

Then we use the routine (6.2.7) to check whether w is identical to a^{k^2}, entering a terminal state if and only if this test has a positive result. The language recognized is precisely

$$\{a^{n^2} : n \geq 1\}.$$

6.11. We give an informal description. In Exercise 6.4 we found a TM which would determine, for a word $a^m b^n$, whether or not m divides n. This is the basis of a subroutine which we shall call the 'divisibility test'. From a typical initial ID ia^n we begin with a computation

$$ia^n \overset{*}{\rightarrow} qa^n \# \beta^2.$$

We test first whether $2 = n$, and if this succeeds enter a terminal state. If not, we apply the divisibility test, to dicover whether 2 divides n, and if this succeeds we arrange for the computation to fail. If the divisibility test fails, we generate the ID

$$qa^n \# \beta^3$$

and then test first whether $3 = n$, and, if this does not hold, whether 3 divides n. If this fails, we go on to

$$qa^n \# \beta^4$$

and repeat the routine. The appearance of success when $n = 4$ is illusory, for if $n = 4$ the computation has already failed from the positive answer to the question 'does 2 divide n'. The language recognized is $\{a^n : n \text{ is prime}\}$.

6.12. Suppose that

$$\Gamma_1 = (V_1, A, \pi_1, \sigma_1), \quad \Gamma_2 = (V_2, A, \pi_2, \sigma_2)$$

are such that $L_1 = L(\Gamma_1)$, $L_2 = L(\Gamma_2)$. We may assume that $V_1 \backslash A$ and $V_2 \backslash A$ are disjoint. Let

$$\Gamma = (V_1 \cup V_2 \cup \{\sigma\}, A, \pi, \sigma),$$

where $\sigma \notin V_1 \cup V_2$, and where π contains $\pi_1 \cup \pi_2$ together with the productions

$$\sigma \rightarrow \sigma_1, \sigma \rightarrow \sigma_2.$$

Then $L(\Gamma) = L_1 \cup L_2$.

If
$$\Gamma' = (V_1 \cup V_2 \cup \{\sigma\}, A, \pi', \sigma),$$

where π' contains $\pi_1 \cup \pi_2$ together with the production $\sigma \to \sigma_1 \sigma_2$, then $L(\Gamma') = L_1 L_2$.

6.13. Suppose that $L = L(\Gamma)$, where $\Gamma = (V, A, \pi, \sigma)$. Let

$$\Gamma'' = (V \cup \{\sigma''\}, A, \pi'', \sigma''),$$

where $\sigma'' \notin V$ and where π'' contains π together with the productions $\sigma'' \to \sigma''\sigma$, $\sigma'' \to 1$. Then $L(\Gamma'') = L^*$.

Exercises 7

7.1. Evidently $M_1 \times \cdots \times M_n \in \mathbf{V}\langle M_1, \ldots, M_n \rangle$, and so

$$\mathbf{V}\langle M_1 \times \cdots \times M_n \rangle \subseteq \mathbf{V}\langle M_1, \ldots, M_n \rangle.$$

Conversely, since each M_i divides $M_1 \times \cdots \times M_n$, we have

$$M_i \in \mathbf{V}\langle M_1 \times \cdots \times M_n \rangle \quad (i = 1, \ldots, n).$$

So $\mathbf{V}\langle M_1, \ldots, M_n \rangle \subseteq \mathbf{V}\langle M_1 \times \cdots \times M_n \rangle$.

7.2. Let $\mathbf{V}$ be an F-variety generated by a monoid M. Then it is ultimately defined by a sequence $u_n = v_n$ $(n \geq 1)$ of equations. In particular M satisfies the equations $u_n = v_n$ $(n \geq m)$ for some sufficiently large m. Hence, since every monoid in $\mathbf{V}$ divides M^k for some k, every monoid in $\mathbf{V}$ satisfies $u_n = v_n$ $(n \geq m)$. Thus

$$\mathbf{V} = \mathbf{V}_F[u_m = v_m, \ u_{m+1} = v_{m+1}, \ldots].$$

7.3. Let $\mathbf{a} = (a_1, a_2, \ldots)$, where $a_n^n = a_n^{n+1}$, but where $a_n, a_n^2, \ldots, a_n^n$ are all distinct. Then the powers

$$1, \ \mathbf{a}, \ \mathbf{a}^2, \ \mathbf{a}^3, \ldots$$

are all distinct, and so the infinite direct product $\langle a_1 \rangle \times \langle a_2 \rangle \times \cdots$ contains an element of infinite order. Thus the (Birkhoff) variety generated by all aperiodic monoids contains the free monoid on one generator, and hence contains all finite cyclic groups.

7.4. If the variety **Fgp** were generated by a single finite group G, then all finite groups would have to satisfy the equation $x^g = 1$, where g is the exponent of G. (The *exponent* g of a finite group G is the least positive integer with the property that $a^g = 1$ for every a in G.) This is clearly not so: simply consider a finite cyclic group of order greater than g. A similar argument applies to **Ap**.

7.5. Let $a \in A$, and define $\varphi : A^* \to \mathbf{Z}_n = \{0, 1, \dots, n-1\}$ by

$$\varphi(a) = 1, \quad \varphi(b) = 0 \; \bigl(b \in A \backslash \{a\}\bigr).$$

Then

$$\varphi^{-1}(k) = \{w \in A^* : |w|_a \equiv k \,(\mathrm{mod}\, n)\} = L(a, k).$$

So $\mathcal{L}_\mathbf{V}(A)$ contains all subsets $L(a, k)$ and hence contains the Boolean algebra generated by the $L(a, k)$.

Conversely, let $\varphi : A^* \to \mathbf{Z}_n$ be an arbitrary morphism. Then

$$w \in \varphi^{-1}(l)$$
   if and only if $\varphi(w) = l$,

i.e., if and only if $\displaystyle\sum_{a \in A} \varphi(a)|w|_a \equiv l \,(\mathrm{mod}\, n)$,

i.e., if and only if $(\forall a \in A)\bigl(\exists k_a \in \{0, 1, \dots n-1\}\bigr)$

$$|w|_a \equiv k_a \,(\mathrm{mod}\, n) \text{ and } \sum_{a \in A} \varphi(a)k_a \equiv l \,(\mathrm{mod}\, n),$$

i.e., if and only if $w \in \displaystyle\bigcap_{a \in A} L(a, k_a)$

for some choice of integers k_a such that

$$\sum_{a \in A} \varphi(a)k_a \equiv l \,(\mathrm{mod}\, n).$$

Thus every $\varphi^{-1}(l)$ is in the Boolean algebra generated by the sets $L(a, k)$, where $a \in A$ and $0 \le k < n$.

7.6. The F-variety **V** of all finite abelian groups is generated by all the cyclic groups $\mathbf{Z}_n$ $(n \ge 2)$. It is the join of the F-varieties $\mathbf{V}_n$ in the lattice of all F-varieties, and so (for every A) $\mathcal{L}_\mathbf{V}(A)$ is the join of the Boolean algebras $\mathcal{L}_{\mathbf{V}_n}(A)$, that is, the Boolean algebra generated by all the sets $L(n, a, k)$ as described.

7.7. (i) If T is a subsemigroup of S and $S^k = 0$ then certainly $T^k = 0$. If $\varphi : S \to T$ is a surjective morphism and $S^k = 0$ then for all

$t_1, \ldots, t_k$ in T we choose $s_1, \ldots, s_k$ in S so that $\varphi(s_i) = t_i$ for $i = 1, \ldots, k$. Then

$$t_1 t_2 \ldots t_k = \varphi(s_1 s_2 \ldots s_k) = \varphi(0) = 0,$$

and so $T^k = 0$. If $S_1, S_2, \ldots, S_n \in \mathbf{Nil}_k$ and if

$$\mathbf{x}^{(1)}, \mathbf{x}^{(2)}, \ldots, \mathbf{x}^{(k)} \in S_1 \times S_2 \times \cdots \times S_n,$$

with

$$\mathbf{x}^{(i)} = \left(x_1^{(i)}, x_2^{(i)}, \ldots, x_n^{(i)}\right) \quad (i = 1, 2, \ldots, k),$$

then

$$\mathbf{x}^{(1)} \mathbf{x}^{(2)} \ldots \mathbf{x}^{(k)}$$
$$= \left(x_1^{(1)} x_1^{(2)} \ldots x_1^{(k)}, x_2^{(1)} x_2^{(2)} \ldots x_2^{(k)}, \ldots, x_n^{(1)} x_n^{(2)} \ldots x_n^{(k)}\right)$$
$$= (0, 0, \ldots, 0) = 0.$$

Thus $S_1 \times S_2 \times \cdots \times S_n \in \mathbf{Nil}_k$. Thus $\mathbf{Nil}_k$ is an F-variety.
(ii) Certainly every semigroup S in $\mathbf{Nil}_k$ satisfies the equation

$$x_1 x_2 \ldots x_k = y_1 y_2 \ldots y_k,$$

since all products of k elements in S are equal to 0. Conversely, if S satisfies the given equation, then all products of k elements in S have a common value, which we may denote by z. Symbolically we have $S^k = \{z\}$. If $s \in S$ then for an arbitrary choice of $a_1, a_2, \ldots, a_k$ in S we have

$$sz = s(a_1 a_2 \ldots a_k) = (sa_1)a_2 \ldots a_k \in S^k = \{z\}.$$

Thus $sz = z$, and similarly $zs = z$, for every s in S. Thus z is a zero element for S, and so $S \in \mathbf{Nil}_k$.
(iii) Let $S = A^+/I_k$, where I_k is the ideal of A^+ consisting of all w in A^+ such that $|w| \geq k$. Then S is a finite semigroup of order

$$|A| + |A|^2 + \cdots |A|^{k-1} + 1$$

and $S \in \mathbf{Nil}_k$. Let φ be the natural morphism from A^+ onto S: thus $\varphi(w) = w$ if $|w| < k$, and $\varphi(w) = 0$ if $|w| \geq k$. Since $\varphi^{-1}(w) = \{w\}$ for all w with $|w| < k$, every language $\{w\}$ (with $|w| < k$) is recognized by a semigroup in $\mathbf{Nil}_k$. Hence the Boolean algebra generated by the languages $\{w\}$ (with $|w| < k$) is contained in $\mathcal{L}_{\mathbf{Nil}_k}(A)$.

Conversely, suppose that $L\ (\subseteq A^+)$ is recognized by a semigroup S in $\mathbf{Nil}_k$, so that there is a morphism $\varphi : A^+ \to S$ and a subset P of S such that $L = \varphi^{-1}(P)$. The morphism φ is determined by its effect on the elements of A, and if $w = a_1 a_2 \ldots a_m \in A^+$ is such that $|w| = m \geq k$ then

$$\varphi(w) = \varphi(a_1)\varphi(a_2)\ldots\varphi(a_m) \in S^m = \{0\}.$$

Hence for all $b \in S \backslash \{0\}$ the set $\varphi^{-1}(b)$ consists entirely of words w in A^+ such that $|w| < k$. Thus $\varphi^{-1}(b)$ is in the Boolean algebra $\mathcal{B}_k$ generated by the singleton sets $\{w\}$, where $|w| < k$. Since

$$\varphi^{-1}(0) = A^+ \backslash \bigcup_{b \in S\backslash\{0\}} \varphi^{-1}(b),$$

we also have $\varphi^{-1}(0) \in \mathcal{B}_k$. Hence

$$L = \varphi^{-1}(P) = \bigcup_{z \in P} \varphi^{-1}(z) \in \mathcal{B}_k.$$

7.8. (i) If $S \in \mathbf{Nil}$ then $S \in \mathbf{Nil}_k$ for some k. Hence, by the argument of the last exercise, every subsemigroup and every quotient semigroup of S is in $\mathbf{Nil}_k$ and so is in $\mathbf{Nil}$. If $S_1, S_2, \ldots, S_n \in \mathbf{Nil}$ then there exist integers $k_1, k_2, \ldots, k_n$ such that

$$S_i \in \mathbf{Nil}_{k_i} \quad (i = 1, 2, \ldots, n).$$

Since
$$\mathbf{Nil}_2 \subset \mathbf{Nil}_3 \subset \mathbf{Nil}_4 \subset \cdots,$$

we deduce that $S_i \in \mathbf{Nil}_k$ for every i, where $k = \max\{k_1, k_2, \ldots k_n\}$. Thus
$$S_1 \times S_2 \times \cdots \times S_n \in \mathbf{Nil}_k \subset \mathbf{Nil}.$$

We have shown that $\mathbf{Nil}$ is an F-variety.

(ii) Every semigroup in $\mathbf{Nil}$ satisfies the equation

$$x_1 x_2 \ldots x_n = y_1 y_2 \ldots y_n$$

for some sufficiently large n. Arguing as in the previous Exercise, we conclude that

$$\mathbf{Nil} = [[x_1 x_2 \ldots x_n = y_1 y_2 \ldots y_n \ (n \geq 2)]].$$

(iii) From the previous Exercise we have that $\mathcal{L}_{\mathbf{Nil}}(A) = \mathcal{B}$, the Boolean algebra generated by all sets $\{w\}$, where $w \in A^+$. Since every finite set $\{w_1, w_2, \ldots w_n\}$ can be written as a finite union

$$\{w_1\} \cup \{w_2\} \cup \cdots \{w_n\},$$

and since every cofinite set is by definition the complement of a finite set, it is clear that the set $\mathcal{FCF}$ of all finite or cofinite sets of A^+ is contained in $\mathcal{B}$.

Notice now that the set $\mathcal{FCF}$ is closed under the Boolean operations $\cup$ and $\setminus$, for we have the implications

$$X, Y \text{ finite} \;\Rightarrow\; X \cup Y \text{ finite},$$
$$X, Y \text{ cofinite} \;\Rightarrow\; X \cup Y \text{ cofinite},$$
$$X \text{ finite}, Y \text{ cofinite} \;\Rightarrow\; X \cup Y \text{ cofinite},$$
$$X \text{ finite}, \;\Rightarrow\; A^+ \setminus X \text{ cofinite},$$
$$X \text{ cofinite}, \;\Rightarrow\; A^+ \setminus X \text{ finite}.$$

Also, $\mathcal{FCF}$ contains all singleton sets $\{w\}$, where $w \in A+$. Hence $\mathcal{FCF}$, being a Boolean algebra containing all the singleton sets, contains the *smallest* Boolean algebra containing these sets. That is, $\mathcal{B} \subseteq \mathcal{FCF}$. Thus

$$\mathcal{L}_{\mathbf{Nil}}(A) = \mathcal{B} = \mathcal{FCF}.$$

7.9. The monoid in question is the syntactic monoid of the star-free language $A^* abaA^*$. (Despite its appearance, the language A^*, being the complement of the finite language $\emptyset$, is star-free.) By the Schutzenberger Theorem the monoid must be aperiodic. The direct verification consists of the observations

$$1^2 = 1, \qquad\qquad 0^2 = 0, \qquad\qquad a^2 = a,$$
$$b^3 = b^2, \qquad (ab)^3 = (ab)^2, \qquad (ba)^3 = (ba)^2,$$
$$(b^2)^2 = b^2, \qquad (ab^2)^2 = ab^2, \qquad (bab)^2 = bab,$$
$$(b^2a)^2 = b^2a, \quad (bab^2)^2 = bab^2, \qquad (b^2ab)^2 = b^2ab.$$

7.10 We compute the syntactic monoid of $(a^2)^*$, using the technique described in Section 3.5. We have

$$L = (a^2)^*, \qquad\qquad L.a = a^{-1}L = a(a^2)^* = L_1,$$
$$L.b = b^{-1}L = \emptyset, \qquad L_1.a = a^{-2}L = L,$$
$$L_1.b = b^{-1}L_1 = \emptyset, \qquad \emptyset.a = \emptyset.b = \emptyset.$$

The minimal automaton has state diagram

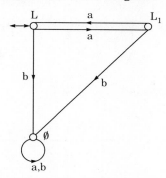

The syntactic monoid is generated by the transformations

$$a = \begin{pmatrix} 0 & 1 & 2 \\ 0 & 2 & 1 \end{pmatrix}, \quad b = \begin{pmatrix} 0 & 1 & 2 \\ 0 & 0 & 0 \end{pmatrix}$$

and is $\{1, a, b\}$, where

$$a^2 = 1, \quad b^2 = ab = ba = b.$$

This is *not* aperiodic, since it contains the non-trivial subgroup $\{1, a\}$.

By contrast, if we carry out the same procedure for $L = (ab)^*$ we obtain the minimal automaton with state diagram

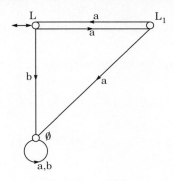

where $L_1 = b(ab)^*$. The syntactic monoid, generated by

$$a = \begin{pmatrix} 0 & 1 & 2 \\ 0 & 2 & 0 \end{pmatrix}, \quad b = \begin{pmatrix} 0 & 1 & 2 \\ 0 & 0 & 1 \end{pmatrix},$$

consists of the six elements $1, a, b, e, f, 0$, where

$$e = \begin{pmatrix} 0 & 1 & 2 \\ 0 & 1 & 0 \end{pmatrix}, \quad f = \begin{pmatrix} 0 & 1 & 2 \\ 0 & 0 & 2 \end{pmatrix}.$$

The Cayley table is

	1	a	b	e	f	0
1	1	a	b	e	f	0
a	a	0	e	0	a	0
b	b	f	0	b	0	0
e	e	a	0	e	0	0
f	f	0	b	0	f	0
0	0	0	0	0	0	0

It is easy to verify that this monoid satisfies the equation $x^3 = x^2$ and so is aperiodic. Hence $(ab)^*$ is star-free, by the Schutzenberger Theorem.

The direct verification consists of the observation that

$$(ab)^* = A^* \backslash (bA^* \cup A^*a \cup A^*a^2A^* \cup A^*b^2A^*).$$

7.11. The state diagram of the minimal automaton of

$$L = \{w \in A^* : |w| \equiv 0 \,(\mathrm{mod}\, n)\}$$

is depicted on the facing page.

$$L_i = \{w \in A^* : |w| \equiv -i \,(\mathrm{mod}\, n)\}$$

and $L_i.a = L_{i+1}$ for every a in A. The syntactic monoid of L contains the cyclic group $\langle a \rangle$ of order n and so is *not* aperiodic. Hence L is not star-free.

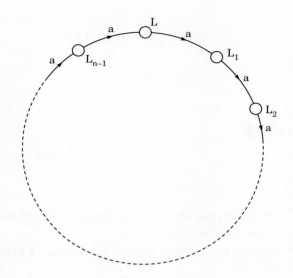

References

Birkhoff, G. (1935). On the structure of abstract algebras. *Proc. Cambridge Phil. Soc.* **31**, 433–454.

Chomsky, N. (1959). On certain formal properties of grammars. *Inf. Control* **2**, 137–167.

Clifford, A. H. and Preston, G. B. (1961). *The algebraic theory of semigroups* Vol. 1. Math. Surveys of the American Math. Soc. 7.

Cohn, P, M. (1965). *Universal algebra.* Harper and Row, New York.

Davis, M. (1958). *Computability and unsolvability.* McGraw-Hill, New York.

Eilenberg, S. (1974). *Automata, languages and machines* Vol. A. Academic Press, New York and London.

— (1976). *Automata, languages and machines* Vol. B. Academic Press, New York and London.

— and Schutzenberger, M. P. (1976). On pseudovarieties of monoids. *Adv. Math.* **19**, 413–448.

Green, J. A. (1951). On the structure of semigroups. *Ann. Math.* **54**, 163–172.

Harary, F. (1969). *Graph theory.* Addison-Wesley, Reading, Mass.

Hopcroft, J. E. and Ullman, J. D. (1979). *Introduction to automata theory, languages and computation.* Addison-Wesley, Reading, Mass.

Howie, J. M. (1976). *An introduction to semigroup theory.* Academic Press, London and New York.

Kleene, S. (1956). Representation of events in nerve sets. *Automata Studies* (ed. C. E. Shannon and J. McCarthy) 3–40. Princeton University Press, Princeton, N. J.

Lallement, G. (1979). *Semigroups and combinatorial applications.* Wiley-Interscience, New York.

288

Pin, J.-E. (1986). *Varieties of formal languages.* North Oxford Academic
Publishers, Oxford (a translation of *Variétés de langages formels,*
Masson, Paris, 1984).

— , Straubing, H. and Thérien, D. (1984). Small varieties of finite
semigroups and extensions. *J. Australian Math. Soc. A* **37**, 269–281.

Rees, D. (1940). On semi-groups. *Proc. Cambridge Phil. Soc.* **36**, 387–
400.

Schutzenberger, M. P. (1965). On finite monoids having only trivial
subgroups. *Inf. Control* **8**, 190–194.

Notations used in the text

$\mathcal{A}$	a finite state automaton
$\mathcal{A}_L$	the minimal automaton of L
A^+	the free semigroup on a set A
A^*	the free monoid on a set A
$A \Diamond B$	the Schutzenberger product of A and B
$\alpha \overset{*}{\to} \beta$	there is a computation from α to β
$\alpha \overset{*}{\Rightarrow} \beta$	there is a derivation of β from α
Ap	the variety of aperiodic semigroups
$\mathrm{Aut}(X)$	the automaton of a transformation monoid X
$C(w)$	the content of w
$\mathcal{D}$	Green's relation
Δ	a blank symbol (on a Turing machine tape)
$\mathcal{F}(A)$	the set of all finite subsets of A
FSA	a finite state automaton
Γ	a phrase structure grammar
ID	an instantaneous description
$\mathcal{J}$	Green's relation
$\mathcal{L}$	Green's relation
$\mathcal{L}(\mathcal{A})$	the language recognized by an automaton
$\mathcal{L}(\Gamma)$	the language generated by a grammar
$\mathcal{L}(\mathcal{P})$	the language recognized by a pushdown automaton
$\mathcal{L}(\mathcal{T})$	the language recognized by a Turing machine
N	the set of natural numbers
$\mathbf{N}^0$	the set of natural numbers together with 0
$\mathcal{P}$	a pushdown automaton
$\mathcal{P}(A)$	the set of all subsets of A
PDA	a pushdown automaton
Q	the set of rational numbers
(Q, S)	a transformation monoid

290

$\mathbf{R}$	the set of real numbers
$\mathcal{R}$	Green's relation
$\operatorname{Rat} A^*$	the set of rational subsets of A^*
$\operatorname{Rec} A^*$	the set of recognizable subsets of A^*
$\rho^\natural$	the natural morphism associated with a congruence ρ
σ_L	the syntactic congruence of L
$S \mid T$	S divides T
$\operatorname{Syn}(L)$	the syntactic monoid of L
$\mathcal{T}$	a Turing machine
$\mathbf{T}^\#$	the congruence generated by $\mathbf{T}$
$\mathbf{TM}$	a Turing machine
$\operatorname{TM}(\mathcal{A})$	the transformation monoid of $\mathcal{A}$
V	the vocabulary of a grammar
$\mathbf{V}$	a variety or F-variety
$\mathbf{V}[u_n = v_n \ (n \geq 1)]$	the variety defined by the equations $u_n = v_n$
$\mathbf{V}_F[u_n = v_n \ (n \geq 1)]$	the F-variety defined by the equations $u_n = v_n$
$\mathbf{V}[[u_n = v_n \ (n \geq 1)]]$	the F-variety ultimately defined by the equations $u_n = v_n$
$\mathbf{V}\langle U \rangle$	the F-variety generated by U
$|w|$	the length of w
$|w|_a$	the number of occurrences of a in w
w^R	the reverse of w

Index

accessible, 51
alphabet, 29
alphabet (of FSA), 40, 47
aperiodic monoid, 213
arithmetical congruence, 5
associative, 9
automaton, 40, 47

band, 199
behaviour (of FSA), 43
bijection, 9
bijective, 9
binary operation, 15
blank symbol, 158
Boolean algebra, 6

cancellative monoid, 32
Cartesian product, 4
Cayley table, 16
Chomsky hierarchy, 188
Chomsky Normal Form, 105
coaccessible, 54
complement, 10
complete automaton, 43
composition, 9
computation (in PDA), 122
computation (in TM), 158
congruence, 23
congruence generated by relation, 193
context-free grammar, 95
context-sensitive grammar, 96

contrapositive, 5

De Morgan laws, 3
derivation, 95
deterministic automaton, 43
diagonal relation, 5
disjoint, 8
distributive, 5
divides, 27
domain, 3

element, 1
empty set, 8
empty word, 30
equationally defined class, 200
equidivisible monoid, 33
equivalence relation, 8
exponent (of group), 201
extension (of function), 31

F-variety, 198
finite state automaton, 40, 47
formal grammar, 93
fractional part, 8
free monoid, 30
free semigroup, 30
FSA, 40, 47
full transformation semigroup, 36
function, 8

generators, 22
grammar, 93
Green's equivalences, 195

group, 18

homomorphism, 23
hyper-regular grammar, 98

ideal, 194
idempotent, 20
identity element, 17
image, 2
initial state
   of FSA, 43, 47
   of PDA, 122
   of TM, 158
injection, 10
injective, 10
inputs (to an FSA), 40, 47
instantaneous description
   of PDA, 122
   of TM, 158
integers, 8
integral part, 8
intersection of sets, 2
inverse (in a group), 19
inverse function, 8
isomorphic, 23
isomorphism
   of automata, 80
   of monoids, 23
   of transformation monoids, 29

kernel equivalence, 9
Kleene's star operation, 57
Kleene's Theorem, 58

language accepted
   by FSA, 43
   by PDA, 123
   by TM, 159
language generated by
   grammar, 95
language recognized
   by FSA, 43
   by PDA, 123

   by TM, 159
left factor (of a word), 32
left regular grammar, 117
leftmost derivation, 115
length (of a word), 32

map, 8
mapping, 8
memory stack alphabet, 122
minimal automaton, 71
monoautomaton, 66
monoid, 17
morphism
   of monoids, 23
   of semigroups, 23

natural numbers, 8
NDTM, 174
non-deterministic
   automaton, 48
   Turing machine, 174

palindrome, 90
partition, 8
path (in a state diagram), 45
PDA, 122
permutation, 35
phrase structure, 93
polyautomaton, 66
power set, 4
product (of sets), 57
production, 93
progressive step
   (in PDA computation), 123
proper left factor (of a word), 32
proper subset, 13
pseudo-variety, 198
Pumping Lemma
   (for FSAs), 46
   (for PDAs), 143
pushdown automaton, 122

quantifiers, 4

quotient set, 9

range, 9
rational language map, 205
rational set, 58
recognition
    by FSA, 43
    by monoid, 68
    by PDA, 123
    by TM, 159
recognizable set, 45
recursively enumerable set, 189
reduced automaton, 76
Rees congruence, 194
Rees quotient, 194
reflexive, 14
regular grammar, 95
regular language, 95
relation, 5
respects, 68
restriction (of function), 31
right linear grammar, 95
RL-map, 205

Schutzenberger product, 215
semigroup, 16
semigroup with identity, 17
singleton, 12
star operation, 57
star-free language, 216
state diagram, 40
states
    of an FSA, 40, 47
    of a PDA, 122
    of a TM, 158
stationary step
    (in PDA computation), 122
submonoid, 18
subsemigroup, 18
subset, 12
successful computation
    in PDA, 123
    in TM, 158

successful path, 45
surjection, 13
surjective, 14
symmetric, 14
symmetric group, 35
symmetric semigroup, 36
syntactic congruence, 67
syntactic monoid, 68

tape symbol, 158
terminal alphabet, 93
terminal state
    of FSA, 43
    of PDA, 122
    of TM, 158
TM, 158
transformation monoid, 29
    of automaton, 82
transitive, 14
transposition, 36
trim automaton, 51
Turing machine, 158

ultimately defined, 200
union of sets, 2

varietal RL-map, 205
variety, 197
Variety Theorem, 208
Venn diagram, 15
vocabulary, 93
VRL-map, 205

word, 30

zero element, 17